$\bar{Y}$	Sample mean of the response Y.
$\hat{\mu}$	Estimate of population mean.
$\hat{\sigma}^2$	Estimate of the population variance.
λ	Expected value of the mean and variance in the Poisson distribution function.
p	Computed probability.
ρ	Population probability or population failure rate.
Σ	Summation in mathematical operations. Also used as the variance-covariance matrix in multivariate analysis.
$\times$	Multiplication in mathematical operations.
$+$	High level of variable in matrix experiment designs.
$-$	Low level of variable in matrix experiment designs.
0	Middle level of variable in matrix experiment designs.
μ_0	A fixed number against which some population mean is compared.
σ_0^2	A fixed number against which some population variance is compared.
T	Number of trials in a Hadamard matrix experiment.
$>$	Mathematical symbol for "greater than."
$<$	Mathematical symbol for "less than."
$\geq$	Mathematical symbol for "equal to or greater than."
$\leq$	[illegible]
$\ngtr$	[illegible]
$\nless$	Mathematical symbol for "not less than."
D_i^*	Decision criterion for sequential random-strategy experiments.
$\lvert\ \rvert$	Mathematical symbol for "absolute value of."
∞	Infinity.
$\sim$	Distributed as; for example, $X \sim N\ (0,\ 10)$ means X is distributed normally with mean 0 and variance 10.
b_0	The intercept of a line in a regression equation.
b_1	The coefficient of the linear effect of the first variable in a regression equation.
b_{11}	The coefficient of the quadratic (squared) term for the first variable in a regression equation.
b_{12}	The coefficient of the cross-product term of the first and second variables in a regression equation.
γ	Percentage of good space within the total experimental space.
Ψ	The value of the star points in central composite rotatable designs.
$\prod_{i=1}^{n}$	Mathematical symbol for multiplication of n terms.

Practical Experiment Designs

for Engineers and Scientists

Practical Experiment Designs

for Engineers and Scientists

Second Edition

WILLIAM J. DIAMOND

VNR VAN NOSTRAND REINHOLD
New York

Library of Congress Catalog Card Number 89-34044
ISBN 0-442-31849-9

Printed in the United States of America

Van Nostrand Reinhold
115 Fifth Avenue
New York, New York 10003

Van Nostrand Reinhold International Company Limited
11 New Fetter Land
London EC4P 4EE, England

Van Nostrand Reinhold
480 La Trobe Street
Melbourne, Victoria 3000, Australia

Nelson Canada
1120 Birchmount Road
Scarborough, Ontario M1K 5G4, Canada

16 15 14 13 12 11 10 9 8 7 6 5 4 3 2 1

Library of Congress Cataloging-in-Publication Data

Diamond, William J., 1919–
Practical experiment designs for engineers and scientists
William J. Diamond—2nd ed.
p. cm.—(Van Nostrand Reinhold competitive manufacturing series)
Includes index.
ISBN 0-442-31849-9
1. Experimental design. I. Title. II. Series.
QA279.D5 1989 89-34044
519.5′3—dc20 CIP

—VNR COMPETITIVE MANUFACTURING SERIES—

Product and Process Design

PRACTICAL EXPERIMENT DESIGN by William J. Diamond
VALUE ANALYSIS IN DESIGN by Theodore C. Fowler (early 1990)
A PRIMER ON THE TAGUCHI METHOD by Ranjit Roy (early 1990)
MANAGING NEW-PRODUCT DEVELOPMENT by Geoff Vincent
ART AND SCIENCE OF INVENTING by Gilbert Kivenson
RELIABILITY ENGINEERING IN SYSTEMS DESIGN AND OPERATION by Balbir S. Dhillon
RELIABILITY AND MAINTAINABILITY MANAGEMENT by Balbir S. Dhillon and Hans Reiche
APPLIED RELIABILITY by Paul A. Tobias and David C. Trindad

Manufacturing (hard)

INDUSTRIAL ROBOT HANDBOOK: CASE HISTORIES OF EFFECTIVE ROBOT USE IN 70 INDUSTRIES by Richard K. Miller
ROBOTIC TECHNOLOGY: PRINCIPLES AND PRACTICE by Werner G. Holzbock
MACHINE VISION by Nello Zuech and Richard K. Miller
DESIGN OF AUTOMATIC MACHINERY by Kendrick W. Lentz, Jr.
TRANSDUCERS FOR AUTOMATION by Michael Hordeski
MICROPROCESSORS IN INDUSTRY by Michael Hordeski
DISTRIBUTED CONTROL SYSTEMS by Michael P. Lukas
BULK MATERIALS HANDLING HANDBOOK by Jacob Fruchtbaum
MICROCOMPUTER SOFTWARE FOR MECHANICAL ENGINEERS by Howard Falk

Manufacturing (soft)

WORKING TOWARDS JUST-IN-TIME by Anthony Dear
GROUP TECHNOLOGY: FOUNDATION FOR COMPETITIVE MANUFACTURING by Charles S. Snead
FROM IDEA TO PROFIT: MANAGING ADVANCED MANUFACTURING TECHNOLOGY by Jule A. Miller
COMPETITIVE MANUFACTURING by Stanley Miller
STRATEGIC PLANNING FOR THE INDUSTRIAL ENGINEERING FUNCTION by Jack Byrd and L. Ted Moore
SUCCESSFUL COST REDUCTION PROGRAMS FOR ENGINEERS AND MANAGERS by E. A. Criner

MATERIAL REQUIREMENTS OF MANUFACTURING by Donald P. Smolik
PRODUCTS LIABILITY by Warren Freedman
LABORATORY MANAGEMENT: PRINCIPLES & PRACTICE by Homer Black, Ronald Hart, Orrin Peterson

Materials Management

TOTAL MATERIALS MANAGEMENT: THE FRONTIER FOR COST-CUTTING IN THE 1990S by Eugene L. Magad and John Amos
MATERIALS HANDLING: PRINCIPLES and PRACTICE by Theodore H. Allegri, Sr.
PRACTICAL STOCK AND INVENTORY TECHNIQUES THAT CUT COSTS AND IMPROVE PROFITS by C. Louis Hohenstein

To Dr. Harry Levin, who introduced me to the concepts of experiment design and started my career in this field. To Bud Beattie, who taught me that engineers are not interested in statistics per se, but are interested in everything that can assist them with their projects. To Dr. William Horne for his many years of support and encouragement during the development of my philosophy, teaching methods, and statistical techniques to better serve practicing engineers and scientists.

Contents

Preface

Audience

This book is for engineers and scientists with little or no statistical background who want to learn to design efficient experiments and to analyze data correctly, but who have no desire to become statisticians. It is an especially useful text for short courses on the fundamentals of experiment design taught by statisticians to practicing engineers and scientists. As one benefit of taking such a course and/or reading this book, the engineer or scientist should be able to use the services of a consulting statistician more knowledgeably and effectively.

This is a tried and proven work that can be used by consulting statisticians in industry. And while not intended as a text for statistics majors in the university, it is a realistic work for graduate, or even undergraduate, engineering and science students who need to learn methods they can use once they are working in industry.

The examples and applications in this book are, perforce, limited to a few fields. Nevertheless, the methods presented apply to virtually every aspect of engineering, as well as to such sciences as chemistry, physics, biomedical research, operations research, psychology, computer science, and others.

Approach

The emphasis is on *practical methods*, rather than on statistical theory. This approach is a result of the author's firm conviction, based on many years of successful consulting and of teaching experimental design to engineers and scientists in industry from all disciplines, that experimenters do not need to know—and in most cases do not want to know—statistical theory or proofs in order to use good experiment designs and statistical analysis of data effectively.

Of course, the author recognizes that there is danger of misuse if the reader does not understand the theory behind the designs. Therefore, the author recommends that the reader supplement this book with a theoretical statistical text or verify, if possible, any design or analysis with a competent statistician. The statistical foundation established in Part One and particularly in Chapters 1 and 2 is sufficient (from this author's experience) to enable the reader to understand the designs and their use, which are described in later chapters.

This book concentrates on methods that meet the needs of engineers and scientists faced with real industrial problems. The author has found that some methods and procedures that have an appeal to statisticians are unpalatable to engineers and scientists and frequently create mental blocks that make it difficult for them to learn the subject and, in many cases, prevent them from using the methods. Accordingly, this book organizes and presents the subject matter differently from more conventional texts on experiment design. The engineer and scientist will be pleased by the resultant simplicity and the ease of using the methods. For example, in Part One the reader is shown how to use the t test or the normal-distribution test to make decisions about means. In Part Two, on two-level multivariable experiments, the same t test or normal-distribution test is used to make decisions about the means. This is in contrast with most texts, which introduce analysis of variance (ANOVA) as soon as they introduce more than one variable; ANOVA is not mentioned in this work until multilevel experiments are presented.

It is the author's belief that a book on experiment design must give the reader a method or methods for the stated problem with all variables included and that, if necessary, the statistics should be modified to meet the user's need. As a result of this view, the present work concentrates on Resolution V, IV, and III designs *with up to 127 variables*, and only briefly mentions complete factorials. With the ready availability of the powerful methods presented in this book, there is no need to disregard some variables simply to fit more conventional methods. And because it is the author's opinion that engineering managers want a firm decision at the end of an experiment (for example, to build a new plant or not to build a new plant), the rigid Neyman-Pearson approach to hypothesis testing is used throughout the book.

Practical Applications

Methods described in this book can be applied to every engineering and scientific problem requiring experimentation. All designs and procedures covered in Chapters 1 through 11 are independent of the scientific discipline of the experiment; they are also independent of the department in a corporation that conducts the experiment. In Chapter 6, for example, an 8-trial Hadamard matrix experiment is described for a product assurance experiment conducted by a mechanical engineer. The very same 8-trial Hadamard matrix can be used by an electronics engineer in product development engineering or by a chemist in a pilot plant. Only the labeling of the variables of the matrix will differ among the three experiments.

Procedures for applying the methods described in Chapters 1 through 11 are summarized in Chapter 12. The experimenter calculates the proper minimum sample size, specifies the *number* of variables to be included, and the *quality* of information required from the experiment, and finds in the list of designs the one and only matrix size that is correct for the experiment. Then by using the computer program described in Chapter 13, the experimenter obtains a printout of the correct matrix, properly labeled with the experiment variables. When the experiment is completed, the computer program performs the analysis of data.

Organization

Part One of this book establishes the fundamentals of experiment design. It defines terms, establishes the logic behind a correct sample size, demonstrates the value of an objective criterion for making decisions from data, and familiarizes the reader with the use of the various statistical tables.

Part Two applies the principles from the first part to the more realistic two-level, multivariable problems typically encountered by the engineer or scientist. The unifying theme in Part Two is the Hadamard matrix. It shows that, for a given Hadamard matrix of size T, the reader can obtain an experiment design for any number of variables from 1 to $(T - 1)$. It further shows that these matrices can be combined, augmented, or otherwise modified to obtain designs that are helpful in special situations. For the engineer or scientist, the learning process is simplified; only one basic experiment design tool, the Hadamard matrix, must be learned. Furthermore, it is easy to compare many designs of different sizes and to weigh the information that will be obtained against the time and cost of the experiments. The process is further simplified by the computer program discussed in Chapter 13. With a few simple statements, this program will generate every two-level experiment design up to 127 variables in 128 trials, and the same program will analyze data from any design that is generated.

Part Three covers the topic of multilevel, multivariable experiments. It shows that for experiments involving quantitative variables, the designs that are available are either augmented or modified Hadamard matrices. Thus, these designs are easy to comprehend simply as an extension of Part Two. For qualitative variables, multilevel complete factorials are used; this is a simple extension of the concept of two-level factorials. Analysis of the data, however, requires introduction of the concepts of ANOVA. Random strategy is introduced because extensive research by the author in this procedure has proven its usefulness to the engineer in many practical situations where a matrix design is not feasible or necessary. Many of the designs of Part Two are applicable to chemical problems; however, when the chemist conducts experiments where the composition of the product is varied, these designs cannot be applied. A number of elegant, efficient designs have been developed for these special problems, and the topic is introduced in Part Three because most of these designs are multilevel.

Part Four covers related topics that don't fit well in the other sections, and *Part Five* contains general references, a list of symbols, tables, and answers to exercises.

As indicated above, some readers will want more detail on the theory behind the topics in various chapters. Therefore, at the end of most chapters, there are references to texts that this author considers the best (for an engineer or scientist) for a fundamental or theoretical discussion of the topic of that chapter. The author hopes that the reader will examine these references after reading each chapter in this book, and then do the exercises at the end of the chapter. The reader is also referred to Part Five, with its examination of important works on experiment design and its coverage of pertinent articles published in *Technometrics*.

Acknowledgments

I am grateful to the Literary Executor of the late Sir Ronald A. Fisher, F.R.S., to Dr. Frank Yates, F.R.S., and to Longmans Group, Ltd., London, for permission to reprint Table V from their book *Statistical Tables for Biological, Agricultural and Medical Research* (sixth edition, 1974).

I am grateful to Marvin Smoak for the example and the computer output in Chapter 15.

William J. Diamond

Practical Experiment Designs

for Engineers and Scientists

Part One

Fundamentals of Experiment Design

1 Introduction to Experiment Design: Fundamental Concepts

The information covered under Introduction to Experiment Design has been divided into two chapters so that the reader unfamiliar with statistics will not be overwhelmed with a huge mass of new concepts. This division into two chapters will allow the reader to easily study and assimilate those concepts that are fundamental to understanding all of the chapters that follow. Chapter 1 covers fundamental concepts; Chapter 2 discusses elements of decision making and other important concepts that the reader will find useful in designing experiments.

By carefully reading and, from time to time, reviewing these two chapters, the reader will quickly gain an insight into what factors distinguish good experiments from bad experiments and an understanding of the principles involved in designing good experiments (i.e., those experiments that will provide correct results in the shortest time at the least cost). A set of exercises and references will be found at the end of each chapter; these exercises and references provide useful additional insights and information to the material presented in the first two chapters.

The topic of Chapter 1—Fundamental Concepts—is discussed under the headings of Origin of Bad Experiments, Philosophy for Good Experiments, Project Strategy, Experiment Strategy, and Perceptions of Reality.

Origin of Bad Experiments

Engineers and scientists who conduct experiments are usually well trained in the mechanics of experimenting, at least in their own disciplines, and have the desire to perform good experiments and to make correct engineering decisions. In larger organizations, a trained statistician is usually available to assist the experimenters in interpreting their data, and experimenters usually have available elegant and sophisticated equipment for preparing samples and measuring the properties of interest. Certainly, management expects, and presumes to be true, that their experimenters are performing at a high level of efficiency in the experiments being conducted and are making correct decisions and reporting accurate results on the bases of these experiments.

In spite of all these efforts, resources, and good intentions directed towards performing good experiments, many experiments are, in fact, either poor experiments, which can result in wrong decisions, or inefficient experiments, which result in excessive cost or time delay in reaching a decision. Among the prime reasons for this state of affairs are the following:

- Engineers and scientists typically are well trained in the mechanics of experimenting but usually receive no training in the strategy of experimenting. The strategy typically used by experimenters is the instinctive one-factor-at-a-time strategy; i.e., hold all variables constant except one, change that one variable until the best setting for that variable is found, then do the same thing with a second variable, a third variable, etc. This is an inefficient strategy in most experiments.
- Experimenters often make what they believe to be correct decisions based on experimental data when, in fact, the decisions are incorrect because the experiment strategy has been incorrect or inefficient. The only clue that an experimenter has made an incorrect decision is, for example, when a product fails in the field. Even then, the development engineer may blame manufacturing, and manufacturing may, in turn, blame the parts suppliers for the problem.
- Management contributes to bad engineering decisions by setting vague objectives for projects; e.g., a latching mechanism that is "as good as possible." Furthermore, management virtually imposes the one-factor-at-a-time strategy on experimenters by almost-daily demands for evidence of "progress" on the project. Finally, management often accepts decisions from their experimenters based on data that have not been verified by a skilled data analyst and that they themselves are not qualified to evaluate.
- Statistical courses taught at the university level—even courses on experimental design—are predominantly theoretical in flavor. Because of this they tend to present techniques not readily applicable in practical settings. Such courses are generally taught to a varied audience and are not specifically aimed at engineers and scientists; consequently they don't present techniques that are useful in industrial and research settings. Finally, and inevitably, such courses use textbooks with oversimplified examples—something that is justifiable pedagogically, but doesn't prepare the student for reality.

All of these reasons contribute heavily towards the performance of poor or inefficient experiments in spite of the best efforts of experimenters and their managers. The following section sets out some basic philosophy necessary for the design of good experiments.

Philosophy for Good Experiments

The best experiment designs result from the combined efforts of a skilled experimenter, who has had basic training in experiment design methods, and of a skilled statistician, who has had training in engineering or science. The statistician alone cannot design good experiments in every possible scientific discipline; neither can the scientist or engineer who is untrained in statistical experiment design be a good experiment designer. This book is designed both to assist teams of statisticians and experimenters in designing good experiments and to provide the necessary guidelines and techniques for good experiment design to those experimenters who, for one reason or another, are unable to call upon the assistance of a statistician.

But there is more to the design of good experiments than the availability of skilled personnel or useful resources. A total project strategy must first be established and then, within that project strategy, a series of experiment strategies must be devised.

Project Strategy

As a necessary first step in any project, management must define the objective of the project as specifically as possible; e.g., to develop a latching mechanism that will have a failure rate of less than one failure per 100,000 operations over a life of 1,000,000 operations. The true objective of the project must be clearly recognized by both the engineer and the statistician. In development projects, the objective of the project is usually a product that meets certain limited specifications. In manufacturing projects, the objective is usually to determine the cause of some defect in the product being produced or to determine if some cost-reduction change in the process can be made without downgrading the product.

The objective of an engineering project is not to find interactions, to define response surfaces, or to estimate variances—subjects of more theoretical interest than practical value. These are only means to an end and are unnecessary unless they contribute to the true project objective. On the other hand, in basic research projects, the objective is usually quite different than in engineering projects: the researcher is frequently trying to establish some law of nature, and such things as response surfaces and interactions are usually vital to the project.

After defining the project objective, the experimenter must define, in writing, the total experimental space in which a satisfactory product or process is expected to be found. The experimenter does this by specifying all the variables of the product or process that could possibly influence the quality or performance of the end product or process and by defining the reasonable range of interest for each variable so specified. If this step in project planning is done adequately, there should be no need to define additional variables at later stages in the project. When a project is still in the research and development phase, the experimenter may have some difficulty in defining all the variables and their ranges of interest; nevertheless, these definitions should be made to the extent that it is possible to do so.

All the responses of interest must also be defined by the experimenter at this stage. It is incorrect and inefficient to conduct one set of experiments to develop a product or process that is satisfactory for one property and then to conduct another set of experiments to improve a second property without degrading the first one, and so on.

Experiment Strategy

Almost certainly, most projects will require several experiments. When devising experiment strategies, the experimenter must bear in mind that it is desirable to establish early in the project whether a product or process with the specified properties is feasible or infeasible in the defined variable space. If the latter is indicated, it is imperative that the project be killed as soon as possible to prevent the waste of organizational resources. On the other hand, if the former is indicated, it is imperative that the project be completed as soon as possible at the least cost.

The strategy for the first experiment will probably be entirely different than the strategy for an experiment in the middle or at the end of a project. The strategy for a development-oriented experiment will be different than the strategy for a product-assurance experiment or for a manufacturing experiment. The correct strategy for any specific experiment can only be determined if the objective of the experiment is stated correctly and completely.

The initial experiment in a development project should have as its objective the identification of a good starting point in the total experimental space and must include all of the identified variables. The second experiment in this type of project should have the objective of identifying the most significant variables and must also include all of the identified variables. The number of variables that will be involved will probably be so large as to preclude the use of complete factorial and multilevel experiments (described later in this book). Breaking down the complete set of variables into smaller subsets on which complete factorial experiment designs are feasible is a completely erroneous approach at this stage in most engineering projects.

Multilevel experiments are time consuming, resource consuming, and, in the early stages of a project, unnecessary. If, in any experiment, the range of the variable is reasonable, the effect of the variable on the response will be approximately linear or at least monotonic, and can be adequately estimated by a two-level experiment. If there is curvature in the response over any extended range of the variable, subsequent experiments will define both the curvature and the approximate location of any maximum or minimum response. In the latter stages of a project, however, multilevel experiments on a limited number of variables are frequently required.

Because most engineering and scientific systems consist of a multitude of variables, some of which interact synergistically to produce either a good effect or a bad effect on the end-product properties, it frequently becomes necessary to investigate these interactions. Sometimes, usually at the end of a project, it is necessary to identify the interactions—a process costly both in time and effort; at other times, usually at the start of a project, it is only necessary that the interactions be permitted to demonstrate their effects—usually at a substantial savings in time and effort.

The engineer must decide for each experiment how much information is needed from that experiment and with what confidence level (described later in this chapter) the final engineering decisions are to be made. Once these facts have been established and the number of variables to be included in the experiment has been determined, the correct experiment design is virtually automatic and will certainly be found among the designs described in subsequent chapters. One thing is certain: as the amount of information required is increased, the amount of experimenting required will also increase.

Finally, before the actual experimenting begins, the test methods and test equipment that will be used must be specified, and the validity and the precision of the test methods must be validated.

Perceptions of Reality

Science does not yet—and may never—allow us to know and understand precisely the real world around us. Scientists and engineers can only perceive reality, with varying degrees of clarity, by means of hypotheses, theories, and experiments. Even the most precise experiments and the most closely argued theories can only be meaningful within certain tolerances, confidence levels, and assumptions.

Fortunately, statistics allows the experimenter to perceive and understand the real world with far greater precision than would otherwise be possible. This section will expound a theology for experimenters (to relate the perceived world to the real world) and will explain the concepts of populations, distributions, and means and variances by means of which we can arrive statistically at a close approximation of the real world.

> *The purpose of an experiment is to better understand the real world, not to understand the experimental data.*

Theology for Experimenters

When God created the universe, He also ordained mathematical relationships between all things then existing and all things that ever would exist. Some of these relationships are very simple, but most of them are very complex. For example, the life of a tire is defined by:

$$
\begin{aligned}
\text{Life of Tire} = {} & f(\text{type of rubber}) + f(\text{curing temperature of rubber}) \\
& + f(\text{additives to the rubber}) + f(\text{groove design}) \\
& + f(\text{mixing procedure for the rubber}) + f(\text{size of tire}) \\
& + f(\text{car on which used}) + f(\text{roadbeds}) \\
& + f(\text{operating speeds}) + f(\text{atmospheric conditions}) \\
& + f(\text{operator of the car}) + f(\text{maintenance}) \\
& + f[(\text{type of rubber})(\text{curing temperature of rubber})]
\end{aligned}
$$

where $f(x)$ is read as "some mathematical function of (x)."

Of course, each term in the equation can be either simple or complex. Furthermore, the terms might interact with each other as shown in the last term above. For example, f[(curing temperature)(type of rubber)] might be:

$$\text{Life of Tire} = 265 + (2.76)(\text{temp}) - 4.82(\text{temp})^2 \quad \text{for rubber A, and}$$

$$\text{Life of Tire} = 487 + (4.87)(\text{temp}) - 1.75(\text{temp})^3 \quad \text{for rubber B.}$$

For the term f(operating speed), the relationship might be:

$$\text{Life of Tire} = 87 + \exp\left[7 - (\text{speed}) + \left(\frac{1}{\text{speed}}\right)^3\right].$$

Suppose a rubber chemist is given the task of using a new polymer to develop an improved tire that would have a better life than the present best product of the company. Obviously, the chemist's task would be simple if God's equation were known: the values of the new polymer would be plugged into f(type of polymer) and the proper level of all other variables would be adjusted so that Life of Tire (new) is substantially greater than Life of Tire (old). Unfortunately, the chemist doesn't know God's equation and probably doesn't even accurately know the life of the present tire or all of the functions that influence the life of a tire.

In practice, the chemist simplifies the problem by conducting tests on sample blocks of material for selected properties, such as abrasion and tensile strength, to determine how these properties are affected by variables such as additives, mixing procedures, temperature of curing, etc. In conducting an experiment to determine the effect of curing temperature, for example, the chemist is, in essence, trying to determine the values that God ordained for f(curing temperature) as it influences f(tensile strength), f(elongation), and other properties. Almost certainly, the right answer will not be obtained. In fact, a different result will be obtained every time an experiment is conducted because of slight differences in all of the other functions that are nominally (but not actually) being held constant and because of all the unknown (to the chemist) functions that God wrote into the equation.

All the experimenter can hope for is that the answer obtained from the experiment will be a good estimate of the value in God's equation. The experimenter must also be aware that the answer obtained applies only at the level of all of the other possible variables that were held constant; the answer does not necessarily apply at different levels of other variables.

An experiment can only produce an estimate of the true state of nature.

EXAMPLE

An experimenter evaluates the effects of several different curing temperatures on samples of a specific polymer formulation produced under "identical" conditions. Results of tensile tests on the polymer samples are shown in figure 1-1. The experimental result is wrong in that the experimenter's best line doesn't coincide with God's line; however, the experimental result is a pretty good estimate of God's equation. The experimenter would probably conclude that a temperature of 170°F is adequate to obtain a maximum tensile strength of about

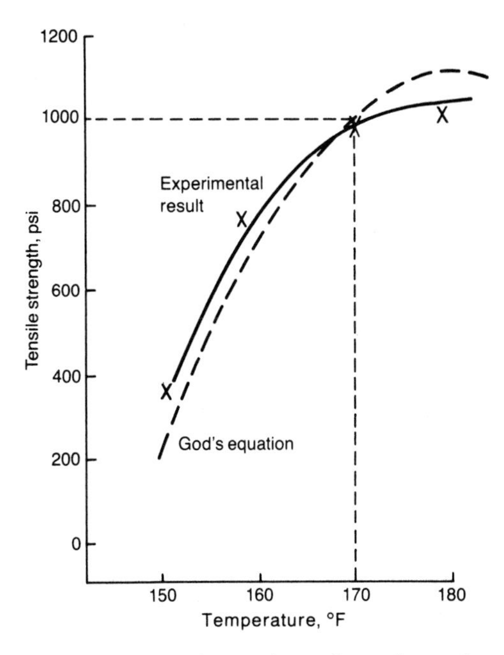

Fig. 1-1. Comparison of experimental results with the truth.

1000 psi. This conclusion, however, only applies to the given formulation and the given procedure. It is quite possible that the tensile strength of polymer samples would be a different function of curing temperature if, for example, an additive were used in the formulation.

If the experimenter now adds an additive, Z, to the polymer and cures it at 170°F, a tensile strength of about 900 psi will be obtained if God's law is as shown in figure 1-2.

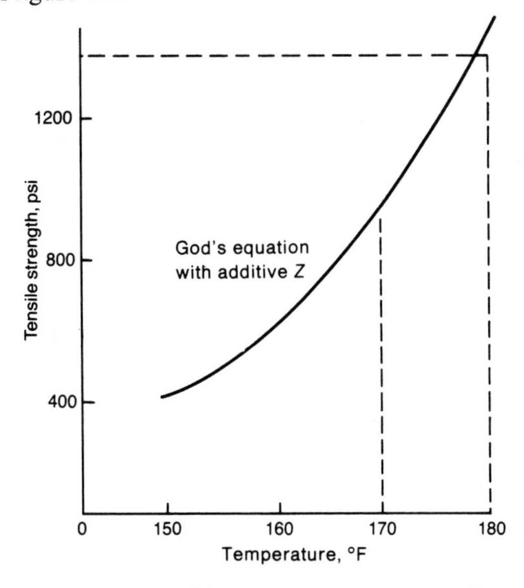

Fig. 1-2. Effect of temperature on tensile strength of polymer formulation with additive Z.

If the chemist first determined, without an additive, that 170°F was the best curing temperature and then used this temperature to evaluate the effect of additive *Z* on the polymer, the chemist would probably conclude, erroneously, that additive *Z* decreased tensile strength. In reality, of course, the tensile strength at 170°F truly is less with the additive in the formulation, but the additive produces a stronger part at 180°F.

Populations

A population is defined as that group or set of things about which a decision is to be made. In some cases, the population is obvious. For example, if the plant manager requests that yesterday's production of motors be tested to determine the average noise level, then the population consists of all motors produced yesterday. If the plant manager is interested in this week's production vs last week's production, there are now two populations: population A consists of all motors made last week, and population B consists of all motors produced this week.

In some cases, the population is not obvious. For example, a process has been used in accord with certain specifications for many years to produce a machined part. It is proposed that the process be speeded up to increase output with the hope that the quality of the part will not decline. The old population, then, is the set of all parts that have been produced and will be produced if the change in speed is not adopted; the new population is the set of all parts that will be produced if the change in speed is adopted.

In development engineering experiments, the populations are defined somewhat differently. For example, a certain mechanism being developed requires a clutch of some type, so the engineer tests the mechanism with both a spring clutch and a friction clutch. There are two populations: the set of all mechanisms that will be produced with a spring clutch if that clutch is adopted; and the set of all mechanisms that will be produced if a friction clutch is adopted.

Product assurance engineers frequently misdefine their population. If they are given four new machines to test prior to production, they think their population size is four. Actually, of course, four is the size of a sample that represents a population of all machines that will be produced if the sample of four machines passes the product-testing process and if the machines that are being tested are representative of the machines that actually will be manufactured. Note that if the machines are not representative of those that will be manufactured, the sample size of the production population is zero, and the testing is a waste of time and money.

The population consists of all items about which a decision is to be made.

It is, of course, vitally important that the sample that is tested not only belongs to the population about which a decision is to be made but also is representative of the entire population. Many apparently unexplainable experimental results are the result of incorrect sampling. For example, the manufacturer of a programmable desk calculator was encountering about 20 percent failures in the field with the motor drive shaft for the tape system. In the laboratory, several dozen motors were tested under extreme conditions to accelerate the failure mode without a single failure occurring. According to the laboratory, there was no problem. It was then discovered that the technician who had secured the motors for test had obtained all of them from the production line just outside the laboratory door; no motors had been secured from the other four production lines, which would have required walking across the plant. When the test was repeated with four motors from each production line, there were four failures in the sample of 20 motors, and all failures were from the third production line. Further experiments on the population of calculators from the third production line established that this entire population was bad due to a faulty assembly of the shaft into a bearing.

Distributions

Most engineers and scientists are at least vaguely acquainted with the normal distribution and are probably aware that there must be other kinds of distributions. Actually, there is an infinity of possible distributions of data; however, for the most part, industrial data will approximate one of a relatively few classical types of distributions: normal, uniform, chi-square, binomial, Poisson, t, F, or lognormal.

The term distribution is really a shortening of the term frequency distribution of the data. Distribution is, along with population mean and population standard deviation (both of which are defined later in this chapter), one of the parameters that define the nature of a population. Specifically, distribution defines the shape of data when the data are plotted in the form of a histogram. Figure 1-3 shows a histogram plot obtained when a wheel with numbers from 0 to 6, in increments of 0.1, was spun 120 times and the frequency with which each interval of numbers came up was recorded. Each interval of numbers could be expected to come up about 20 times out of the 120 spins, and the data, indeed, are shown to be clustered around the expected frequency line.

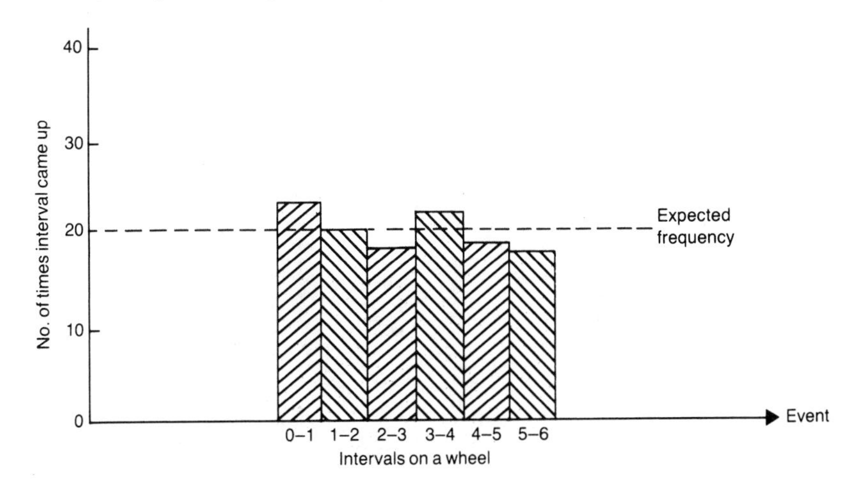

Fig. 1-3. Frequency with which each interval of numbers on a wheel comes up out of 120 spins.

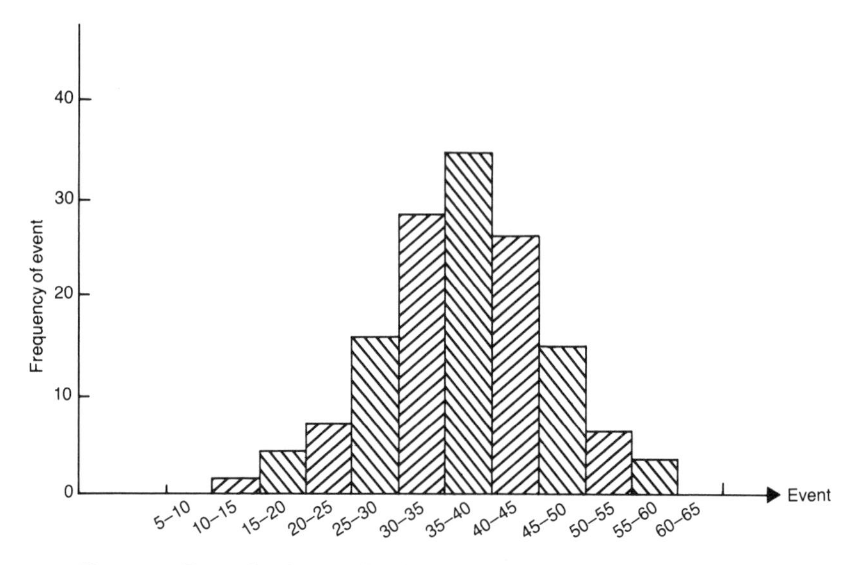

Fig. 1-4. Example of normally distributed data.

The data distribution shown in figure 1-3 is called a uniform distribution. By contrast, if a set of data was obtained from a normally distributed population, and the data were grouped into classes, a histogram of the type shown in figure 1-4 would be obtained.

From any given histogram of sample results, predictions can be made concerning the frequency of occurrence or nonoccurrence of any possible event in that population. More important, however, is the inverse: given that a population should have a certain distribution, if the experimental data do not correspond to that distribution or to parameters of that distribution, then the experimenter must reject the hypothesis that the sample came from the proposed population.

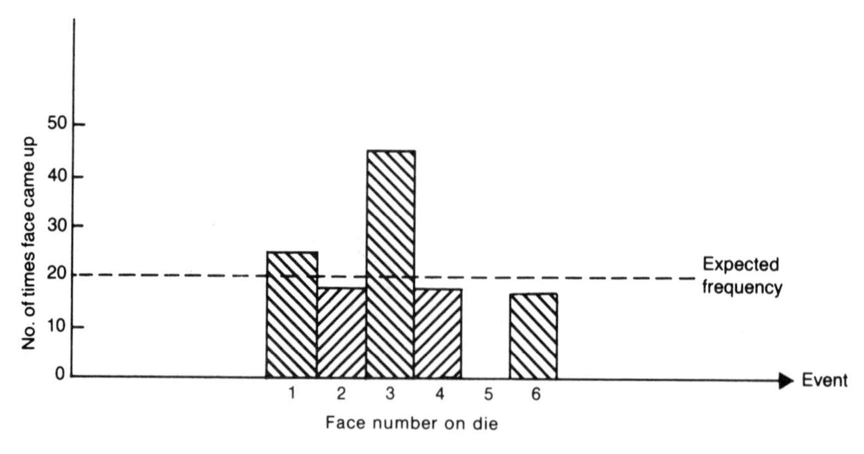

Fig. 1-5. Frequency with which each numbered face comes up—nonstandard die

For instance, if a die is thrown 120 times and a histogram such as that shown in figure 1-5 is obtained, a strong inference can be drawn either that the die is loaded or that there are two faces with the number 3 and none with the number 5. The data could not have come from the population of fair dice.

Similarly, suppose that the histogram of the normal distribution (figure 1-4) is the result of samples taken over a three-month period for some production process. If, at some later date, a sample of five items was tested and results of 52, 57, 54, 59, and 55 were obtained, it is highly unlikely that the process during the second sampling period is the same as that during the first sampling period.

Though histograms are nice, they are not as convenient as distribution functions and tables for determining probabilities. A distribution function is a cumulative probability and can be plotted as shown in figure 1-6 for the uniform distribution of a variable that can take on any value between 2 and 5.

Any probability (p) can immediately be obtained from the graph shown in figure 1-6 in the following manner:

$$p(>5) = (1.0 - 1.0) = 0$$
$$p(<2) = (0 - 0) = 0$$
$$p(3 - 4) = (0.667 - 0.333) = 0.333$$
$$p(>4) = (1.000 - 0.667) = 0.333$$
$$p(<5) = (1.000 - 0) = 1.000$$

If a mean result (see below for definition of mean) of 5.5 was obtained with a sample of 10 items from some population, it would be absolutely certain that the sample did not come from the above population. Likewise, if the mean result was 4.9, it would be a good bet that this sample, too, didn't come from this population.

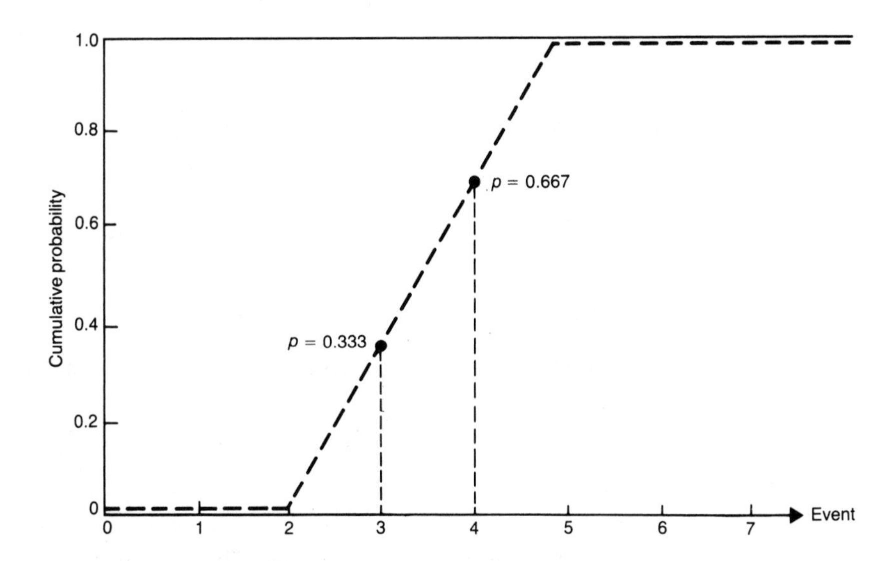

Fig. 1-6. Example of a distribution function.

The situations that most experimenters encounter, and with which this book is primarily concerned, are experiments that require decisions to be made about the population mean (described below), not about the individual items in the population. In these cases, the underlying distribution of the individual items is unimportant with any reasonable sample size. This curious phenomenon is explained by the central limit theorem, which states that the sample mean ($\bar{X}$) (described below) is approximately normally distributed about the population mean (μ) (described below) if the sample size is relatively large—which, for all practical purposes, is almost always the case in engineering and scientific work. Therefore, if the variance (σ^2) of the population (described below) is known, one can compute sample size and all the other factors necessary for a good experiment from the normal distribution tables, even though the distribution of the population may not be normally distributed or the distribution may be unknown. If the variance of the population is not known, the sample size can be calculated from t tables (described later in this book) regardless of the distribution of the population.

Means and Variances

In conducting experiments, the experimenter is trying to find some information regarding a particular population of interest. Since it is almost always impossible, impractical, or uneconomical to test an entire population, the experimenter compromises and tests a number of items selected from the population in such a manner as to be representative of the population as a whole. This group of items is called a random sample.

When a relatively large amount of data is obtained from experiments, as is usually the case, it is very difficult for the human mind to correlate all the numbers and make a valid interpretation of their meaning. Even if one experimenter should arrive at one conclusion from an examination of the data, another experimenter might very well arrive at a totally different conclusion from the same set of data. This problem is simply resolved by the use of statistics: every set of data obtained from a sample produced under a fixed set of conditions can, for most practical engineering problems, be adequately defined by three numbers calculated from the data:

- The number of test results (N).
- The mean of the test results ($\bar{X}$).
- The variance of the test results (S^2).

After obtaining the above numbers (statistics) from the test data, and after verifying the accuracy of the data, the experimenter could then throw away the individual results for all practical purposes: all the important information in the data would be in the values of N, $\bar{X}$, and S^2.

The value $\bar{X}$ is referred to as the sample mean, and the value S^2 is referred to as the sample variance. Both of these sample values are related to the equivalent population values as described below. The sample mean ($\bar{X}$) is an arithmetic average computed by adding all the test results together and dividing by the number of test samples (N):

$$\bar{X} = \frac{\sum X_i}{N},$$

where X = test result, $i = 1, 2, 3, \ldots, N$, and $\sum$ is the symbol for summation.

The sample mean is an estimate of the true central value around which all the data in the population congregate. This true central value is the population mean. Thus, the sample mean ($\bar{X}$) is a point estimate of the population mean, which is denoted by the Greek letter mu (μ). (Note: sample values are denoted by Roman letters, and population values are denoted by Greek letters.) The population mean is itself an arithmetic average derived by adding together the appropriate values for all units in the population and dividing the result by the number of units in the population. It should be obvious that the population mean can, in most cases, only be computed by God and is usually unknown to the experimenter.

The sample variance (S^2) is the square of the sample standard deviation (S), and both values represent the extent of scatter of the data around the sample mean: the larger the scatter, the greater the variance and standard deviation; the less the scatter, the smaller the variance and standard deviation. Sample variance (S^2) and sample standard deviation (S) are each point estimates of, respectively, population variance (σ^2) (sigma squared) and population standard deviation (σ) (sigma).

Sample variance can be calculated by either of the following two formulas:

$$S^2 = \frac{\sum (X_i - \bar{X})^2}{N - 1}, \quad \text{or}$$

$$S^2 = \frac{\sum X_i^2 - (\sum X_i)^2/N}{N - 1},$$

where $i = 1, 2, 3, \ldots, N$. The term in the denominator ($N - 1$) is called degrees of freedom. It is an important and much used parameter of any experiment, as will be seen in later chapters. The number of degrees of freedom is designated either by the Greek letter phi (ϕ) or by the abbreviation df.

The relationship between sample mean ($\bar{X}$) and population mean (μ) is illustrated in the following example. A similar relationship would exist between sample variance and population variance.

EXAMPLE

A metallurgist wanted to determine the change in tensile strength that would result from changing the carbon content from 5 percent to 10 percent in a given cast-iron formulation made with a specified process.

God's equation (unknown) is (tensile strength) = α_1(% carbon) + $\sum(f_i)$, where α_1 is the effect of each 1 percent carbon (say, 1000 psi), and $\sum(f_i)$ is the effect of all other potential variables that can influence the tensile strength of a sample. These potential variables, however, are presumed to be constant in the experiment, so $\sum(f_i)$ is presumed to be a constant (say 40,000 psi).

The right answers are easily computed by God:

$$\mu_{5\%\,C} = \text{mean tensile strength at } 5\%\,C = 1000(5) + 40{,}000 = 45{,}000 \text{ psi.}$$

$$\mu_{10\%\,C} = \text{mean tensile strength at } 10\%\,C = 1000(10) + 40{,}000 = 50{,}000 \text{ psi.}$$

Therefore, the true effect of changing from 5 percent carbon to 10 percent carbon, as computed by God, would be an increase of 5000 psi in tensile strength.

If the metallurgist produced a sample of 20 ingots with 10 percent carbon, the first thing that would be noticed is that every sample result would be different. This would be positive proof that all the potential variables were not, in fact, kept absolutely the same and, probably, that the carbon content was not exactly at 10 percent in all the samples. The same thing would apply to the sample results obtained from ingots made with 5 percent carbon. The metallurgist might then calculate the following based on test results:

$$\bar{X}_{5\%\,C} = 44{,}010 \text{ psi.}$$

$$\bar{X}_{10\%\,C} = 50{,}040 \text{ psi.}$$

$$a_1 = \frac{\bar{X}_{10\%\,C} - \bar{X}_{5\%\,C}}{5} = 1206 \text{ psi}/\%\text{C},$$

where a_1 is a point estimate of α_1.

The metallurgist would conclude that the effect of changing the carbon level, a_1, is 1206 psi for each 1 percent increase in carbon content. This answer is obviously wrong, since it disagrees with God's number; however, any experiment can only estimate God's numbers, not determine these numbers precisely.

To summarize:

- $\bar{X}_{10\%\,C}$ (50,040 psi) is a point estimate of $\mu_{10\%\,C}$ (50,000 psi).
- $\bar{X}_{5\%\,C}$ (44,010 psi) is a point estimate of $\mu_{5\%\,C}$ (45,000 psi).
- a_1 (1206 psi/%) is a point estimate of α_1 (1000 psi/%).

The experimenter, not knowing God's numbers, could not know how good these estimates were: the mere fact that $\bar{X}_{10\%\,C}$ was bigger than $\bar{X}_{5\%\,C}$ would not justify a statement that $\mu_{10\%\,C} > \mu_{5\%\,C}$. The experimenter could, however, by using statistics and an estimate of the variance, relate the difference between the sample means ($\bar{X}$) to precise limits on the difference between the population means (μ). Statistics is the necessary connecting link between sample data and population parameters:

- $(\bar{X}_{10\%\,C} - \bar{X}_{5\%\,C})$ + statistics → knowledge of $(\mu_{10\%\,C} - \mu_{5\%\,C})$.
- a_1 + statistics → knowledge of α_1.

$\bar{X}$ is not equal to μ; it is only an estimate of μ.
S^2 is not equal to σ^2; it is only an estimate of σ^2.

EXERCISES

1. Take 10 coins, shake them, drop them on a table, count the number of heads, and record the results. Repeat this process about 30 times. Plot the results on a histogram.
2. Take six dice, toss them, and count the number of spots on the up faces. Divide by 6 to obtain $\bar{X}$, the sample mean number of spots on a die with a sample size of six. Repeat about 30 times, and plot the results on a histogram. Now repeat the experiment using a single die. What is the appearance of the distribution with a single die? What is the appearance of the distribution of the mean of six dice?
3. The following test results (power output in mW) were obtained from a sample of ten lasers: 5.8, 6.1, 5.9, 6.1, 6.4, 5.8, 5.2, 5.9, 5.7, and 5.6. Calculate $\bar{X}$, S^2, S, and ϕ.

REFERENCES

Davies, O. L. 1978. *Design and analysis of industrial experiments.* 2nd ed. New York: Longmans, Inc.

Ott, L. 1977. *An introduction to statistical methods and data analysis.* North Scituate, Mass.: Duxbury Press.

2 Introduction to Experiment Design: Elements of Decision Making

Virtually all industrial experiments are comparative experiments. For example, experiments may be designed and conducted to evaluate the yield of an established process vs the yield of a modified or new process; to ascertain the change in tensile strength of an alloy if its formulation is changed; to determine the change in production costs if a proposed production-line change should be made; to compare test results on an entirely new product with its engineering specifications; and so on.

In all such cases, after the testing has been completed and the data have been reduced, *the experimenter must make a decision:* do the results indicate that the old product or process is equal to, better than, or worse than the new product or process? Test results alone, without any preestablished guidelines, are frequently ambiguous to the point where the experimenter can make serious errors of judgment.

Statistics can greatly assist the experimenter in the decision-making process. Four major steps are involved in establishing guidelines for decision making before the experiment is conducted:

- Stating the two alternative decisions.
- Defining the acceptable risks for selecting the wrong alternative.
- Establishing an objective criterion for selecting between the alternative decisions.
- Computing the requisite sample size.

Each of these steps is discussed in detail below.

Stating the Alternative Decisions

The experimenter, in selecting and testing samples and statistically analyzing test results, is usually making a comparison between two populations; e.g., one population could comprise all units that have been made or will be made under certain circumstances, and a second population could comprise all similar units that have been made or will be made under different circumstances. In the metallurgical example described in Chapter 1, one population was all of the units that could be made with cast iron containing 5 percent carbon, and the second population was all of the units that could be made with cast iron containing 10 percent carbon.

When comparing properties of interest between two populations, the experimenter will be confronted with two alternative possibilities: either the properties of interest are essentially the same in both populations; or the properties of interest are significantly different between the two populations. In statistical terms, these alternatives are stated as:

- Null hypothesis (H_0)—no essential difference exists between the properties of interest in the two populations.
- Alternative hypothesis (H_a)—a significant difference does exist between the properties of interest in the two populations.

When statistically analyzing test results, the experimenter basically is concerned either with the population means or with the population variances. Thus, the null and alternative hypotheses may be stated mathematically as follows:

- Null hypothesis
  $H_0{:}\mu_1 = \mu_2$ for population means.
  $H_0{:}\sigma_1^2 = \sigma_2^2$ for population variances.
- Alternative hypothesis
  $H_a{:}\mu_1 \neq \mu_2$ for population means.
  $H_a{:}\sigma_1^2 \neq \sigma_2^2$ for population variances.

The alternative hypotheses shown above are termed double-sided alternative hypotheses. This means that the experimenter has no commitment to either population and is only interested in knowing which of the two is the better population. In some cases, however, the experimenter may be primarily focused on one population or the other and is only interested in results where the population of interest is better than the other population. In these cases, the experimenter would state a single-sided alternative hypothesis; e.g.,

- $H_a{:}\mu_2 > \mu_1$ (primary focus on population 2), or
- $H_a{:}\mu_1 > \mu_2$ (primary focus on population 1)

In the metallurgical example cited above, the alternatives would have been stated as follows:

$$H_0{:}\mu_{10\%\,\mathrm{C}} = \mu_{5\%\,\mathrm{C}}$$

$$H_a{:}\mu_{10\%\,\mathrm{C}} > \mu_{5\%\,\mathrm{C}}.$$

In this statement, the alternative hypothesis is stated as a single-sided hypothesis: the experimenter is interested in the population with 10 percent carbon only if this population has a mean tensile strength higher than that of the population with 5 percent carbon. The experimenter has no interest in making any decision about the population with 5 percent carbon; the population with 5 percent carbon is only the point of reference against which the population with 10 percent carbon is to be compared.

Sometimes the experimenter is making a comparison between a population and some fixed number, e.g., a value in a quality-control specification. Such a fixed number is designated by the symbol mu zero (μ_0). Thus, the null hypothesis in this case would be

$$H_0: \mu_1 = \mu_0,$$

and the alternative hypothesis could be

$$H_a: \mu_1 > \mu_0,$$

$$H_a: \mu_1 < \mu_0,$$

or

$$H_a: \mu_1 \neq \mu_0.$$

Likewise, in some experiments concerning variance, the purpose is to compare the variance of a population with some fixed number, such as a specification value, which is designated by the symbol sigma zero squared (σ_0^2).

It is impossible to prove a null hypothesis correct, but it is relatively easy to prove a null hypothesis incorrect. Accordingly, the proper decision-making procedure is to try to prove with high probability that the null hypothesis is false. If, with statistics, the null hypothesis is proved false, the experimenter can accept the alternative hypothesis to be true. On the other hand, if the null hypothesis cannot be proved false, it will be accepted as true.

A null hypothesis is written, for example, $H_0: \mu_1 = 10$, with an alternative hypothesis, $H_a: \mu_1 > 10$, for statistical correctness. The experimenter should recognize that these hypotheses correspond, respectively, to the following two decisions:

- The true mean of population 1 is not appreciably better than, or is even lower than, 10 if H_0 is true.
- The true mean of population 1 is higher than 10 if H_a is true.

Defining the Risks

In every experiment, there is a risk that the experimenter will infer the wrong decision from the test data. However, the experimenter can control this risk by selecting the proper sample size to be used in the experiment. (The relationship between sample size and risk of error is described later in this chapter.)

The experimenter runs the risk of making either of two errors in any experiment:

- Alpha error (α)—the experimenter accepts the alternative hypothesis (H_a) as being true when the null hypothesis (H_0) is actually true.
- Beta error (β)—the experimenter accepts the null hypothesis as being true when the alternative hypothesis is actually true.

Note that an experimenter cannot make both an alpha error and a beta error in the same experiment. This is because, at the conclusion of the experiment, the experimenter must decide which of the two alternatives to select: either the null hypothesis, or the alternative hypothesis. If the alternative hypothesis is selected, only an alpha error can be made (accepting the alternative hypothesis as true when the null hypothesis is actually true); conversely, if the null hypothesis is selected, only a beta error can be made (accepting the null hypothesis as true when the alternative hypothesis is actually true).

The risk of making either an alpha error or a beta error is expressed as a probability of making that error; e.g., an experimenter, by proper sample-size selection (described below), may establish an alpha-error probability of 0.05 and a beta-error probability of 0.10. This means that the experimenter will have 5 chances in 100 (a 5 percent chance) of making an alpha error or 10 chances in 100 (a 10 percent chance) of making a beta error when the experimental data are used to choose one of the alternative decisions. Expressed another way, the experimenter will have a probability of $(1 - \alpha)(100)$, a 95 percent chance, of obtaining a set of data that will lead to accepting the null hypothesis, if that is the true state of nature; similarly, the experimenter will have a probability of $(1 - \beta)(100)$, a 90 percent chance, of obtaining a set of data that will lead to accepting the alternative hypothesis, if that is the true state of nature.

Selection of probability values for alpha and beta errors in any given experiment is difficult and somewhat subjective. The experimenter must consider the specific objectives of the experiment, the relative criticalness of the decisions, the amount of money and time available, and, among whatever other factors may be important, personal intuition and judgment regarding the particular situation. An example of a typical procedure for selecting alpha and beta values is given later in this chapter.

A probability matrix of the possible choices and errors that can be made in an experiment with a single-sided alternative hypothesis is shown in figure 2-1.

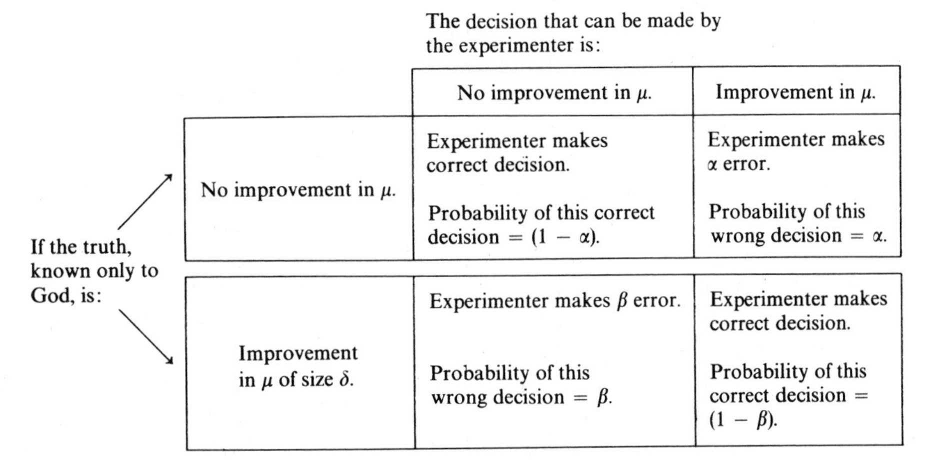

Fig. 2-1. Probability matrix for typical experiment before the experiment is conducted (single-sided alternative hypothesis).

Note that this probability matrix is *applicable only before the experiment is conducted*. (The term delta used in figure 2-1 is explained in the following subsection.)

The probability matrix is applicable only before the experiment is conducted.

Establishing an Objective Criterion

In many experiments, the experimenter will want to determine whether a proposed change to a product or process will produce a desired degree of improvement in a particular property of interest. The minimum acceptable degree of improvement must be determined before the experiment is begun, and this minimum degree of improvement must also be related to an acceptable β risk (the risk of accepting the null hypothesis when, in fact, the alternative hypothesis is true) to minimize the possibility of rejecting the improvement. (The minimum degree of improvement is related to the β risk through selection of the proper sample size, as will be described later in this chapter.)

For example, suppose that an experimenter wants to evaluate a proposed reformulation of a particular alloy. The existing alloy has a population mean tensile strength (μ_0) of 30,000 psi. The experimenter has determined that if the reformulation results in a population mean tensile strength (μ_1) of at least 31,500 psi, the reformulation will be cost effective, provided that the experiment is conducted so that there is only a β risk of not reformulating and, thereby, incurring the costs of not improving the alloy. (The β risk in this instance would, of course, be subjectively determined on the basis of the particular situation confronting the experimenter.)

The difference between μ_0 and the potential value of μ_1 at which the β risk applies is termed delta and is denoted by the Greek letter delta (δ).

In reformulating the alloy, then, the experimenter hopes that the new population mean of the property of interest (μ_1) will be at least equal to ($\mu_0 + \delta$). However, the actual value of μ_1 could fall anywhere: it could be equal to or less than μ_0, in which case the reformulation would not be cost-effective; it could be equal to or greater than ($\mu_0 + \delta$), in which case the reformulation definitely would be cost-effective; or it could fall somewhere between μ_0 and ($\mu_0 + \delta$), in which case the cost-effectiveness would be diminished in the proportion of the proximity of μ_1 to μ_0. If the experimenter could directly measure the population mean of the reformulated alloy, there would be no problem in deciding whether or not to reformulate: the experimenter would simply compare the value of μ_1 with the value of ($\mu_0 + \delta$) and make the appropriate decision. The experimenter cannot, of course, directly measure the population mean but must, instead, make a decision on the basis of the sample mean ($\bar{X}$), which only approximates the population mean.

If $\bar{X}$ is equal to or less than μ_0, it is highly probable that the reformulation did not improve the alloy; in this case, intuition (which could be confirmed by the laws of statistics) would cause the experimenter to accept the null hypothesis. If

$\bar{X}$ is equal to or greater than $(\mu_0 + \delta)$, intuition (which again could be confirmed by the laws of statistics) would cause the experimenter to accept the alternative hypothesis. The experimenter has a problem only when $\bar{X}$ falls between μ_0 and $(\mu_0 + \delta)$.

Statisticians have solved this problem by deriving formulas that can be used to compute a value for an objective criterion against which the experimenter can compare the value for $\bar{X}$ and decide which of the two hypotheses to accept. Two types of cases are considered for computing an objective criterion: the case where only one population is sampled; and the case where two populations are sampled.

One Population Sampled

The one-population-sampled situation arises when an experimenter wishes to compare a population with a fixed number (μ_0). (See the discussion above on stating the alternative decisions.) The following procedure applies if σ^2 is known; the procedure to follow if the variance is unknown is described in Chapter 3.

Three alternative hypotheses are possible, depending upon the experimenter's objective for the experiment:

- $H_a: \mu_1 > \mu_0$
- $H_a: \mu_1 < \mu_0$
- $H_a: \mu_1 \neq \mu_0$

The procedure to follow for each of these cases is described below.

When the alternative hypothesis is $H_a: \mu_1 > \mu_0$, the objective criterion is calculated by the following formula:

$$\bar{X}^* = \mu_0 + \frac{\sigma U_\alpha}{\sqrt{N}},$$

where $\bar{X}^*$ = criterion value for sample mean, μ_0 = population mean of original population, σ = population standard deviation (square root of population variance) of population being studied, U_α = number obtained from table 1 for related α probability,† and N = number of items tested.

> *Note: all criterion values in this book are denoted by asterisks.*

It should be noted that $\bar{X}^*$ can also be computed with the formula

$$\bar{X}^* = (\mu_0 + \delta) - \frac{\sigma U_\beta}{\sqrt{N}},$$

† All tables are contained in Part Five.

where U_β = number obtained from table 1 for related β probability and δ = minimum acceptable degree of improvement at β risk. The formula for computing N, described later, is the simultaneous solution of these two formulas for computing $\bar{X}^*$.

Once the objective criterion has been calculated and the tests have been made, the decision-making procedure is virtually automatic:

- If the sample mean ($\bar{X}$) is equal to or greater than the criterion value ($\bar{X}^*$), the experimenter can accept the alternative hypothesis ($\mu_1 > \mu_0$) with at least $(1 - \alpha)(100)$ percent confidence that the engineering decision is correct.
- If $\bar{X}$ is less than $\bar{X}^*$, the experimenter can accept the null hypothesis ($\mu_1 = \mu_0$) with at least $(1 - \beta)(100)$ percent confidence that the engineering decision is correct. Note that experimental data leading to acceptance of the null hypothesis do not prove that the old and new population means are exactly equal; rather, the data merely indicate that the new population mean, whatever it may be, is not so large as $(\mu_0 + \delta)$, and the right decision is to not change the process.

Figure 2-2 illustrates the situation when $\bar{X}$ is greater than $\bar{X}^*$.

In the situation shown, if $\bar{X}$ were exactly equal to $\bar{X}^*$, there would be exactly a 0.05 probability (5 chances in 100) that the sample mean could have come from a population with a mean of μ_0 and a variance of σ^2. Conversely, there is a 95 percent chance that the sample could not have come from the population with a mean of μ_0 and a variance of σ^2. If $\bar{X}$ is greater than $\bar{X}^*$, as shown, then the odds that the sample did not come from a population with a mean of μ_0 and a variance of σ^2 is even greater than 95 percent.

Thus, the experimenter can conclude, with at least $(1 - \alpha)(100)$ percent confidence, that the true population mean of the sampled population is not μ_0 but some value greater than μ_0. Note carefully that the experimenter cannot conclude that the true mean is $(\mu_0 + \delta)$, only that the true mean is greater than μ_0.

Figure 2-3 illustrates the situation when $\bar{X}$ is less than $\bar{X}^*$.

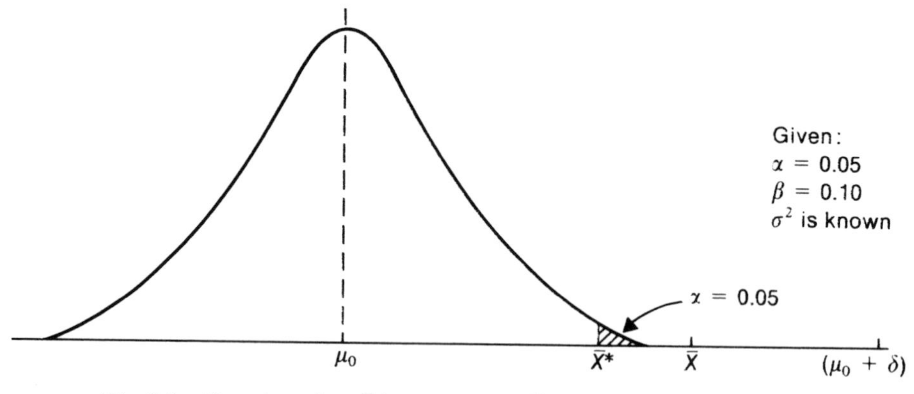

Fig. 2-2. Situation when $\bar{X}$ is greater than $\bar{X}^*$.

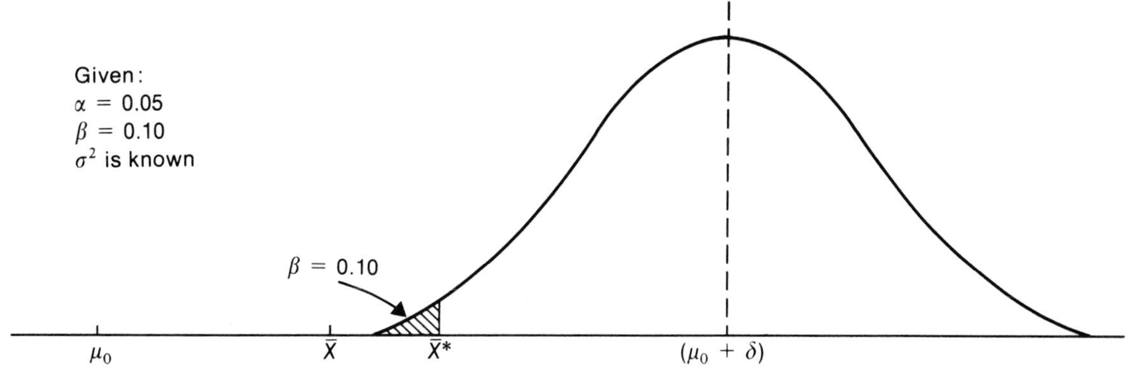

Fig. 2-3. Situation when $\bar{X}$ is less than $\bar{X}^*$.

In the situation shown, if $\bar{X}$ were exactly equal to $\bar{X}^*$, there would be exactly a 0.10 probability (10 chances in 100) that the sample could have come from a population with a mean of $(\mu_0 + \delta)$ and a variance of σ^2. Conversely, there would be a 90 percent chance that the sample could not have come from a population with a mean of $(\mu_0 + \delta)$ and a variance of σ^2. If $\bar{X}$ is less than $\bar{X}^*$, as shown, then the odds that the sample did not come from a population with a mean of $(\mu_0 + \delta)$ and a variance of σ^2 are even greater than 90 percent.

Thus, the experimenter can conclude with at least $(1 - \beta)(100)$ percent confidence that the true population mean of the sampled population is not $(\mu_0 + \delta)$ but some value less than $(\mu_0 + \delta)$. Again, note carefully that the experimenter cannot conclude that the true mean is μ_0, only that the true mean is less than $(\mu_0 + \delta)$.

To summarize: when the alternative hypothesis is $H_a: \mu_1 > \mu_0$, the decision-making process basically consists of choosing between two negative statements:

- When $\bar{X}$ is equal to or greater than $\bar{X}^*$ (figure 2-2), the mean is not μ_0; therefore, accept the alternative hypothesis $H_a: \mu_1 > \mu_0$.
- When $\bar{X}$ is less than $\bar{X}^*$ (figure 2-3), the mean is not $(\mu_0 + \delta)$; therefore, accept the null hypothesis $H_0: \mu_1 = \mu_0$.

When the alternative hypothesis is $H_a: \mu_1 < \mu_0$, the plus sign in the formula for calculating the objective criterion, above, is changed to a minus sign:

$$\bar{X}^* = \mu_0 - \frac{\sigma U_\alpha}{\sqrt{N}}.$$

The terms are defined as above.

Following a logic pattern similar to that used above, the decision-making process for the case when the alternative hypothesis is $H_a: \mu_1 < \mu_0$ becomes the following:

- When $\bar{X}$ is equal to or less than $\bar{X}^*$, the mean is not μ_0; therefore, accept the alternative hypothesis $H_a: \mu_1 < \mu_0$.
- When $\bar{X}$ is greater than $\bar{X}^*$, the mean is not $(\mu_0 - \delta)$; therefore, accept the null hypothesis $H_0: \mu_1 = \mu_0$.

Note that the term $(\mu_0 - \delta)$ is used in the second statement above rather than the term $(\mu_0 + \delta)$ that has previously been used. This is because, when the alternative hypothesis is $H_a{:}\mu_1 < \mu_0$, the experimenter is interested in an important decrease from μ_0 rather than, as in the previous case, an important increase above μ_0.

For the case when the alternative hypothesis is $H_a{:}\mu_1 \neq \mu_0$, the double-sided alternative hypothesis is redefined as two single-sided hypotheses: $H_{a1}{:}\mu_1 > \mu_0$ and $H_{a2}{:}\mu_1 < \mu_0$. Each of these single-sided hypotheses is then treated exactly as described above: an objective criterion is calculated for each hypothesis, $\bar{X}$ from the experiment is compared with each $\bar{X}^*$, and the appropriate decision is made regarding each alternative hypothesis or the null hypothesis.

Two Populations Sampled

For the case where there are two populations, the variances of the populations are known and equal, and the alternative hypothesis is $H_a{:}\mu_A \neq \mu_B$, the objective criterion is calculated by the following formula:

$$|\bar{X}_A - \bar{X}_B|^* = U_\alpha \sigma \sqrt{\frac{1}{N_A} + \frac{1}{N_B}},$$

where $|\bar{X}_A - \bar{X}_B|^*$ = symbol for criterion for the difference between sample means, σ = population standard deviation (square root of variance), U_α = number obtained from table 2 for related α probability, N_A = number of samples tested from population A, and N_B = number of samples tested from population B.

Note carefully that values for U_α are obtained from table 1 for single-sided alternative hypotheses and from table 2 for double-sided alternative hypotheses.

The double-sided alternative hypothesis really consists of two complementary single-sided hypotheses. Thus, instead of being written as $H_a{:}\mu_A \neq \mu_B$, the double-sided alternative hypothesis could be written as

$$H_{a1}{:}\mu_A > \mu_B$$

$$H_{a2}{:}\mu_B > \mu_A.$$

When the double-sided alternative hypothesis is expressed in this fashion, the alpha risk associated with the double-sided hypothesis is divided between the two single-sided hypotheses. Thus, if there is an alpha risk of 0.05 for H_a, there would be an $\alpha/2$ risk, or 0.025, for each of H_{a1} and H_{a2}.

In a similar manner, the formula for calculating an objective criterion for a double-sided alternative hypothesis can be broken down into two separate criteria:

$$(\bar{X}_A - \bar{X}_B)^* = U_{\alpha/2}\sigma \sqrt{\frac{1}{N_A} + \frac{1}{N_B}} \qquad \text{when } \bar{X}_A > \bar{X}_B, \text{ and}$$

$$(\bar{X}_A - \bar{X}_B)^* = -U_{\alpha/2}\sigma \sqrt{\frac{1}{N_A} + \frac{1}{N_B}} \qquad \text{when } \bar{X}_B > \bar{X}_A.$$

In these two formulas, the term $(\bar{X}_A - \bar{X}_B)^*$ is not an absolute value; consequently, when $\bar{X}_B > \bar{X}_A$, the criterion value will be a negative number. Note, too, that the value for $U_{\alpha/2}$ is obtained from table 1 when a double-sided alternative hypothesis is expressed as two single-sided hypotheses.

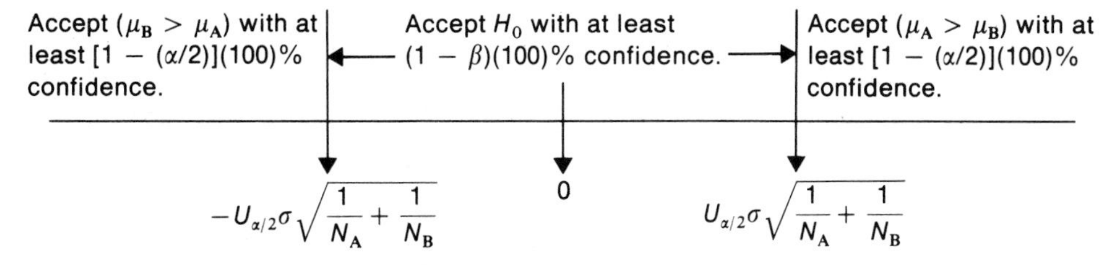

Fig. 2-4. Decision areas available when double-sided alternative hypothesis is expressed as two complementary single-sided hypotheses.

Figure 2-4 illustrates the three decision areas that exist when a double-sided alternative hypothesis is expressed as two single-sided hypotheses with two related criterion values.

Chapter 3 will describe the procedures for obtaining criteria when the variance is unknown or when the variances are unequal.

The objective criterion is used to make a good engineering decision after the experiment is complete.

Computing the Sample Size

Statisticians are frequently asked the question, "How many samples do I need if I want to test a thus-and-so?" Unfortunately, there is no correct sample size that can be determined without additional information. The size of the sample required for a given experiment is influenced by the values selected for alpha and beta risks, by the selection of an important increment of test response (δ), and by the value, or values, of population variance (σ^2).

If the population variance is known for each population of concern in the experiment, calculation of the required number of samples is straightforward; if the variance is not known, calculation of the sample size is somewhat more complex. This section will describe the calculation of sample sizes when population variances are known; Chapter 3 will deal with the calculation of samples sizes when the variances are unknown.

With population variances known, three experimental situations exist, each of which requires a somewhat different approach to calculating sample size:

- $H_0: \mu_1 = \mu_0$; σ_1^2 is known.
- $H_0: \mu_1 = \mu_2$; $\sigma_1^2 = \sigma_2^2$.
- $H_0: \mu_1 = \mu_2$; $\sigma_1^2 \neq \sigma_2^2$.

For each of these situations, the null hypothesis is used as the most convenient descriptor of that situation.

For the first situation ($H_0: \mu_1 = \mu_0$; σ_1^2 is known), the null hypothesis states that the mean of the new population (μ_1) is to be compared against some fixed number (μ_0). By definition, the variance of the population is known. Given this situation, the experimenter calculates the required sample size by the following equation:

$$N = (U_\alpha + U_\beta)^2 \frac{\sigma^2}{\delta^2} \text{ or } N = (U_\alpha + U_\beta)^2 \left(\frac{\sigma}{\delta}\right)^2$$

where N = number of sample items required from the new population; σ^2 = population variance; δ = important engineering increment; U_α = normal distribution number for alpha risk (obtained from table 1 if alternative hypothesis is single-sided, and obtained from table 2 if alternative hypothesis is double-sided); and U_β = normal distribution number for beta risk (obtained from table 1 in all cases).

EXAMPLE

A photoconductor film has been manufactured for some time according to a set of specifications. Based on much test data, the experimenter knows that the mean speed of the photoconductor (μ_0) being produced is 1.1 μJ/in.^{2}(μJ = microjoule). Engineering wishes to improve the speed, i.e., to lower the value of the term μJ/in.^{2}. The experimenter believes that a desirable speed increase might be obtained by making the film thinner, reducing its present 20-mil thickness to an 18-mil thickness. It is anticipated that the decrease in thickness will not change the population variance of the film.

The experimenter knows that the present population mean film speed ($\mu_{20\text{ mils}}$) is 1.1 μJ/in.^{2} and the population variance of the film speed (σ^2) is 0.01. The experimenter can state the purpose of the experiment as:

$$H_0: \mu_{18\text{ mils}} = \mu_{20\text{ mils}} = 1.1\ \mu\text{J/in.}^2$$
$$H_a: \mu_{18\text{ mils}} < 1.1\ \mu\text{J/in.}^2$$

The alternative hypothesis is single-sided since the experimenter is interested only in decreasing the energy requirements, not in proving that the change in thickness makes the film slower, if that is true. The experimenter only wishes to choose between two possibilities: either the film is better or it is not better at an 18-mil thickness.

The experimenter decides that good values for alpha, beta, and delta are $\alpha = 0.05$, $\beta = 0.10$, and $\delta = 0.1$. U_α and U_β are obtained from table 1 by going down the first column (probability) and then across to the number in the second column (U): $U_{0.05} = 1.645$, and $U_{0.10} = 1.282$.

N can now be calculated:

$$N = (U_\alpha + U_\beta)^2 \frac{\sigma^2}{\delta^2}$$

$$N = (1.645 + 1.282)^2 \frac{0.01}{0.01} = 8.567$$

The experimenter has determined that nine sample runs of the photoconductor must be made and tested. Everything in the process is specified exactly the same as it always has been except that the thickness must be controlled at 18 mils instead of 20 mils. Based on the mean sample speed ($\bar{X}_{18\text{ mils}}$) of these nine results, the experimenter will make a decision (using the methods described in Chapter 3) whether or not to permanently change the thickness to 18 mils. In any case, when the experiment is complete, there are at most 5 chances in 100 (α) of making the decision that 18 mils is better than 20 mils if that decision should, in fact, be wrong, and there are at most 10 chances in 100 (β) of making the decision that 18 mils is equivalent to 20 mils if that decision should, in fact, be wrong.

For the second situation ($H_0:\mu_1 = \mu_2;\ \sigma_1^2 = \sigma_2^2$), the null hypothesis states that neither population mean is known; however, the two population variances are known (by definition) and are equal. Given this second situation, the experimenter calculates the required sample size for each population by the following equation:

$$N_A = N_B = 2(U_\alpha + U_\beta)^2 \frac{\sigma^2}{\delta^2} \text{ or } 2(U_\alpha + U_\beta)^2 \left(\frac{\sigma}{\delta}\right)^2$$

where N = sample size that must be tested from each population, and all other terms are defined as in the previous equation. Note carefully that the factor 2 is in this formula because there are two populations. Thus, if identical products from two different vendors are to be tested, and if the values for α, β, and δ are the same as in the foregoing example, the experiment would be set up as follows:

$$H_0:\mu_1 = \mu_2$$
$$H_a:\mu_1 \neq \mu_2$$

By definition, $\alpha = 0.05$, $\beta = 0.10$, $\delta = 0.1$, and $\sigma_1^2 = \sigma_2^2 = 0.01$. From table 2, $U_\alpha = U_{0.05} = 1.960$. (Table 2 is used because H_a is double-sided.) From table 1, $U_\beta = U_{0.10} = 1.282$. (Remember, table 1 is always used to obtain values for U_β.) The experimenter then calculates N:

$$N_A = N_B = 2(1.960 + 1.282)^2 \frac{0.01}{0.01} = 21.02$$

The experimenter must obtain and test a sample of 21 parts from vendor A and 21 parts from vendor B.

For the third situation ($H_0:\mu_1 = \mu_2;\ \sigma_1^2 \neq \sigma_2^2$), the null hypothesis states that two populations are being compared and that the two population variances, although known, are not equal. In this situation, the experimenter must compute two sample sizes:

$$N_1 = (U_\alpha + U_\beta)^2 \frac{\sigma_1(\sigma_1 + \sigma_2)}{\delta^2}, \qquad \text{and}$$

$$N_2 = (U_\alpha + U_\beta)^2 \frac{\sigma_2(\sigma_1 + \sigma_2)}{\delta^2},$$

where N_1 = sample size from first population that must be tested, N_2 = sample size from second population that must be tested, σ_1 = standard deviation of first population, and σ_2 = standard deviation of second population. All other terms are defined as before.

Importance of *N*

The formulas for calculating N are simple and easy to use, and, more importantly, they are the primary criteria for the validity of experiments. If, for a given experiment, the values for α, β, and δ are carefully chosen, and if σ^2 is accurately known, then the computed value for N is the only correct sample size. With even one sample less than N, the experiment would be invalid, since either α or β would then be greater than the α or β that was specified as absolutely necessary. On the other hand, if even one sample more than N is tested, the experimenter would waste time and money, since either α or β would then be smaller than the minimum α or β that was specified as necessary.

An experiment would be suspect if an experimenter properly calculated an N of, say, 20 samples but then decided that this sample size was excessive for the time or money available and proceeded to test a sample of 10 parts instead of the 20 required. Nevertheless, it is always necessary to be realistic, and sometimes it will be necessary to choose an N based on time available, or money available, or parts available, rather than on a calculation from the proper formula. In this type of situation, it is recommended that the efficiency of the experiment be evaluated before it is started. If the α error is critical, the way to determine the efficiency of a proposed experiment is to compute the value of the β error that would be obtained by using the correct α and δ and the proposed sample size; i.e., the experimenter would predetermine the risk of missing an improvement of size δ if a substandard sample size should be used. On the other hand, if the β error is critical, the efficiency of a proposed experiment is determined by computing the value of α with the proposed sample size and the correct β and δ.

EXAMPLE

If the correct N is 20, the desired α is 0.05, $\sigma_1^2 = \sigma_2^2 = 0.01$, and the desired δ is 0.1, and if the experimenter is considering using only 10 samples, the resultant probability of a beta error is computed by the following technique:

$$H_0: \mu_1 = \mu_2$$
$$H_a: \mu_1 \neq \mu_2$$

By definition, $\alpha = 0.05$, $\delta = 0.1$, $\sigma_1^2 = \sigma_2^2 = 0.01$, and $N = 10$. From table 2, $U_\alpha = U_{0.05} = 1.96$. Thus,

$$N = 2(U_\alpha + U_\beta)^2 \frac{\sigma^2}{\delta^2}$$

$$10 = 2(1.96 + U_\beta)^2 \frac{0.01}{0.01}$$

$$\frac{10}{2} = (1.96 + U_\beta)^2$$

$$\sqrt{\frac{10}{2}} = 1.96 + U_\beta$$

$$U_\beta = \sqrt{5} - 1.96 = 2.24 - 1.96 = 0.28$$

From table 1, $\beta = 0.39$.

Unless the experimenter is willing to accept a 39 percent chance of missing an important improvement of 0.1 (δ), the experiment could be a waste of time and money with a sample size of 10.

An identical procedure is used for computing the probability of an α error when β is critical.

Each formula for N is also a connecting link between statistics and engineering in an experiment. The size of N is a function of U_α and U_β, which reflect the laws of statistics and the normal distribution; of δ, which reflects the engineering requirements of the experiment; and of σ^2, which reflects the ability of the experimenter to control the variability of the experiment (e.g., experimental techniques, measuring equipment, etc.). Since U_α and U_β are in the numerator of each formula, the size of the experiment must increase as the desired risk of error becomes smaller. Likewise, any decrease in variance (e.g., using better measuring devices) causes a decrease in the size of an experiment. Finally, the smaller the size of an effect that is to be detected, the greater the sample size required.

The following example gives a clearer picture of the effect of sample size on the decision-making process.

EXAMPLE

Suppose that there are several bins, each filled with parts produced by one of two different processes. Population A (the parts produced by one process) is normally distributed and is described by the parameters $\mu_A = 100$ and $\sigma_A^2 = 100$. Population B (the parts produced by the other process) is also normally distributed and is described by the parameters $\mu_B = 105$ and $\sigma_B^2 = 81$. The normal distributions of these two populations are depicted in figure 2-5.

Unfortunately, one of the bins is not labeled as to which process produced the parts. The experimenter must determine, by sampling the bin and testing the parts, which process produced the parts in the unmarked bin.

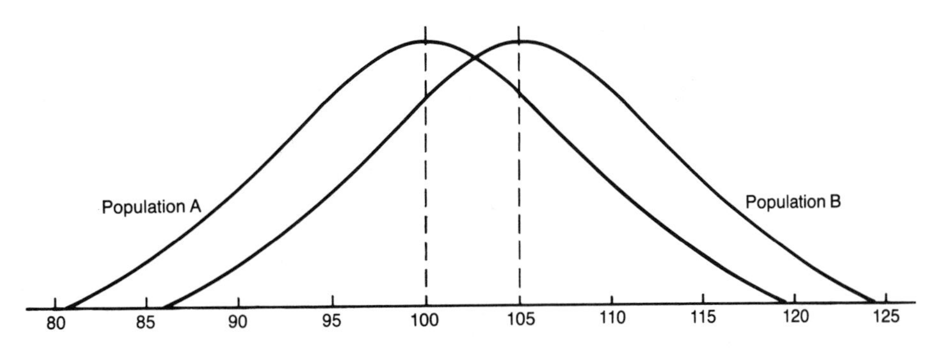

Fig. 2-5. Normal distributions of populations A and B.

If the experimenter tests a single item, obtaining a result of 105, little more than guesswork would be involved if that sample (and the bin) was assigned to either population. From figure 2-5, it is apparent that a result of 105 is not an unexpected result for either population.

The experimenter can calculate the distribution of $\bar{X}$ and the interval within which $\bar{X}$ will fall for each population for any given sample size and any given α risk by the formula

$$\mu \pm \frac{U_{\alpha}\sigma}{\sqrt{N}}.$$

If an α risk of 0.05 is assigned to the experiment, then, with a sample of only one item, computations from the above formula would lead one to expect any result between 80.4 and 119.6 (100 ± 19.6) if the part came from population A and any result between 87.4 and 122.6 (105 ± 17.6) if the part came from population B.

If the experimenter tests a sample of two parts from the bin, obtaining a mean result of 105, it would still not be possible to make a decision with α risk, since the expected intervals for $\bar{X}$ would be 100 ± 13.9 for population A and 105 ± 12.5 for population B.

The distribution of $\bar{X}$ with a sample size of two if $\mu_A = 100$ and $\mu_B = 105$ can be drawn as shown in figure 2-6.

If the experimenter tested a sample of 16 parts from the bin, $\bar{X}$ would be in the interval of 100 ± 4.9 for population A and 105 ± 4.41 for population B. The distribution of $\bar{X}$ for a sample of 16 parts can be drawn as shown in figure 2-7. With a sample mean ($\bar{X}$) result of 105, there would be less than a 2.5 percent chance that the sample of 16 units came from population A. Therefore, it is a good bet that the sample did not come from population A. The only alternative is that it came from population B.

Note carefully that the decision can be made with at least 97.5 percent confidence that the bin was not produced by process A.

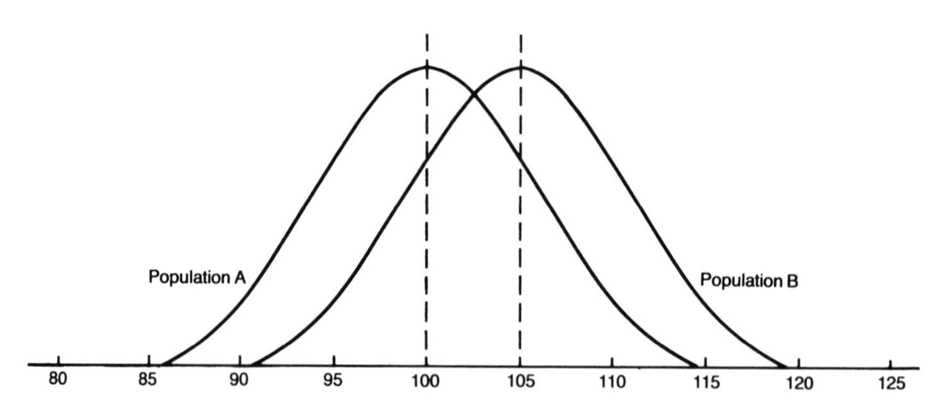

Fig. 2-6. Distribution of $\bar{X}$ for a sample size of two.

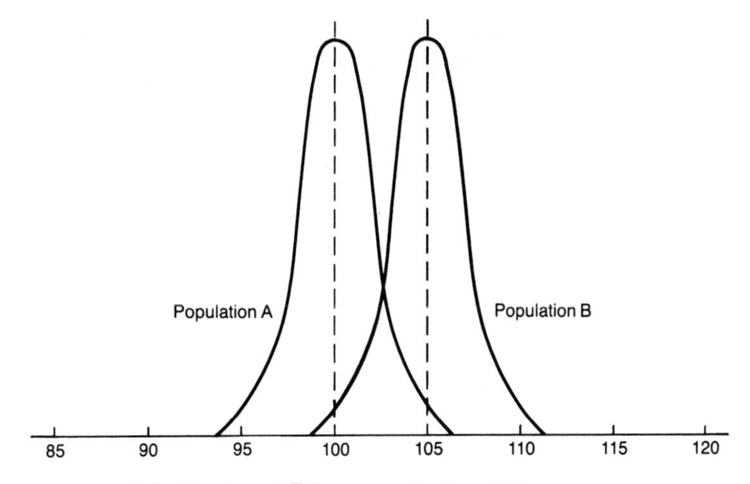

Fig. 2-7. Distribution of $\bar{X}$ for a sample size of 16.

The correct sample size assures the experimenter that the risks of error will be equal to or less than α and β when the experiment is completed.

Choosing α, β, and δ

If an experiment is designed so that the probability of making an α error is 0.50 and the probability of making a β error is also 0.50, the experimenter obviously could do just as well by tossing a coin; there is a 50 percent chance of being right just by tossing a coin, and it is certainly inexpensive. In a like manner, it is equally useless to try to design an experiment with zero risk of making either an α or a β error; such an experiment would require testing the entire population. The correct experiment design is one that will reduce the possibilities of α and β errors to some levels that are both technically acceptable and economically feasible.

It is impossible to give a general procedure for choosing values for α, β, and δ that would be applicable to every possible project. The only truly correct values for α, β, and δ are those arrived at as the product of the experimenter's thought process for a specific situation. This type of thought process is illustrated in the following example.

EXAMPLE

A production engineer is responsible for a chrome-plating operation. One of the steps in the process is the application of a copper-flash undercoating followed by a rinse. The specification for the rinse operation is that it shall be 6 to 8 minutes long. The specification on the final plating is that the adhesion of the plating shall be greater than 72 pounds when measured on a special testing device. Ten parts are tested every day.

The process has been in operation for several months. The population mean adhesion of the plating is 75 pounds, and the standard deviation is 3 pounds; the data appear to be normally distributed. Obviously, a substantial number of parts are below specification, and there have been a significant number of failures of the part in the field.

The process mean rinse time is 7 minutes. It is thought that the cause of adhesion failures is a too-short rinse time, and an experiment is proposed to determine if increasing the rinse time will increase the mean adhesion of the plating.

The first step in designing the experiment is to define the experimental space. Suppose that the entire variation in the plating adhesion is due to variation in rinse time. Obviously, this is not strictly true, since some of the variation is due to testing variations and to lack of other precise controls; however, it is a good starting assumption. To date, 95 percent of the parts produced have adhesion values of $\mu_0 \pm 2\sigma$ (i.e., 75 ± 6 pounds); assume that the best parts (adhesion value of 81 pounds) are produced by a rinse time of 8 minutes and that the worst parts (adhesion value of 69 pounds) are produced by a rinse time of 6 minutes. The anticipated experimental situation can then be plotted as shown in figure 2-8.

Note that the experimenter has proposed a linear relationship between adhesion and rinse time. It might be that God's law is a curved line; even if true, the experimenter should encounter no difficulty if the relationship is monotonic.

If the mean rinse time were to be increased to 8 minutes, with the same tolerance being maintained on the rinse time, the experimenter could expect a mean plating adhesion of 81 pounds with a tolerance of ± 6 pounds (a range of 75 to 87 pounds) on the resultant parts.

The next step is to determine the cost of increasing the rinse time and the profits that would result if the plating adhesion should be increased. For example, suppose that each minute added to the processing time would cost \$1500 per month and that every 1-pound increase in plating adhesion would increase profits

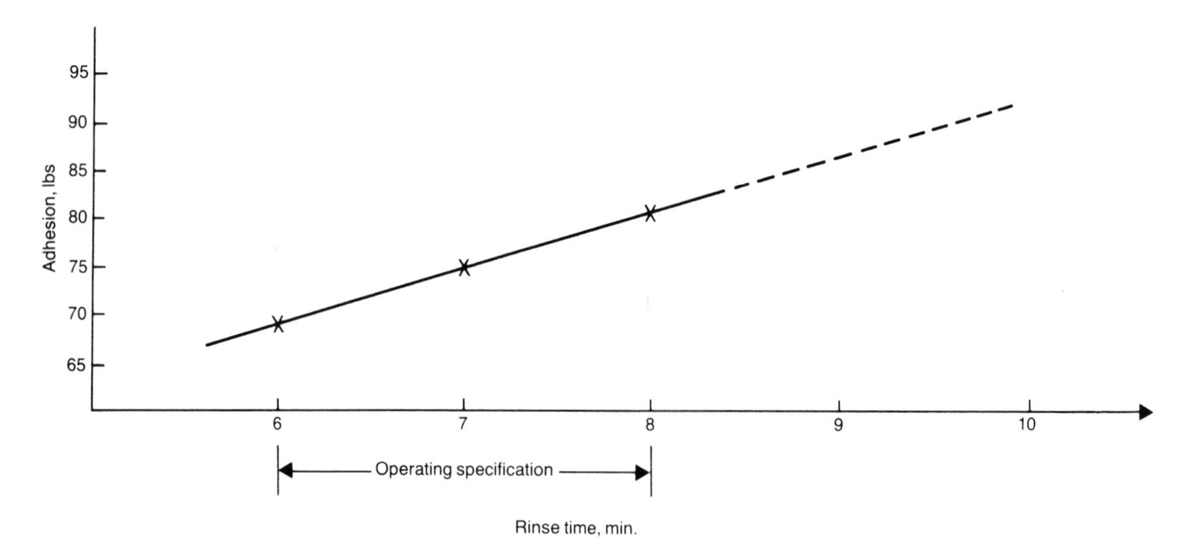

Fig. 2-8. Anticipated experimental situation.

by \$1000 per month. If the experimenter decides to increase the mean rinse time to 8 minutes, from 7 minutes, the increase in processing costs will be \$1500 per month. The increased gross and net profits that would result for various increments in plating adhesion due to the 1-minute increase in rinse time are shown below:

Increase in adhesion, lbs.	Gross profits, \$/mo	Cost of increased rinse time, \$/mo	Net profit, \$/mo
0	0	1500	(1500)
1	1000	1500	(500)
1.5	1500	1500	0
2	2000	1500	500
3	3000	1500	1500
4	4000	1500	2500
5	5000	1500	3500
6	6000	1500	4500

Note: Brackets ( ) indicate net loss.

The experiment will be conducted on the production line; the cost of the experiment will therefore be considered as negligible, since all the parts produced (except those tested) will be sold.

Next, the experiment is described in terms of the null and alternative hypotheses.

$$H_0: \mu_1 = \mu_0 = 75 \text{ pounds}$$
$$H_a: \mu_1 > \mu_0$$

A single-sided alternative hypothesis is used because the experimenter is interested in the increased rinse time only if it increases the plating adhesion.

If the adhesion strength should be increased exactly 1.5 pounds, it would not make any difference, economically, whether the rinse time was changed or not. Therefore, $\mu_0 + 1.5 = 76.5$ pounds is a good tentative value of $\bar{X}^*$. With $\bar{X}^* = 76.5$, any sample mean ($\bar{X}$) less than 76.5 pounds would cause rejection of the increased rinse time; any value of $\bar{X}$ equal to or greater than 76.5 pounds would lead to acceptance of the increased rinse time. If the true population mean (μ_1) is exactly 76.5 pounds, there is a 50 percent chance of accepting H_0 and a 50 percent chance of accepting H_a. This tentative value of $\bar{X}^*$ may be changed slightly, depending upon the final values selected for α and β.

With reference again to the economic data, the risk of accepting the alternative hypothesis if $\mu_1 = \mu_0$ (75 pounds) should be set equal to the risk of accepting the null hypothesis at $\mu_1 = \mu_0 + 3$ (78 pounds), since the economic losses are equal at these two values. In the first case, acceptance of the alternative hypothesis would lead to a loss of \$1500 per month; in the second case, acceptance of the null hypothesis would mean forgoing a potential profit of \$1500 per month. Expressed another way, the alpha risk if the population mean is 75 pounds should equal the beta risk if the population mean is 78 pounds.

If $\mu_0 + 3$ is chosen, as above, for the value of adhesion at which β will apply, the value of δ is then established as 3 pounds if the loss of \$1500 per month is important to the economics of the process.

		Experimental decision based on $\bar{X}^* = 76.5$ lbs	
		No improvement	Improvement
Truth	No improvement	No gain, No loss.	Loss of \$1500/month.
	Improvement of 3 lbs	Potential profit of \$1500/month not realized.	Gain of \$1500/month.

Fig. 2-9. Risk matrix for plating experiment.

In summary, the experimenter has now defined the experiment in the following manner:

$$H_0: \mu_{8\text{ min}} = \mu_0 = 75 \text{ pounds}$$
$$H_a: \mu_{8\text{ min}} > \mu_0$$
$$\bar{X}^* = 76.5 \text{ pounds} \quad \text{(tentative)}$$
$$\sigma = 3 \text{ pounds}$$
$$\alpha \text{ at 75 pounds} = \beta \text{ at 78 pounds}$$
$$\delta = 3 \text{ pounds}$$

The risk matrix for the experiment is shown in figure 2-9.

The experimenter must now choose specific values for α and β. Four different situations will be considered, and the influence of each situation on selecting values for α and β, and the resultant influence on values for N and $\bar{X}^*$, will be described.

In the first situation, a potential gain or loss of \$1500 per month is important but not vital in comparison to the total cost of the process. The experimenter has good reason to believe that extending the rinse cycle by 1 minute will improve the mean coating adhesion to about 81 pounds. In this situation, the experimenter would be justified in selecting fairly high risk factors for both α and β, say 0.10 for each. With $\alpha = 0.10$, $\beta = 0.10$, $\delta = 3$, and $\sigma = 3$, the experimenter calculates that $N = 7$ and $\bar{X}^* = 76.5$ pounds. Note that, if the improvement in adhesion should actually be 81 pounds, the probability of accepting H_a is virtually a certainty (>99.9 percent).

In the second situation, the loss of \$1500 per month if H_0 is true is relatively minor; however, the experimenter suspects that the new value for coating adhesion may be closer to 78 pounds than to 81 pounds, and the potential gain of \$1500 per month is important if H_a is true. In this situation, the experimenter will want greater assurance than before that a beta error (accepting the null hypothesis as true when the alternative hypothesis is actually true) is not made; thus, β may be set at 0.05, or even 0.01, while keeping α at 0.10. For $\beta = 0.05$, the experimenter calculates that $N = 9$ and $\bar{X}^* = 76.3$ pounds; for $\beta = 0.01$, $N = 13$ and $\bar{X}^* =$ 76.1 pounds.

In the third situation, the experimenter suspects that the new adhesion strength will be substantially above 78 pounds, but the economics of the process are such that a loss of \$1500 per month would be intolerable if the true adhesion strength is actually 75 pounds. In this situation, an alpha error (accepting the alternative hypothesis as true when the null hypothesis is actually true) would be more critical than a beta error. Thus, the experimenter probably would want to lower the value of α to 0.05, or even to 0.01, while retaining β at 0.10. For $\alpha = 0.05$, the experimenter calculates $N = 9$ and $\bar{X}^* = 76.7$ pounds; for $\alpha = 0.01$, $N = 13$ and $\bar{X}^* = 76.9$ pounds.

The fourth situation would arise when either a gain or a loss of $1500 per month would be a major consideration in the total process cost. In this situation, the experimenter would pick relatively low values for both α and β. If a value of 0.05 is selected for both α and β (a reasonable value), the experimenter would then calculate $N = 11$ and $\bar{X}^* = 76.5$. In this situation, as in the first situation, the calculated value for $\bar{X}^*$ is the same as the tentative value assigned to $\bar{X}^*$ earlier in the design of this experiment.

The above example is strictly one person's view of the problem, using only the information given. Certainly, if additional information were available, the entire view of the problem and the choice or computation of α, β, δ, N, and $\bar{X}^*$ could be different. For example, if the failure of the part was causing a safety hazard to the user of the machine, the economics would become a minor consideration, and α might be set at 0.20 and β might be set at 0.01, or even 0.001, based on the potential cost of lawsuits that might result from failure to improve the part.

It is also possible that another individual, given the same problem and the same information as in the example, would come up with an entirely different null and alternative hypothesis and would select entirely different values for α, β, and δ. Under certain circumstances, one might choose to test $H_0: \mu_1 = 76.5$ and $H_a: \mu_1 > 76.5$. With these hypotheses, one would have little chance of losing $1500 per month because of an α error. However, with a sample size of 11 and an α of 0.05, the criterion for accepting H_a would be 78.3 pounds. Thus, there is a probability of approximately 0.50 of missing an improvement of 3 pounds and a potential gain of $1500 per month. In fact, there is even a high probability of missing a gain of $2500, or even $3500, per month with this plan.

Operating Characteristic Curves

An operating characteristic curve is a useful visual aid that completely describes an experiment. When an experimenter chooses values for α, β, and δ, knows μ_0 and σ_0^2, and calculates N, the outcome probabilities of the experiment have been completely defined, regardless of what the true value may be for μ_1, the mean of the population under study. By constructing an operating characteristic curve, the experimenter can predict the probability of making a good decision for every possible value of μ_1 if the experiment is conducted.

The initial construction of an operating characteristic curve is illustrated in figure 2-10.

Two points on the curve can be located immediately, given the information

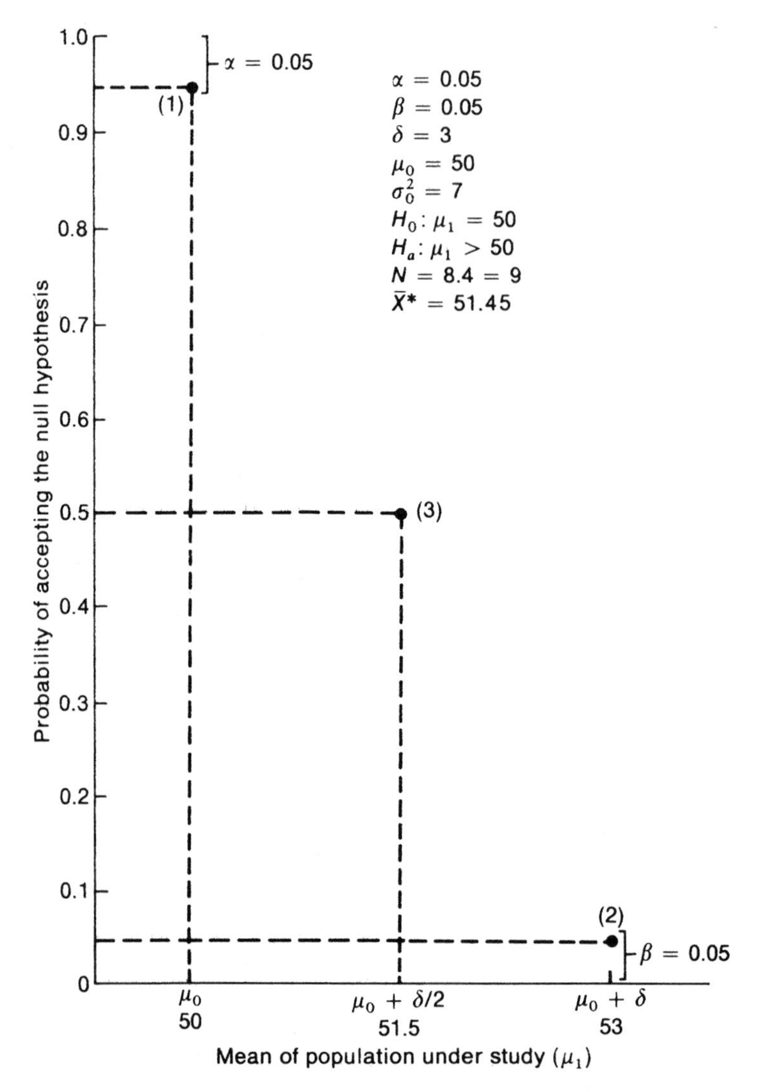

Fig. 2-10. Initial construction of an operating characteristic curve.

shown in figure 2-10. With $\alpha = 0.05$, there are only 5 chances in 100 that the experimenter will reject the null hypothesis if, in fact, $\mu_1 = 50$. Stated another way, there are 95 chances in 100 that the experimenter will accept the null hypothesis if $\mu_1 = 50$. Therefore, the first point on the curve can be plotted at the intersection of $\mu_1 = 50$ and $p = 0.95$.

The value of β chosen by the experimenter applies at only one value for μ_1; i.e., when $\mu_1 = \mu_0 + \delta$. In this example, $\beta = 0.05$, $\delta = 3$, and $\mu_0 = 50$; thus, if $\mu_1 = \mu_0 + \delta = 53$, there is only β probability (0.05) that the experimenter will accept the null hypothesis if, in fact, $\mu_1 = 53$. Therefore, a second point can be plotted on the curve at the intersection of $(\mu_0 + \delta) = 53$ and $p = 0.05$.

When $\alpha = \beta$, as is the case shown in figure 2-10, a third point can be plotted without difficulty because, at $\mu_0 + \delta/2$, the probability of accepting the null hypothesis will always be approximately 0.50. Other points on the plot can be

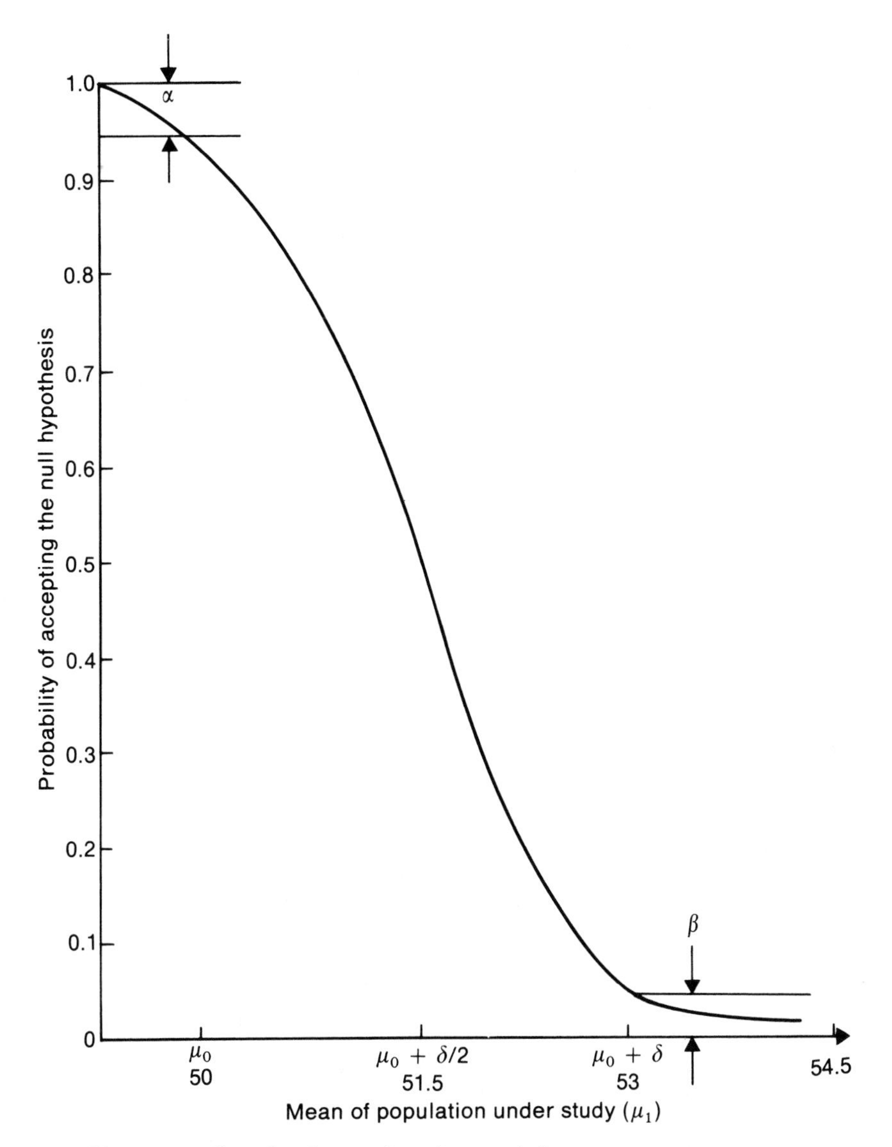

Fig. 2-11. Completed operating characteristic curve.

obtained by rearranging the formula for N [in this case, $N = (U_\alpha + U_\beta)^2(\sigma^2/\delta^2)$] to $U_\beta = (\sqrt{N}\delta/\sigma) - U_\alpha$ and by calculating the value of U_β for various values of δ. By referring to table 1, the value of β for various deltas can then be obtained and plotted on the chart.

The completed operating characteristic curve is shown in figure 2-11. According to the curve shown in figure 2-11, if μ_1 were actually 50, and an experimenter conducted an experiment with $N = 9$ trials at the given set of operating conditions, then 95 times out of 100 data ($\bar{X}$) would be obtained such that the experimenter would conclude that μ_1 is not larger than 50.

If μ_1 were actually 51.5 ($\mu_0 + \delta/2$), and the experimenter conducted $N = 9$ trials under the specified conditions, then 50 percent of the time data ($\bar{X}$) would be obtained that would lead to a decision that $\mu_1 = 50$, and 50 percent of the time data would be obtained that would lead to a decision that $\mu_1 > 50$. Note that the population mean value of 51.5, which leads to a 50 percent probability of either accepting or rejecting the null hypothesis, is almost precisely the value of the objective criterion ($\overline{X}^* = 51.5$).

If μ_1 were actually 54.5, and the experimenter conducted $N = 9$ trials under the specified conditions, then data would almost certainly be obtained that would cause rejection of the null hypothesis. The probability of accepting the null hypothesis would be considerably less than 0.05 in this case.

The operating characteristic curve makes precise what most people know intuitively: the bigger the effect of a variable on some response, the easier it is to make a correct decision that the variable is important; the smaller the effect of a variable, the easier it is to make the correct decision that the variable is not important. On the other hand, if the effect of a variable on the response is halfway between an important difference and essentially no difference, it matters little which decision is made.

> *The operating characteristic curve defines the probable outcome of an experiment for every possible value of μ.*

Power Curves

The inverse of an operating characteristic curve is called a power curve. Some experimenters will find this curve more to their liking than the operating

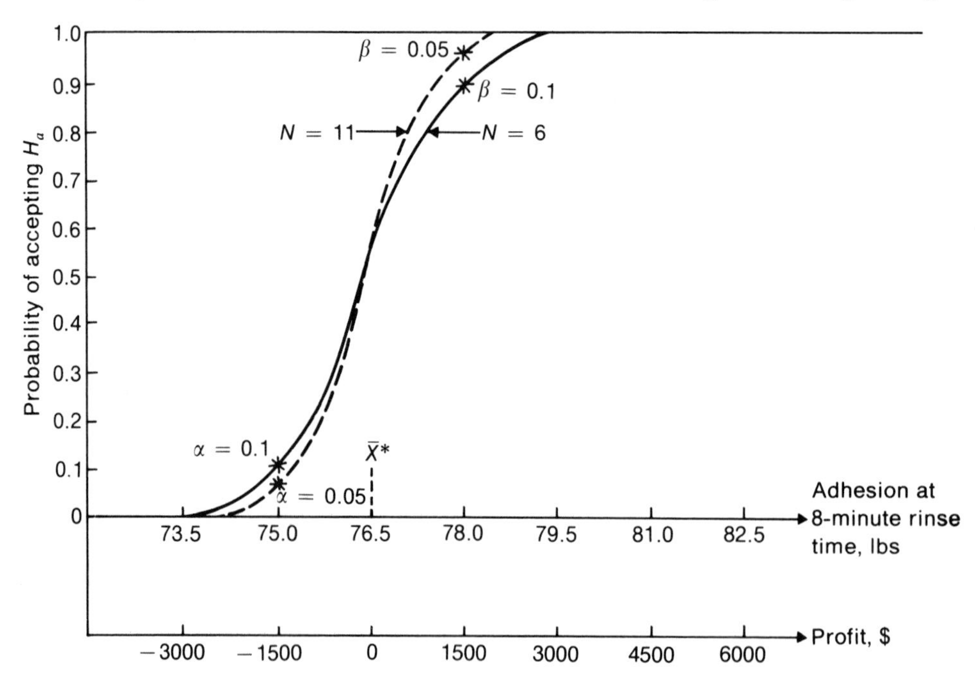

Fig. 2-12. Power curves for plating experiment.

characteristic curve, even though exactly the same information is contained in both curves. For the power curve, various values of δ and given values for α and N are used to calculate the probability of accepting H_a. The following formula is used to calculate U_β:

$$U_\beta = \frac{\sqrt{N}\delta}{\sigma} - U_\alpha.$$

The value of β is obtained from table 1, and the probability of accepting H_a is obtained by subtracting this value of β from 1. If, for a given δ, the value of $\beta = 0.17$, then the probability of accepting H_a is $1.0 - 0.17 = 0.83$.

Figure 2-12 is the power curve for two of the cases described in the plating experiment described earlier:

- $N = 6, \ \alpha = 0.1, \ \beta = 0.1, \ \delta = 3.$
- $N = 11, \alpha = 0.05, \beta = 0.05, \delta = 3.$

EXERCISES

1. Given: $\alpha = 0.10$, $\beta = 0.10$, $\mu_0 = 30$, $\delta = 3$, $\sigma^2 = 9$, $H_0{:}\mu_1 = 30$, and $H_a{:}\mu_1 > 30$. Calculate the sample size and construct an operating characteristic curve between $\mu_1 = 25$ and $\mu_1 = 55$ for the calculated sample size. Calculate the objective criterion for making a decision.

2. The following results are obtained in an experiment:

Experimental conditions	Result (pounds)
5 minutes, 200°	85
10 minutes, 200°	95
5 minutes, 300°	120
10 minutes, 300°	75

   Plot the results. Is there an interaction of time and temperature?

3. It is known that vendor A is supplying a plastic gear that has a normally distributed impact strength with a population mean of 280 foot-pounds and a standard deviation of 6. Vendor B is supplying the same part with a population mean of 290 foot-pounds and a standard deviation of 10, also normally distributed. What percentage of each vendor's parts will be stronger than 272 foot-pounds?

4. Which of the following statements are true?

   a. $H_0{:}\mu_A = \mu_0$ is an alternative hypothesis about a fixed number.
   b. $H_a{:}\mu_A \neq \mu_0$ is a double-sided alternative hypothesis about a population mean as compared to a fixed number.
   c. An experimenter commits an α error when the alternative hypothesis is accepted but H_0 is actually true.
   d. If the alternative hypothesis is accepted after the experiment is completed, and if the experiment was double-sided, the confidence statement is made with $(1 - \alpha/2)(100)$ percent confidence.
   e. σ is the symbol for sample population variance.
   f. If a sample mean $\overline{X}$ is greater than criterion $\overline{X}^*$, we conclude that $\mu > \mu_0$.
   g. If a sample mean $\overline{X}$ is greater than the criterion $\overline{X}^*$, we conclude that $\mu = \mu_0 + \delta$.
   h. If the null hypothesis is $H_0{:}\mu_A = \mu_B$, the experimenter has two populations.

j. If the sample size is correctly calculated to be 10, a better experiment will result if a sample size of 20 is used.

REFERENCES

Bowker, A. H. and Lieberman, G. L. 1972. *Engineering statistics*. 2nd ed. Englewood Cliffs, N.J.: Prentice-Hall, Inc.

Guenther, W. C. 1973. *Concepts of statistical inference*. 2nd ed. New York: McGraw-Hill Book Co.

Ott, L. 1977. *An introduction to statistical methods and data analysis*. North Scituate, Mass.: Duxbury Press.

3 Simple Comparative Experiments: Decisions About Population Means

A simple comparative experiment is a set of trials conducted to determine if changing a single variable from one condition to another, while holding all other potential variables constant, has any effect on a response or responses of interest to the experimenter.

Simple comparative experiments are appropriate for engineering situations such as the following:

- The plating on a part is evidently not adequate because 20 percent of the parts are failing in the field. It is proposed that the toughness of the plating be increased by adding a new ingredient to the plating bath that should increase the abrasion resistance of the parts. A good simple comparative experiment will answer the engineering question: Does the additive increase the abrasion resistance of the population of parts that will be produced if the additive is used in the process?
- A certain device or machine is performing satisfactorily in all respects except one. It is proposed that this one problem can be corrected by replacing an existing spring with one having more compression. A good simple comparative experiment can answer the questions: Does the spring with greater compression solve the problem? Does the spring with greater compression create any new problems?
- An electronic circuit in a developmental machine is unstable. It is proposed that the stability will be enhanced if a certain resistor is changed from 5 ohms to 15 ohms. A good simple comparative experiment can answer the question: Is a 15-ohm resistor better than a 5-ohm resistor for stabilizing the current?

In this chapter, experiment designs will be presented for simple comparative experiments in which the experimenter is interested in determining information about population means. The following chapter will present similar designs for simple comparative experiments in which the experimenter is interested in population variances.

		$H_0{:}\mu_A = \mu_B$	
	$H_0{:}\mu_A = \mu_0$	$\sigma_A^2 = \sigma_B^2$	$\sigma_A^2 \neq \sigma_B^2$
σ^2 Known	Case 1 $N = (U_\alpha + U_\beta)^2 \frac{\sigma^2}{\delta^2}$ $\bar{X}^* = \mu_0 + \frac{U_\alpha \sigma}{\sqrt{N}}$	Case 2 $N_A = N_B = 2(U_\alpha + U_\beta)^2 \frac{\sigma^2}{\delta^2}$ $\lvert\bar{X}_A - \bar{X}_B\rvert^* =$ $U_\alpha \sigma \sqrt{\frac{1}{N_A} + \frac{1}{N_B}}$	Case 3 $N_A = (U_\alpha + U_\beta)^2 \frac{\sigma_A (\sigma_A + \sigma_B)}{\delta^2}$ $N_B = (U_\alpha + U_\beta)^2 \frac{\sigma_B (\sigma_A + \sigma_B)}{\delta^2}$ $\lvert\bar{X}_A - \bar{X}_B\rvert^* = U_\alpha \sqrt{\frac{\sigma_A^2}{N_A^2} + \frac{\sigma_B^2}{N_B}}$
σ^2 and δ^2 Unknown	Case 4 $N = (U_\alpha + U_\beta)^2 \frac{\sigma^2}{\delta^2}$ $\phi = N - 1$ $N_t = (t_\alpha + t_\beta)^2 \frac{\sigma^2}{\delta^2}$ $\bar{X}^* = \mu_0 + \frac{t_\alpha S}{\sqrt{N}}$	Case 5 $N_A = N_B = 2(U_\beta + U_\beta)^2 \frac{\sigma^2}{\delta^2}$ $\phi = N_A + N_B - 2$ $N_{At} = N_{Bt} = 2(t_\alpha + t_\beta)^2 \frac{\sigma^2}{\delta^2}$ $\lvert\bar{X}_A - \bar{X}_B\rvert^* =$ $t_\alpha S \sqrt{\frac{1}{N_A} + \frac{1}{N_B}}$	Case 6 $N_A = (U_\alpha + U_\beta)^2 \frac{\sigma_A (\sigma_A + \sigma_B)}{\delta^2}$ $N_{At} = (t_\alpha + t_\beta)^2 \frac{\sigma_A (\sigma_A + \sigma_B)}{\delta^2}$ $\phi = N_A + N_B - 2$ $\lvert\bar{X}_A - \bar{X}_B\rvert^* = t_\alpha \sqrt{\frac{S_A^2}{N_A} + \frac{S_B^2}{N_B}}$
S^2 Known	Case 7 $N = (t_\alpha + t_\beta)^2 \frac{S^2}{\delta^2}$ $\bar{X}^* = \mu_0 + \frac{t_\alpha S}{\sqrt{N}}$	Case 8 $N_A = N_B = 2(t_\alpha + t_\beta)^2 \frac{S^2}{\delta^2}$ $\lvert\bar{X}_A - \bar{X}_B\rvert^*$ $= t_\alpha S \sqrt{\frac{1}{N_A} + \frac{1}{N_B}}$	Case 9 $N_A = (t_\alpha + t_\beta)^2 \frac{S_A(S_A + S_B)}{\delta^2}$ $N_B = (t_\alpha + t_\beta)^2 \frac{S_B(S_A + S_B)}{\delta^2}$ $\lvert\bar{X}_A - \bar{X}_B\rvert^* = t_\alpha \sqrt{\frac{S_A^2}{N_A} + \frac{S_B^2}{N_B}}$

Fig. 3-1. Summary of cases for simple comparative experiments related to population means.

There are nine possible problem types concerning population means that are of interest to the experimenter:

- Case 1. $H_0{:}\mu_1 = \mu_0$; variance (σ_1^2) is known.
- Case 2. $H_0{:}\mu_1 = \mu_2$; $\sigma_1^2 = \sigma_2^2$ and both are known.
- Case 3. $H_0{:}\mu_1 = \mu_2$; $\sigma_1^2 \neq \sigma_2^2$ and both are known.
- Case 4. $H_0{:}\mu_1 = \mu_0$; σ_1^2 is unknown.
- Case 5. $H_0{:}\mu_1 = \mu_2$; $\sigma_1^2 = \sigma_2^2$ but both σ_1^2 and σ_2^2 are unknown.
- Case 6. $H_0{:}\mu_1 = \mu_2$; $\sigma_1^2 \neq \sigma_2^2$ and both σ_1^2 and σ_2^2 are unknown.
- Case 7. $H_0{:}\mu_1 = \mu_0$; S^2 known.
- Case 8. $H_0{:}\mu_1 = \mu_2$; S^2 known and $\sigma_1^2 = \sigma_2^2$.
- Case 9. $H_0{:}\mu_1 = \mu_2$; S_1^2 and S_2^2 known and $\sigma_1^2 \neq \sigma_2^2$.

Figure 3-1 presents a summary of these nine cases and the formulas that will be needed by the experimenter for each case. Experiment designs will be presented in this chapter for nine cases; in addition, two special cases will be considered:

- Special Case 1. (Paired Comparisons)
- Special Case 2. (Some Data Lost)

Cases 1, 2, and 3. Variance or Variances Are Known

CASE 1

$H_0: \mu_1 = \mu_0$; Variance Is Known

A solid-state welding process is producing parts that have a population mean impact strength (μ_0) of 300 foot-pounds, and the variance (σ_0^2) is known to be 100. The part is encountering a 10 percent failure rate in the field. It is thought that increasing the welding temperature from the present 500° to 600° will increase the impact strength and, presumably, decrease the failure rate in the field. Unless the change really would increase the impact strength, however, it is undesirable to increase the temperature since the higher temperature would increase the cost of the part.

Case 1 Experiment Design

A. State the object of the experiment.

$H_0: \mu_{600} = \mu_{500} = 300$ foot-pounds
$H_a: \mu_{600} > 300$ foot-pounds
(single-sided)

B. Choose α, β, and δ based on economics, etc., and state the value of σ^2.

$\alpha = 0.05$
$\beta = 0.10$
$\delta = 12$ foot-pounds $\quad \delta^2 = 144$
$\sigma^2 = 100 \quad \sigma = 10$ foot-pounds

C. Look up the value of U_α in table 1 (single-sided normal distribution table).

$U_\alpha = 1.645$

Look up the value of U_β in table 1.

$U_\beta = 1.282$

Compute $N = (U_\alpha + U_\beta)^2 \dfrac{\sigma^2}{\delta^2}$.

$$N = (1.645 + 1.282)^2 \frac{100}{144} = 5.95$$

D. Calculate criterion.

$$\bar{X}^* = \mu_0 + \frac{\sigma U_\alpha}{\sqrt{N}}$$

$$\bar{X}^* = 300 + \frac{(10)(1.645)}{\sqrt{6}} = 306.7$$

E. Make six parts on the welder with the temperature set at 600°, and test for impact strength.
Calculate $\bar{X}$.

Results: 310, 311,
300, 290,
320, and 306 foot-pounds
$\bar{X} = 306.16$ foot-pounds

F. Compare $\bar{X}$ with $\bar{X}^*$. $(\bar{X} = 306.16) < (\bar{X}^* = 306.7)$

G. Make the decision. Accept H_0. With at least 90 percent confidence, it can be stated that the correct decision is: don't increase the temperature.

The above experiment was designed to test a hypothesis: that μ_{600} would be larger than μ_{500}. In some instances, the experimenter may wish only to establish a confidence interval for μ_{600} or may simply wish to estimate a value for μ_{600}.

Any $(1 - \alpha)(100)$ percent confidence interval can be calculated for μ_{600} in the forgoing experiment by using the following formula:

$$\left(\bar{X} - \frac{\sigma U_{\alpha/2}}{\sqrt{N}}\right) \leq \mu_{600} \leq \left(\bar{X} + \frac{\sigma U_{\alpha/2}}{\sqrt{N}}\right)$$

If a 90 percent confidence is wanted, the U value for $\alpha/2 = 0.05$ from the single-sided table (table 1) is used in the calculation. Note, however, that this is the same U value that would be obtained by using $\alpha = 0.10$ and the double-sided table (table 2).

The 90 percent confidence interval for μ_{600} is then calculated:

$$\left(306.16 - \frac{(10)(1.645)}{\sqrt{6}}\right) \leq \mu_{600} \leq \left(306.16 + \frac{(10)(1.645)}{\sqrt{6}}\right)$$

$$299.44 \leq \mu_{600} \leq 312.87$$

The above inequality states that the population mean (μ_{600}) is, with 90 percent confidence, between the values of 299.44 and 312.87 foot-pounds.

One-sided confidence intervals can also be calculated by using the expressions from the above formula:

$$-\infty \leq \mu_{600} \leq \bar{X} + \frac{\sigma U_\alpha}{\sqrt{N}} \quad \text{or} \quad \bar{X} - \frac{\sigma U_\alpha}{\sqrt{N}} \leq \mu \leq \infty$$

For one-sided confidence intervals, U_α would be selected from table 1. Thus, for a 99 percent one-sided confidence interval ($\alpha = 0.01$), $U_\alpha = 2.326$, and the confidence interval would be:

$$-\infty \leq \mu_{600} \leq 306.16 + \frac{10(2.326)}{\sqrt{6}}$$

$\mu_{600} \leq 315.66$ with 99% confidence.

If the experimenter wishes to conduct an experiment with the sole object of estimating a value for μ_{600}, the following formula can be used to determine N:

$$\frac{\sigma U_{\alpha/2}}{\sqrt{N}} = \text{magnitude of error on } \mu \text{ that can be tolerated.}$$

In addition to selecting a desired value for α, the experimenter must select the magnitude of error for μ that can be tolerated. The above formula is rearranged in terms of the sample size, N, to the following formula:

$$N = \frac{\sigma^2(U_{\alpha/2})^2}{(\text{magnitude of error})^2}$$

If $\sigma = 10$, $\alpha/2 = 0.05$, and the tolerable magnitude of error $= \pm 10$, the sample size would be:

$$N = \frac{(10)^2(1.645)^2}{(\pm 10)^2} = (1.645)^2 = 2.70, \quad \text{or } 3.$$

If three samples are tested at 600°, and the sample mean ($\bar{X}$) is 306, then the estimated value of μ_{600} is:

$$\mu_{600} = \bar{X} \pm \frac{\sigma U_{\alpha/2}}{\sqrt{N}} = 306 \pm \frac{(10)(1.645)}{\sqrt{3}}$$

$$\mu_{600} = 306 \pm 9.50 \text{ with 90 percent confidence.}$$

CASE 2 $H_0: \mu_1 = \mu_2$; Variances Are Equal and Known

Two vendors, A and B, are approached as potential suppliers of a rubber part that will be subjected to abrasion in a test machine until the part fails. Both vendors state that their parts will meet the engineering specification for the part, which is a maximum wear of 20 mg/1000 cycles on the test, and both state that the variance (σ^2) is 16. Since each vendor quotes the same price, it is advantageous to select the vendor with the smallest population mean abrasion (μ).

Case 2 Experiment Design

A. State the object of the experiment.

$H_0: \mu_A = \mu_B$
$H_a: \mu_A \neq \mu_B$ (double-sided)
The alternative hypothesis can also be stated:

$H_{a1}: \mu_B < \mu_A$ with $\alpha/2$ risk
$H_{a2}: \mu_A < \mu_B$ with $\alpha/2$ risk

B. Choose α, β, and δ and state the values of σ^2.

$\alpha = 0.01$
$\beta = 0.02$
$\delta = 2$ mg/1000 cycles $\quad \delta^2 = 4$
$\sigma_A^2 = 16 \quad \sigma_A = 4$ mg/1000 cycles
$\sigma_B^2 = 16 \quad \sigma_B = 4$ mg/1000 cycles

C. Look up the value of U_α in the double-sided table, table 2. Look up the value of U_β in table 1. Compute N_A and N_B.

$U_\alpha = 2.576$

$U_\beta = 2.054$

$$N_A = N_B = 2(U_\alpha + U_\beta)^2 \frac{\sigma^2}{\delta^2}$$

$N_A = N_B = 2(2.576 + 2.054)^2 \frac{16}{4} = 171.5$

D. Calculate decision-making criterion.

$$|\bar{X}_A - \bar{X}_B|^* = U_\alpha \sigma \sqrt{\frac{1}{N_A} + \frac{1}{N_B}}$$

$|\bar{X}_A - \bar{X}_B|^* = (2.576)(4)\sqrt{\frac{1}{172} + \frac{1}{172}} = 1.11$ mg

E. Have each vendor supply 172 parts, test the parts, and calculate $\bar{X}_A$ and $\bar{X}_B$.

$\bar{X}_A = 18.8$ mg/1000 cycles
$\bar{X}_B = 16.6$ mg/1000 cycles
$|\bar{X}_A - \bar{X}_B| = 2.2$ mg/1000 cycles

F. Compare $|\bar{X}_A - \bar{X}_B|$ with $|\bar{X}_A - \bar{X}_B|^*$.

$(|\bar{X}_A - \bar{X}_B| = 2.2) > (|\bar{X}_A - \bar{X}_B|^* = 1.11)$

G. Make the decision.

Since $\bar{X}_B < \bar{X}_A$ and $|\bar{X}_A - \bar{X}_B| > |\bar{X}_A - \bar{X}_B|^*$, accept H_{a1} with $[1 - (\alpha/2)](100)$ percent confidence. Accept $H_{a1}: \mu_{\text{vendor B}} < \mu_{\text{vendor A}}$ with 99.5 percent confidence.

Conclusion: the mean wear on the parts produced by vendor B is less than the mean wear produced on the parts by vendor A; therefore, purchase from vendor B.

In this example, the experimenter might also want to answer the question: With what confidence can it be stated that vendor B is producing parts with a population mean value better than the minimum specifications? This is equivalent to asking the question: With what confidence can it be stated that μ_B is less than 20 mg/1000 cycles? The upper one-sided confidence interval formula is used:

$$-\infty \leq \mu_\beta \leq \bar{X} + \sigma U_\alpha \sqrt{\frac{1}{N_B}} = 20$$

$$\mu_\beta \leq 16.6 + 4U_\alpha \sqrt{\frac{1}{172}} = 20$$

Solving the equality for U_α,

$$U_\alpha = \frac{20 - 16.6}{4\sqrt{\frac{1}{172}}} = \frac{3.4}{0.305} = 11.1$$

From table 1, $U_\alpha = 3.09$ for 99.9 percent confidence. Therefore, since the value of U_α for the data is much greater than 3.09, the experimenter can state with greater than 99.9 percent confidence that vendor B *can* produce parts which will have a mean abrasion of less than 20 mg/1000 cycles. The experimenter cannot say that vendor B *will* produce parts that have abrasion of less than 20 mg/1000 cycles. All that can be said is that, if the vendor uses material equivalent to that used in the sample, and if the vendor processes the material in a manner equivalent to the process used for the samples, vendor B will unquestionably produce parts with a population mean abrasion less than 20 mg/1000 cycles.

A confidence interval estimate of the difference between μ_A and μ_B can also be calculated. The $(1 - \alpha)$ confidence interval for $(\mu_A - \mu_B)$ is:

$$\left[(\bar{X}_A - \bar{X}_B) - U_{\alpha/2}\sigma\sqrt{\frac{1}{N_A} + \frac{1}{N_B}}\right] \leq (\mu_A - \mu_B) \leq \left[(\bar{X}_A - \bar{X}_B) + U_{\alpha/2}\sigma\sqrt{\frac{1}{N_A} + \frac{1}{N_B}}\right]$$

For the example above, the 98 percent confidence interval is:

$$\left[(18.8 - 16.6) - (2.326)(4)\sqrt{\frac{1}{172} + \frac{1}{172}}\right] \leq (\mu_A - \mu_B) \leq \left[(18.8 - 16.6) + (2.326)(4)\sqrt{\frac{1}{172} + \frac{1}{172}}\right]$$

$1.20 \leq (\mu_A - \mu_B) \leq 3.20$ with 98 percent confidence.

CASE 3

$H_0: \mu_1 = \mu_2$; Variances Are Not Equal and Both Are Known

For case 3, the example used is the same as that used for case 2 except that, for case 3, vendor A states that the variance (σ^2) is 9.

Case 3 Experiment Design

A. State the object of the experiment.

$H_0: \mu_A = \mu_B$
$H_a: \mu_A \neq \mu_B$

B. Choose α, β, and δ and state the values of σ_A^2 and σ_B^2.

$\alpha = 0.01$
$\beta = 0.02$
$\delta = 2$ mg/1000 cycles $\quad \delta^2 = 4$
$\sigma_A^2 = 9 \quad \sigma_A = 3$ mg/1000 cycles
$\sigma_B^2 = 16 \quad \sigma_B = 4$ mg/1000 cycles

C. Look up the value of U_α in the double-sided table (table 2). Look up the value of U_β in the single-sided table (table 1).

$U_\alpha = 2.576$
$U_\beta = 2.054$

Compute N_A and N_B.

$$N_A = (U_\alpha + U_\beta)^2 \frac{\sigma_A(\sigma_A + \sigma_B)}{\delta^2} \qquad N_A = (2.576 + 2.054)^2 \frac{3(3 + 4)}{4}$$

$$= 112.54$$

$$N_B = (U_\alpha + U_\beta)^2 \frac{\sigma_B(\sigma_A + \sigma_B)}{\delta^2} \qquad N_B = (2.576 + 2.054)^2 \frac{4(3 + 4)}{4}$$

$$= 150.1$$

D. Calculate criterion.

$$|\bar{X}_A - \bar{X}_B|^* = U_\alpha \sqrt{\frac{\sigma_A^2}{N_A} + \frac{\sigma_B^2}{N_B}} \qquad |\bar{X}_A - \bar{X}_B|^* = 2.576\sqrt{\tfrac{9}{113} + \tfrac{16}{150}}$$

$$= 1.112$$

E. Have vendor A submit 113 parts and vendor B supply 150 parts. Test parts, and compute $\bar{X}_A$ and $\bar{X}_B$ from the data.

$\bar{X}_A = 18.8$ mg/1000 cycles
$\bar{X}_B = 16.6$ mg/1000 cycles
$\bar{X}_A - \bar{X}_B = 2.2$

F. Compare $|\bar{X}_A - \bar{X}_B|$ with $|\bar{X}_A - \bar{X}_B|^*$.

$(|\bar{X}_A - \bar{X}_B| = 2.2) > (|\bar{X}_A - \bar{X}_B|^* = 1.112)$

G. Make the decision.

Accept $H_a: \mu_B < \mu_A$. Conclusion: purchase the parts from vendor B.

For case 3, the confidence interval for $(\mu_1 - \mu_2)$ can be calculated from the following formula:

$$\left[(\bar{X}_A - \bar{X}_B) - U_{\alpha/2}\sqrt{\frac{\sigma_A^2}{N_A} + \frac{\sigma_B^2}{N_B}}\right] \le (\mu_A - \mu_B)$$

$$\le \left[(\bar{X}_A - \bar{X}_B) + U_{\alpha/2}\sqrt{\frac{\sigma_A^2}{N_A} + \frac{\sigma_B^2}{N_B}}\right]$$

Cases 4, 5, and 6. Variance or Variances Are Unknown

For the previous three cases, the experimenter calculated N and the criterion using the normal-distribution tables (tables 1 and 2) because the value of σ^2 was known. For the three cases described in this section, σ^2 is unknown. When σ^2 is unknown, the experimenter can obtain only an estimate of σ^2 from the experiment; this estimate is termed S^2.

Using an estimate of σ^2 introduces an additional element of uncertainty into the experiment and the decision-making process. This additional uncertainty is taken into account by using what is called the t distribution in place of the normal distribution when the variance (σ^2) is only estimated. The t distribution has the same shape as the normal distribution, but the width of the curve of the distribution

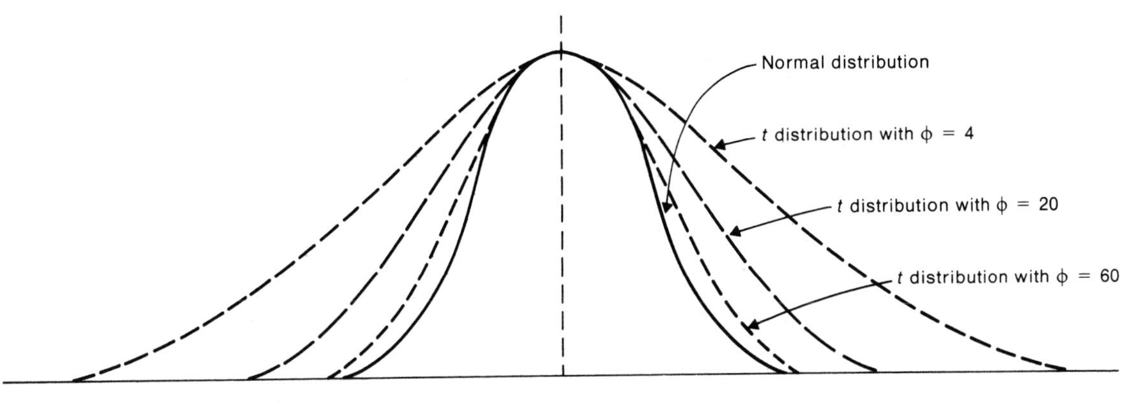

Fig. 3-2. Comparison of t distribution with normal distribution.

is wider. The width of the curve of the t distribution is also a function of the statistical term "degrees of freedom" (ϕ) as shown in figure 3-2. As noted in Chapter 1, $\phi = N - 1$.

Several important observations can be made from figure 3-2:

- ϕ is a measure of the precision with which σ^2 has been estimated.
- As the value of ϕ increases, the width of the t distribution decreases.
- When ϕ becomes very large (about 60), the t distribution curve almost coincides with the normal-distribution curve.
- When an experimenter has an estimate of σ^2 with at least 60 degrees of freedom ($\phi \geq 60$), the experimenter can say that σ^2 is known for all practical purposes; i.e.:

  S^2 (with $\phi = 60$) $= \sigma^2$.

  Therefore, when $\phi > 60$, the experimenter can use the normal-distribution tables. When ϕ is less than 60, the experimenter must use t-distribution tables, as described below.

CASE 4

$H_0{:}\mu_1 = \mu_0$; Variance Is Unknown

This type of problem is equivalent to case 1, the only difference being that σ^2 is unknown. This creates a problem since the formula to calculate a precise value of N cannot be used without knowing σ^2. However, a good estimate of N can be obtained by the following procedure, where δ is specified in terms of σ rather than as a specific number. For example, suppose $\alpha = 0.05$, $\beta = 0.10$, and δ is specified as $\delta = 1\sigma$. Then, for a single-sided alternative hypothesis:

$$N = (U_\alpha + U_\beta)^2 \frac{\sigma^2}{\delta^2} = (1.645 + 1.282)^2 \frac{\sigma^2}{(1\sigma)^2} = 8.5$$

This estimate of N is, however, too low since the decision-making criterion will be based on the t distribution rather than the normal distribution. Therefore, the experimenter must go one step further. In this case, go to table 3 to obtain the values of t_α and t_β with $\phi = N - 1 = 7.5$:

$$t_{0.05} = 1.87 \qquad \text{for } \phi = 7.5$$

$$t_{0.10} = 1.41 \qquad \text{for } \phi = 7.5$$

Substitute these values in place of U_α and U_β and compute:

$$N_t = (t_\alpha + t_\beta)^2 \frac{\sigma^2}{\delta^2}$$

$$N_t = (1.87 + 1.41)^2 \frac{\sigma^2}{\sigma^2} = 10.76$$

But with $N = 10.76$, ϕ will be 9.76 and $t_\alpha = 1.82$ and $t_\beta = 1.38$. Since the values of t_α and t_β are too high, this value of N_t is slightly high, and a still better estimate of N could be obtained by additional iterations; however, this is usually not justified. In this case, it makes little difference whether the experimenter chooses 10 or 11 samples.

A second procedure can also be used to compute an estimated N. If the experimenter has some idea of the value of σ^2 and wishes to compute N_t for a very definite δ, the same procedure as above can be used except that the estimated value of σ^2 and the value of δ will be used in the first step. For example, suppose the experimenter thinks σ^2 is about 10 and wants a δ of 2. For $\alpha = 0.05$, $\beta = 0.10$, and a single-sided alternative hypothesis:

$$N = (U_\alpha + U_\beta)^2 \frac{\hat{\sigma}^2}{\delta^2} = (1.645 + 1.282)^2 \frac{10}{4} = 21.4 \doteq 21$$

$$\phi = 21 - 1 = 20$$

The term $\hat{\sigma}^2$ is the estimate made without data of the true σ^2. For $\phi = 20$, $t_{0.05} = 1.73$, $t_{0.10} = 1.33$, and:

$$N_t = (t_\alpha + t_\beta)^2 \frac{\hat{\sigma}^2}{\delta^2} = (1.73 + 1.33)^2 \frac{10}{4} = 23.4$$

If the estimated value of the variance is too large, the sample size will be larger than necessary; if the estimated variance is too small, the sample size will be too small. In any case, the next chapter will discuss how the experimenter can test the goodness of his estimate of σ^2 at the end of the experiment.

To illustrate case 4, assume that a new welder is purchased and that a set of proposed operating conditions is included in the instructions. The specifications require that the parts produced by the welder have a population mean impact strength of greater than 300 foot-pounds. The process engineer decides to conduct an experiment to prove that the proposed operating conditions give parts with $\mu > 300$ foot-pounds.

Case 4 Experiment Design

A. State the object of the experiment. $H_0: \mu_{\text{new welder}} = \mu_0 = 300$ foot-pounds
$H_a: \mu_{\text{new welder}} > 300$

B. Choose the values of α, β, and δ.

$\alpha = 0.05$
$\beta = 0.01$
$\delta = 0.5\sigma \quad \delta^2 = 0.25\sigma^2$

C. Look up the value of U_α in table 1. $U_\alpha = 1.645$
Look up the value of U_β in table 1. $U_\beta = 2.326$

Compute N and ϕ.

$$N = (U_\alpha + U_\beta)^2 \frac{\sigma^2}{\delta^2} \qquad N = (1.645 + 2.326)^2 \frac{\sigma^2}{0.25\sigma^2} = 63.1$$

$\phi = N - 1$ $\qquad \phi = 62$

Look up t_α from table 3. $\quad t_\alpha = 1.67$ for $\alpha = 0.05$, $\phi = 62$.
Look up t_β from table 3. $\quad t_\beta = 2.39$ for $\beta = 0.01$, $\phi = 62$.

Calculate N_t.

$$N_t = (t_\alpha + t_\beta)^2 \frac{\sigma^2}{\delta^2} \qquad N_t = (1.67 + 2.39)^2 \frac{\sigma^2}{0.25\sigma^2} = 65.9$$

D. Make 66 parts with the welder set at standard conditions.
Calculate $\bar{X}$ and S^2 from the sample data.

$$\bar{X} = \frac{\sum X_i}{N_t} \qquad \bar{X} = 305$$

$$S^2 = \frac{\sum (X_i - \bar{X})^2}{N_t - 1} \qquad S^2 = 25 \quad S = 5$$

E. Compute criterion. $\quad t_\alpha = 1.67$ for $\phi = 65$.

$$\bar{X}^* = \mu_0 + \frac{t_\alpha S}{\sqrt{N_t}} \qquad \bar{X}^* = 300 + \frac{1.67(5)}{\sqrt{66}} = 301$$

F. Compare $\bar{X}$ with $\bar{X}^*$.

$(\bar{X} = 305) > (\bar{X}^* = 301)$
Accept $H_a\!:\mu_{\text{new welder}} > 300$ with at least 95 percent confidence. Conclusion: the new welder will produce parts with a mean impact strength greater than 300 foot-pounds if there are no substantial changes in the material and machine in future operations.

A confidence interval on the population mean for the new welder can be computed by using a formula similar to that used in case 1:

$$\left(\bar{X} - \frac{t_{\alpha/2}S}{\sqrt{N_t}}\right) \leq \mu \leq \left(\bar{X} + \frac{t_{\alpha/2}S}{\sqrt{N_t}}\right)$$

From the above data, for example, the 95 percent confidence interval is computed:

$$\alpha = 0.05, \quad \alpha/2 = 0.025, \quad t_{0.025} = 2.00, \quad N = 66$$

$$\left[305 - \frac{(2.00)(5)}{\sqrt{66}}\right] \leq \mu \leq \left[305 + \frac{(2.00)(5)}{\sqrt{66}}\right]$$

$$303.77 \leq \mu \leq 306.23$$

A single-sided confidence interval with $(1 - \alpha)(100)$ percent confidence is given for this case by:

$$\bar{X} - \frac{t_\alpha S}{\sqrt{N_t}} \leq \mu \leq \infty$$

For the above experiment, the 95 percent one-sided confidence interval is:

$$305 - \frac{(1.67)(5)}{\sqrt{66}} \leq \mu \leq \infty$$

$$303.97 \leq \mu \leq \infty, \quad \text{or}$$

$\mu > 303.97$ with 95 percent confidence.

After obtaining the above results, the experimenter might then want to estimate the percentage of parts that would have an impact strength of less than 299 foot-pounds. To determine this percentage, the experimenter first must assume the distribution is normal and then find a value for K by the following formula:

$$K = \frac{\bar{X} - 299}{S}$$

$$K = \frac{305 - 299}{5} = 1.20$$

To determine the implication of the value of K, go to table 12. Note that this table only goes to $N = 50$, while $N = 66$ in this problem. In a situation like this, use the row for $N = 50$, go across this row to the confidence level that is desired (say, 0.90), and find the value of K. In this case, $K = 1.20$ lies between 0.894 for $\gamma = 0.25$ and 1.560 for $\gamma = 0.10$. By interpolation, $K = 1.20$ has a $\gamma = 0.18$. Therefore, with 90 percent confidence, it can be stated that at least $(1 - \gamma)(100)$ percent of the population will have an impact strength equal to, or greater than, 299 foot-pounds. In this case, it can be stated with 90 percent confidence that at least 82 percent of the parts will have an impact strength of at least 299 foot-pounds. Conversely, it can also be stated with 90 percent confidence that, at most, 18 percent of the parts will have an impact strength less than 299 foot-pounds.

If the experimenter wanted to know the percentage of parts with an impact strength between two different numbers, the same procedure as above would be used, except that the γ value would be obtained from table 11.

CASE 5

$H_0: \mu_1 = \mu_2$; Variances Are Equal and Unknown

Case 5 is similar to case 2 except that the variance is unknown in case 5. Using an example similar to that used in case 2, assume that a vendor proposes to supply a rubber part made from either one of two different formulas, depending upon which formula produces parts with better abrasion resistance. The vendor is certain that the variance will be the same for both formulas.

Case 5 Experiment Design

A. State the object of the experiment.

$H_0: \mu_{\text{formula A}} = \mu_{\text{formula B}}$
$H_a: \mu_{\text{formula A}} \neq \mu_{\text{formula B}}$

B. Choose the values of α, β, and δ.

$\alpha = 0.01$
$\beta = 0.20$
$\delta = 0.5\sigma \quad \delta^2 = 0.25\sigma^2$

C. Look up the value of U_α in table 2 (double-sided). Look up the value of U_β in table 1 (single-sided).

$U_\alpha = 2.576$

$U_\beta = 0.842$

Compute $N_A = N_B$.

$$N = 2(U_\alpha + U_\beta)^2 \frac{\sigma^2}{\delta^2}$$

$$N_A = N_B = 2(2.576 + 0.842)^2 \times \frac{\sigma^2}{0.25\sigma^2} = 93.5$$

Compute ϕ using

$$\phi = N_A + N_B - 2$$

$$\phi = 93.5 + 93.5 - 2 = 185$$

Compute N_t using

$$N_t = 2(t_\alpha + t_\beta)^2 \frac{\sigma^2}{\delta^2}$$

$$N_t = 2(2.61 + 0.85)^2 \frac{\sigma^2}{0.25\sigma^2} = 95.8$$

D. Have the vendor supply 96 samples with formula A and 96 samples with formula B.

Compute $\bar{X}_A$, $\bar{X}_B$, and S^2.

$$\bar{X}_A = \frac{\sum X_{iA}}{N_A}$$

$\bar{X}_A = 15.7$ mg/1000 cycles
$\bar{X}_B = 13.6$ mg/1000 cycles

$$\bar{X}_B = \frac{\sum X_{jB}}{N_B} \qquad (\bar{X}_A - \bar{X}_B) = 2.1$$

$$S^2 = \frac{\sum(X_{iA} - \bar{X}_A)^2 + \sum(X_{jB} - \bar{X}_B)^2}{N_{At} + N_{Bt} - 2} \qquad \text{Suppose } S^2 = 16, S = 4, \phi = 190$$

E. Compute criterion.

$$|\bar{X}_A - \bar{X}_B|^* = t_\alpha S\sqrt{\frac{1}{N_{At}} + \frac{1}{N_{Bt}}} \qquad |\bar{X}_A - \bar{X}_B|^* = (2.61)(4)\sqrt{\frac{1}{96} + \frac{1}{96}} = 1.52$$

F. Compare $|\bar{X}_A - \bar{X}_B|$ with $|\bar{X}_A - \bar{X}_B|^*$

$$(|\bar{X}_A - \bar{X}_B| = 2.1) > (|\bar{X}_A - \bar{X}_B|^* = 1.52)$$

G. Make the decision.

Accept H_a:$\mu_{\text{formula B}} < \mu_{\text{formula A}}$ with at least 99.5 percent confidence. Conclusion: use formula *B* for the part.

For this case, the confidence interval estimate of the difference in means is given by either

$$\left[(\bar{X}_A - \bar{X}_B) - t_{\alpha/2}S\sqrt{\frac{1}{N_A} + \frac{1}{N_B}}\right] \leq (\mu_A - \mu_B) \leq \left[(\bar{X}_A - \bar{X}_B) + t_{\alpha/2}S\sqrt{\frac{1}{N_A} + \frac{1}{N_B}}\right]$$

or

$$\left[(\bar{X}_B - \bar{X}_A) - t_{\alpha/2}S\sqrt{\frac{1}{N_A} + \frac{1}{N_B}}\right] \leq (\mu_B - \mu_A) \leq \left[(\bar{X}_B - \bar{X}_A) + t_{\alpha/2}S\sqrt{\frac{1}{N_A} + \frac{1}{N_B}}\right]$$

CASE 6

H_0:$\mu_1 = \mu_2$; Variances Are Not Equal and Are Unknown

This case causes a rather complex statistical problem and has been the subject of so much controversy that it has been dignified with the name Behrens-Fisher Problem. Under most experimental conditions, the best procedure (certainly the easiest) is to ignore the problem: when the sample size is adequate (larger than 8), the problem essentially disappears. In this book, the problem will be ignored.

Assume that a vendor proposes sending samples from material A and samples from material B for evaluation. The samples will be tested for Izod impact strength. The vendor estimates that material A will have a variance of about 25 and that material B will have a variance of about 9. The specification on the part calls for a population mean Izod impact strength of 150 foot-pounds or greater.

Case 6 Experiment Design

A. State the object of the experiment.

$H_0: \mu_A = \mu_B$
$H_a: \mu_A \neq \mu_B$

B. Choose α, β, and δ and state $\hat{\sigma}_A^2$ and $\hat{\sigma}_B^2$, where $\hat{\sigma}^2$ means an estimate of σ^2 not based on data.

$\alpha = 0.05$
$\beta = 0.10$
$\delta = 5$ foot-pounds
$\hat{\sigma}_A^2 = 25$
$\hat{\sigma}_B^2 = 9$

C. Look up U_α in table 2 (double-sided). Look up U_β in table 1 (single-sided).

$U_\alpha = 1.960$
$U_\beta = 1.282$

Compute:

$$N_A = (U_\alpha + U_\beta)^2 \frac{\hat{\sigma}_A(\hat{\sigma}_A + \hat{\sigma}_B)}{\delta^2} \qquad N_A = (1.960 + 1.282)^2 \frac{5(5 + 3)}{25}$$

$$= 16.82 = 17$$

$$N_B = (U_\alpha + U_\beta)^2 \frac{\hat{\sigma}_B(\hat{\sigma}_A + \hat{\sigma}_B)}{\delta^2} \qquad N_B = (1.960 + 1.282)^2 \frac{3(5 + 3)}{25}$$

$$= 10.09 = 10$$

$\phi_A = 16, \quad \phi_B = 9$

Corrected:

$$N_A = (t_\alpha + t_\beta)^2 \frac{\hat{\sigma}_A(\hat{\sigma}_A + \hat{\sigma}_B)}{\delta^2} \qquad N_A = (2.12 + 1.34)^2 \frac{5(5 + 3)}{25} = 19.2$$

$$N_B = (t_\alpha + t_\beta)^2 \frac{\hat{\sigma}_B(\hat{\sigma}_A + \hat{\sigma}_B)}{\delta^2} \qquad N_B = (2.26 + 1.38)^2 \frac{3(5 + 3)}{25} = 12.7$$

D. Obtain 19 samples of material A and 13 samples of material B and test for Izod impact strength.
Compute $\bar{X}_A$, $\bar{X}_B$, S_A^2, and S_B^2.

$$\bar{X}_A = \frac{\sum X_{iA}}{N_A}$$

$$\bar{X}_B = \frac{\sum X_{jB}}{N_B}$$

$$S_A^2 = \frac{\sum(X_{iA} - \bar{X}_A)^2}{N_A - 1}$$

$$S_B^2 = \frac{\sum(X_{jB} - \bar{X}_B)^2}{N_B - 1}$$

$\bar{X}_A = 150.2, \quad S_A^2 = 30$
$\bar{X}_B = 153.6, \quad S_B^2 = 8$
$(\bar{X}_B - \bar{X}_A) = 3.4$ foot-pounds

E. Compute criterion.

$$\phi = N_A + N_B - 2 \qquad \phi = 19 + 13 - 2 = 30$$

$$(\bar{X}_B - \bar{X}_A)^* = t_\alpha \sqrt{\frac{S_A^2}{N_A} + \frac{S_B^2}{N_B}} \qquad (\bar{X}_B - \bar{X}_A)^* = 2.04\sqrt{\tfrac{30}{19} + \tfrac{8}{13}} = 3.02$$

F. Compare $(\bar{X}_B - \bar{X}_A)$ with $(\bar{X}_B - \bar{X}_A)^*$. $[(\bar{X}_B - \bar{X}_A) = 3.4] > [(\bar{X}_B - \bar{X}_A)^* = 3.02]$

G. Make decision. Accept $H_a: \mu_B > \mu_A$ with at least 97.5 percent confidence. Conclusion: use material B for the part.

For case 6, the $(1 - \alpha)(100)$ percent confidence interval estimate of $(\mu_B - \mu_A)$ can be computed by the formula:

$$\left[(\bar{X}_B - \bar{X}_A) - t_{\alpha/2}\sqrt{\frac{S_A^2}{N_A} + \frac{S_B^2}{N_B}}\right] \le (\mu_B - \mu_A) \le \left[(\bar{X}_B - \bar{X}_A) + t_{\alpha/2}\sqrt{\frac{S_A^2}{N_A} + \frac{S_B^2}{N_B}}\right]$$

$$\phi = N_A + N_B - 2$$

For example, the 99 percent confidence interval estimate of $\mu_B - \mu_A$ in this experiment is:

$$[3.4 - (2.75)\sqrt{\tfrac{30}{19} + \tfrac{8}{13}}] \le (\mu_B - \mu_A) \le [3.4 + (2.75)\sqrt{\tfrac{30}{19} + \tfrac{8}{13}}]$$

$$-0.67 \le (\mu_B - \mu_A) \le 7.47$$

Cases 7, 8, and 9. Sample Estimate of Variance Is Known

For the previous three cases, σ_0^2 and S^2 unknown, the experimenter calculated N by using the subterfuge of setting $\delta = f(\sigma)$ or guessing at the six σ limits of the product or process. The initially calculated value of N, using table 1 or 2, was then corrected using the t tables. This procedure is only required for the first experiment on a product or process.

After the first experiment has been conducted, an estimate of the variance (S^2) can be calculated from the data of the first experiment. The experimenter will assume that the variance will not change between the first and second experiment. Therefore, the S^2 calculated from the data of the first experiment can be used to calculate the proper N for the second experiment using tables 3 and 4. After the second experiment is conducted, a second estimate of the variance (S_2^2) will be calculated from the data of the second experiment. In Chapter 5, a procedure will be described for testing whether the variance has changed between experiment 1 and experiment 2. In Chapter 5, it will also be shown how these two estimates of the variance can, in most cases, be combined for a best estimate of S^2.

CASE 7 $H_0: \mu_A = \mu_0$; S^2 Is Known

A case 5 type experiment was conducted on a photoconductor. Six trials were conducted at each of two thicknesses. $\bar{X}_{thin} = 1.4\ \mu_{J/in^2}$, $\bar{X}_{thick} = 1.8\ \mu_{J/in^2}$, $S^2 = 0.01$, with $\phi = 10$. Several weeks later, it was proposed that a better solution to the problem might be the inclusion of an additive to the thick photoconductor to make a faster-speed photoconductor (lower μ_{J/in^2} is faster).

CASE 7 Experiment Design

A. Define the object of the experiment.

$H_0{:}\mu_2 = \mu_0 = 1.80\ \mu_{g/in^2}$
$H_a{:}\mu_2 < \mu_0$

B. Choose α, β, δ; Specify S^2 and ϕ.

$\alpha = 0.05 \quad \beta = 0.05$
$\delta = 0.2 \quad \mu_{j/in^2}$
$S^2 = 0.01,\ \phi = 10$

C. Look up t_α in table 3; $\alpha = 0.05$, $\phi = 10$.

$t_{0.05} = 1.81$

Look up t_β in table 3; $\beta = 0.05$, $\phi = 10$.

$t_{0.05} = 1.81$

Calculate $N = (t_\alpha + t_\beta)^2 \dfrac{S^2}{\delta^2}$.

$$N = (1.81 + 1.81)^2 \frac{0.01}{0.04} = 3.4 \doteq 4$$

D. Make up four samples with the additive and test; Calculate $\overline{X}_2$ and S_2^2.

$\overline{X}$ additive $= 1.71$
$S_2^2 = .0064,\ \phi = 3$

E. Calculate weighted average of S^2, and S_2^2.

$$S_{avg}^2 = \frac{S_1^2\,\phi_1 + S_2^2\,\phi_1}{\phi_1 + \phi_1}$$

$$S_{avg}^2 = \frac{(0.01)(10) + (.0064)(3)}{10 + 3}$$

.0092

$$\phi_{avg} = \phi_1 + \phi_2$$

$$\phi_{Avg} = 10 + 3 = 13$$

F. Compute criterion.

$$\overline{X}^* = \mu_0 - \frac{t_\alpha S}{\sqrt{N}}$$

$t_\alpha = 1.77,\ \phi = 13$

$$\overline{X}^* = 1.80 \frac{(1.77)(.096)}{\sqrt{4}} = 1.715$$

G. Compare $\overline{X}$ additive with $\overline{X}^*$.

$[(\overline{X} = 1.71)] < [\overline{X}^* = 1.72]$

H. Make the decision.

Accept H_a. Conclusion: With at least 95 percent confidence, it can be stated that the additive increases the speed of the photoconductor. *Note:* it does not increase the speed as much as the thin photoconductor increased the speed.

CASE 8

$H_0{:}\mu_A = \mu_B$; S^2 Is Known and $\sigma_A^2 = \sigma_B^2$

A case 4 type experiment was conducted on a new process producing a resin. The primary property of the product was viscosity. After the experiment was completed, X was calculated to be 2000 cps and S^2 was calculated to be 200, with $\phi = 8$. It was expected that the mean of the process was operator sensitive, so an experiment was planned to check this.

Case 8 Experiment Design

A. State object of experiment.

$H_0{:}\mu_{\text{operator A}} = \mu_{\text{operator B}}$
$H_a{:}\mu_{\text{operator A}} \neq \mu_{\text{operator B}}$

B. Choose α, β, and δ; State S^2 and ϕ.

$\alpha = .05$
$\beta = .05$
$\delta = 20$ cps
$S^2 = 200$
$\phi = 8$

C. Look up t_α in table 4 (double-sided); Look up t_β in table 3 (t_β always single-sided); Compute $N_A = N_B = 2(t_\alpha + t_\beta)^2 \dfrac{S^2}{\delta^2}$.

$t_\alpha = 2.31$
$t_\beta = 1.86$

$N_A = N_B = 2(2.31 + 1.86)^2 \dfrac{200}{400} = 17.4$

D. Have each operator make 17 samples; Compute $\overline{X}_A$, $\overline{X}_B$, S^2.

$$S^2 = \frac{\Sigma(X_{iA} - \overline{X}_A)^2 + \Sigma(X_{jB} - \overline{X}_B)^2}{N_A + N_B - 2}$$

Calculate $(\overline{X}_A - \overline{X}_B)$.

$\overline{X}_A = 2004$ ps
$\overline{X}_B = 1980$ cps
$S^2 = 180$, $\phi = 17 + 17 - 2 = 32$
$\overline{X}_A - \overline{X}_B = (2004 - 1980) = 24$

E. Calculated weighted average S^2.

$$S^2_{\text{avg}} = \frac{(S_1^2)(\phi_1) + (S_2^2)(\phi_2)}{\phi_1 + \phi_2}$$

$$S^2_{\text{avg}} = \frac{(200)(8) + (180)(32)}{8 + 32} = 184$$

F. Compute criterion.

$$|\overline{X}_A - \overline{X}_B|^* = t_\alpha S \sqrt{\frac{1}{N_A} + \frac{1}{N_B}}$$

t from table 4 with $\phi = 40$

$$|\overline{X}_A = \overline{X}_B|^* = (2.03)(13.56) \sqrt{\frac{1}{17} + \frac{1}{17}} = 9.49$$

G. Make the decision. — Accept H_a: conclusion: Operator B obtains a lower viscosity product than operator A with at least 97.5 percent confidence.

CASE 9

$H_0{:}\mu_A = \mu_B$; S_A^2 and S_B^2 Are Known and $\sigma_A^2 \neq \sigma_B^2$

Two electronic chip fabrication lines were operating with a mean etch bias of 40, which was much too high a value. The variance of the two lines was significantly different $S_A^2 = 8$, $S_B^2 = 2$, $\phi_A = \phi_B = 10$. Two new coaters were proposed to reduce the etch bias. To save time, it was proposed that new coater A be installed and tested on production line A and new coater B be installed and tested on production line B. It was not expected that the new coaters would have any effect on the variances.

Case 9 Experiment Design

A. State the object of the experiment.

$H_0{:}\mu_A = \mu_B$
$H_a{:}\mu_A \neq \mu_B$

B. Choose α, β, δ;
State S_A^2, S_B^2, ϕ_A, and ϕ_B.

$\alpha = 0.05$
$\beta = 0.10$
$\delta = 3$
$S_A^2 = 8$, $\phi_A = 10$
$S_B^2 = 2$, $\phi_B = 10$

C. Look up t_α in table 4;
Look up t_β in table 3;

$t_\alpha = 2.23$
$t_\beta = 1.37$

$$N_A = (t_\alpha + t_\beta)^2 \frac{(S_A)(S_A + S_B)}{\delta^2}$$

$$N_A = (2.23 + 1.37)^2 \frac{(2.83)(2.83 + 1.41)}{9} = 17.3$$

$$N_B = (t_\alpha + t_\beta)^2 \frac{(S_B)(S_A + S_B)}{\delta^2}$$

$$N_B = (2.23 + 1.37)^2 \frac{(1.41)(2.83 + 1.41)}{9} = 8.61$$

D. Obtain 17 samples with coater A on production line A, and 9 samples with coater B on production line B;
Calculate $\overline{X}_A$, $\overline{X}_B$, S_A^2 and S_B^2.

$\overline{X}_A = 32$ $\quad S_A^2 = 6$, $\phi_A = 16$
$\overline{X}_B = 29$ $\quad S_B^2 = 1.5$, $\phi_B = 8$

E. Calculate weighted average of S_A^2 and the weighted average of S_B^2.

$$S_A^2 \text{ avg} = \frac{(8)(10) + (6)(16)}{10 + 16} = 6.77$$

$$S^2_{avg} = \frac{(S_1^2)(\phi_1) + (S_2^2)(\phi_2)}{\phi_1 + \phi_2}$$

$\phi_A = 26$

$$S_B^2 \text{ avg} = \frac{(2)(10) + (1.5)(8)}{10 + 8} = 1.78$$

Calculate $\phi_A + \phi_B$.

$\phi_B = 18$

$\phi_A + \phi_B = 26 + 18 = 44$

F. Compute criterion.

$$|\bar{X}_A - \bar{X}_B|^* = t_\alpha \sqrt{\frac{S_A^2}{N_A} + \frac{S_B^2}{N_B}}$$

$$|\bar{X}_A - \bar{X}_B|^* = 2.02 \sqrt{\frac{6.77}{17} + \frac{1.78}{9}} = 1.55$$

G. Compare $(\bar{X}_A - \bar{X}_B)$ with $|\bar{X}_A - \bar{X}_B|^*$.

$[(\bar{X}_A - \bar{X}_B) = 3] > [|\bar{X}_A - \bar{X}_B|^* = 1.55]$

H. Make the decision.

Accept H_a: Conclusion: Coater B is better than coater A, with at least 97.5 percent confidence. Both coaters are better than the present coater.

SPECIAL CASE 1 (Paired Comparisons)

In some experiments, it is possible that the data are obtained in such fashion that the results are paired; i.e., on the same sample item, a result is obtained from both population 1 and population 2. The following are two examples of what are called pair-comparison experiments:

- An experimenter is interested in determining the effect of annealing on the hardness of some heat-treated part. The best procedure is to measure the hardness of each heat-treated part, anneal it, and remeasure the hardness. The difference between the hardness before and after annealing on each part is the effect of annealing on that part.
- An experiment is conducted to determine the effect of humidity on the blackness of printwork produced by typewriter ribbons. The best experiment would consist of taking each typewriter ribbon and determining the blackness at normal humidity, then using a different part of the same ribbon to determine the blackness at high humidity. The difference between the two blacknesses is the effect of high humidity.

In both of the above examples, the differences for all the items tested then become the data, which are analyzed as in case 4.

For the most part, it is impossible to know the variance, or even to estimate the variance, of the differences. Therefore, the only valid estimate of sample size that can be computed is by letting $\delta = f(\sigma)$; e.g., with $\alpha = 0.05$, $\beta = 0.10$, and $\delta_{diff} = \sigma_{diff}$:

$$N_{pairs} = (U_\alpha + U_\beta)^2 \frac{\sigma^2_{diff}}{\delta^2_{diff}}$$

$$= (1.645 + 1.282)^2 \frac{\sigma^2}{\sigma^2} = 8.6$$

Corrected for t:

$$N_{\text{pairs}} = (1.87 + 1.41)^2 \frac{\sigma^2_{\text{diff}}}{\sigma^2_{\text{diff}}} = 10.8$$

To illustrate the design of experiments for paired comparisons, assume that an electronic circuit is supposed to have a current output of 2.5 amps and that the present circuits actually have an average output of 2.4 amps. It is proposed that the circuits can be improved if they are coated with a polyester film.

Special Case 1 Experiment Design

A. State the object of the experiment.

$H_0: \mu_{\text{diff}} = 0$
$H_a: \mu_{\text{diff}} = (\mu_{\text{with film}} - \mu_{\text{without film}}) > 0$

B. Choose α, β, and δ.

$\alpha = 0.10$
$\beta = 0.05$
$\delta_{\text{diff}} = 1\sigma_{\text{diff}}$

C. Look up U_α in table 1.
Look up U_β in table 1.

$U_\alpha = 1.282$
$U_\beta = 1.645$

Compute:

$$N_{\text{pairs}} = (U_\alpha + U_\beta)^2 \frac{\sigma^2}{\delta^2} \qquad N_{\text{pairs}} = (1.282 + 1.645)^2 \frac{\sigma^2}{\sigma^2} = 8.6$$

Correct for t with $\phi = 7.6$.

$$N_t = (t_\alpha + t_\beta)^2 \frac{\sigma^2}{\delta^2} \qquad N_{\text{pairs}} = (1.41 + 1.88)^2 \frac{\sigma^2}{\sigma^2} = 10.8$$

D. Select 11 circuits at random, number each circuit, and test.

$X_1 = 2.5$ $X_7 = 2.4$
$X_2 = 2.0$ $X_8 = 2.3$
$X_3 = 2.4$ $X_9 = 3.0$
$X_4 = 2.9$ $X_{10} = 2.3$
$X_5 = 3.1$ $X_{11} = 2.8$
$X_6 = 2.0$

E. Take each circuit above, coat with polyester film, and retest.

$X_{1a} = 2.8$ $X_{7a} = 2.5$
$X_{2a} = 2.1$ $X_{8a} = 2.7$
$X_{3a} = 2.7$ $X_{9a} = 3.0$
$X_{4a} = 3.4$ $X_{10a} = 2.5$
$X_{5a} = 3.1$ $X_{11a} = 3.1$
$X_{6a} = 2.2$

F. Find the difference for each pair. $X_{ia} - X_i$

Pair	Difference
1	0.3
2	0.1
3	0.3
4	0.5
5	0
6	0.2

Pair	Difference
7	0.1
8	0.4
9	0
10	0.2
11	0.3

Compute $\bar{X}_{\text{diff}}$. — $\bar{X}_{\text{diff}} = 0.218$

Compute S^2_{diff}. — $S^2_{\text{diff}} = 0.0256 \quad S_{\text{diff}} = 0.160$

$\phi = N_{\text{diff}} - 1 = 11 - 1 = 10$

G. Compute $\bar{X}^*_{\text{diff}} = \dfrac{t_\alpha S_{\text{diff}}}{\sqrt{N_{\text{diff}}}}$ — $\bar{X}^*_{\text{diff}} = \dfrac{(1.37)(0.16)}{\sqrt{11}} = 0.068$

H. Compare $\bar{X}_{\text{diff}}$ with $\bar{X}^*_{\text{diff}}$ — $(\bar{X}_{\text{diff}} = 0.218) > (\bar{X}^*_{\text{diff}} = 0.068)$

I. Make the decision. — Accept $H_a\!:\mu_{\text{diff}} > 0$ with at least 90 percent confidence. Conclusion: change the process to include the polyester film.

The reader should note that if the original data are used, but treated erroneously as case 5, the alternative hypothesis will not be accepted at $\alpha = 0.10$ because the variance between parts is so large. Treatment of the data as case 5 is incorrect; treatment as paired comparison is correct.

SPECIAL CASE 2
(Some Data Lost)

In case 5, it was computed that 96 samples each were required for formula A and formula B from a vendor. Suppose that: (1) the vendor was able to supply only 80 samples of formula A and 60 samples of formula B; or (2) the vendor supplied 96 samples of each formulation, which were tested, but some of the results either were lost or were obviously bad data so that only 80 results were available on formula A and 60 results were available on formula B.

The correct procedure in this situation is to analyze the data as if the sample size this experimenter has is the proper size. The criterion, $|\bar{X}_A - \bar{X}_B|^*$, will then be different:

$$|\bar{X}_A - \bar{X}_B|^* = (2.61)(4)\sqrt{\tfrac{1}{80} + \tfrac{1}{60}} = 1.79,$$

which is larger than the 1.52 calculated with 96 samples of each formulation. If $(\bar{X}_A - \bar{X}_B) = 2.1$, the decision will still be to accept $H_a\!:\mu_B < \mu_A$; furthermore, this decision will be made with an α risk equal to or less than 0.01, as required.

If, however, in the above case, the difference between $\bar{X}_A$ and $\bar{X}_B$ had been 1.67, for example, the null hypothesis $(H_0\!:\mu_A = \mu_B)$ would be accepted, but the confidence in this decision would be less than the $(1 - \beta)(100)$ (i.e., 80 percent) confidence that was stipulated.

In the case 7 example, the value of ϕ from the previous experiment $(\phi = 10)$ was large compared to the value of N that was calculated for the present experiment $(N = 4)$. However, if the calculated value of N had been much larger than the value of ϕ from the previous experiment, this value of N would probably be larger than is actually required. For example, assume that new experimental conditions are $S_1^2 = 0.01$, with $\phi = 10$. $\alpha = 0.01$, $\beta = 0.01$, and $\delta = 0.1$. For these conditions, a first estimate of N would be computed.

$$N = (t_\alpha + t_\beta)^2 \frac{S_1^2}{\delta^2} = (2.76 + 2.76)^2 \frac{0.01}{(0.1)^2} = 30.5 = 31$$

After the 31 sample items have been tested, a new estimate of σ^2 (i.e., S_2^2) will be computed from the new data. When an experimenter has two estimates of variance, they are combined as follows:

$$\frac{\phi_1 S_1^2 + \phi_2 S_2^2}{\phi_1 + \phi_2} = S_{avg}^2 = \phi_{avg} = \phi_1 + \phi_2$$

The new value of S_2^2 will have $\phi = 30$ (i.e., $N - 1$); when combined with the previous value of S_1^2, with $\phi = 10$, the latest estimate of σ^2 will have $\phi = 40$. When $\phi = 40$, $t_{0.01} = 2.44$, which is smaller than the 2.76 used to calculate $N = 31$. Using this value of t_α, a new N can be computed.

$$N = (2.44 + 2.44)^2 \frac{0.01}{0.01} = 23.8 = 24$$

Of course, the value of $N = 24$ is based on the assumption that the value of S^2 from both experiments will be exactly the same. This is unlikely; therefore, the value of $N = 31$ is probably too high because t_α and t_β are too high, and the value of $N = 24$ is possibly too low because the value of S_2^2 can be higher than the initial S_1^2. The recommended procedure, therefore, is to split the difference and take $N = 27$.

The criterion should be calculated using the new S_{avg}^2 and t_α should be looked up with the new ϕ_{avg}.

EXERCISES

1. An electronic circuit has a mean output of 7.5 mW, and the standard deviation is known to be 1.2 mW. It is proposed that the output power might be increased substantially if an additional chip is added to the circuit. The present circuit costs $20, and the additional chip will cost $1. Ten thousand chips will be manufactured per month and each additional mW of power will increase profits by $30,000 per month. A cost increase of 5 percent would make the circuit noncompetitive unless the quality were improved.
   - Design an experiment for the purpose of making a business decision on whether to add or not to add the additional chip to the circuit. Justify your decision for selecting α, β, and δ.
   - If the sample mean ($\overline{X}$) for the experiment you design is 7.9 mW, state your engineering decision and your recommended business decision.
   - What is the 95 percent confidence limit on μ?
2. A stud, which is welded to a plate, can be manufactured from either material A or material B. The important criterion is the tensile strength of the weld. The variance is

unknown, but it is presumed to be the same for both materials. Suppose the following values have been chosen for the experiment: $\alpha = 0.05$; $\beta = 0.10$; $\delta = 1.50$.

- Calculate N.
- State the hypotheses.
- Suppose you order the proper number of studs from the vendor but you end up with only 7 welds with material A and 10 welds with material B. You obtain the following tensile-test results (in psi) on the 17 welds:

Material A	Material B
2035	2510
2160	1980
1910	2400
1850	2165
2195	2020
2220	2315
1740	2080
	1910
	2240
	2040

Based on the 17 test results, state your engineering decision and recommend a business decision.

3. You are the development engineer on a project to develop an improved dishwasher. Cleanliness of the dishes is measured on a device that gives a rating from 0 (very bad) to 10 (very good). You propose testing the angle of the water jet both at 40° and at 30° to find out which angle gives the better cleaning. From previous work, you have an estimate of the variance (S^2) of 1.4, with $\phi = 10$. It costs nothing to use either the 40° or 30° angle; however, one unit of improvement is worth $100,000 per year additional profit to the company. Each test costs $1000.

- Design the experiment. Justify all statistical decisions.
- Compute N.
- If $\overline{X}_{40} = 5.2$, $\overline{X}_{30} = 5.7$, $S^2_{40} = 1.3$, and $S^2_{30} = 1.5$, what is your engineering decision?

REFERENCES

Anderson, T. W., and Sclove, S. L. 1974. *Introductory statistical analysis*. Boston: Houghton Mifflin Co.

Guenther, W. C. 1973. *Concepts of statistical inference*. 2nd ed. New York: McGraw-Hill Book Co.

4 Simple Comparative Experiments: Decisions About Population Variances

The previous chapter discussed experiments conducted to make decisions about population means. In many cases, the experimenter is also interested in making decisions about population variances. For example, in case 5 of the previous chapter, the assumption was made that the variances were equal for computation of the sample size and the decision-making criterion. After the experiment was completed, most experimenters would be interested in determining whether or not their assumption was correct. In other problems, the experimenter might have no interest whatsoever in the population mean but a great interest in the population variance.

The philosophy of simple comparative experiments for making decisions about population variances is the same as that for making decisions about population means. The experimenter must choose values for α and β risks, for null and alternative hypotheses, and for the difference (δ) that is important in an engineering sense and must calculate sample size and criterion values. For population variances, however, the formulas used to calculate sample size and criterion values are different than those used in determining population means.

The next two sections of this chapter will describe the techniques for calculating sample size and criterion values when the experimenter is interested in population variances. Following these two sections will be experiment designs for the three possible variance problems that can be encountered and for a problem combining concerns about population mean and population variance.

Sample Size

To determine sample sizes for experiments used to make decisions about population variances, a new term, R, must be calculated. Once a value for R has been determined, the experimenter looks up a value for ϕ in table 5, table 6, or table 9, depending upon the alternative hypothesis, as will be explained below.

When the value for ϕ has been determined, the experimenter adds 1 to this value to obtain the sample size (i.e., $N = \phi + 1$).

For $H_a\!:\sigma^2_{\text{new}} < \sigma^2_0$, R is calculated as follows:

$$R = \frac{\sigma_0^2 - \delta}{\sigma_0^2} < 1.0,$$

where δ is the difference between σ^2_{new} and σ^2_0 at which the experimenter wants to have at most β probability of accepting the null hypothesis. If, for example, $\sigma^2_0 = 100$ and $\delta = 55$, then,

$$R = \frac{100 - 55}{100} = 0.45$$

For $R = 0.45$, and assuming $\alpha = 0.05$ and $\beta = 0.01$, the experimenter determines from table 5 that $\phi = 49$. Thus, $N = 49 + 1 = 50$ is the correct sample size for the experiment.

For $H_a\!:\sigma^2_{\text{new}} > \sigma^2_0$, R is calculated as follows:

$$R = \frac{\sigma_0^2 + \delta}{\sigma_0^2} > 1.0$$

If $\sigma^2_0 = 100$ and $\delta = 50$, then

$$R = \frac{100 + 50}{100} = 1.5$$

For $R = 1.5$, and assuming that $\alpha = 0.05$ and $\beta = 0.05$, the experimenter determines from table 6 that $\phi = 120$. Thus, $N = 120 + 1 = 121$ is the correct sample size for this experiment.

It should be noted at this point that there is one other possible case:

$$H_0\!:\sigma^2_{\text{A}} = \sigma^2_0$$
$$H_a\!:\sigma^2_{\text{A}} \neq \sigma^2_0$$

This case will not be discussed in detail since it is rarely encountered in practice. However, one procedure for calculating N is given for reference:

- Choose a δ_1 for $H_{a1}\!:\sigma^2_{\text{A}} > \sigma^2_0$ and calculate

  $$R = \frac{\sigma_0^2 + \delta_1}{\sigma_0^2},$$ then look up ϕ_1 in table 6.
- Choose a δ_2 for $H_{a2}\!:\sigma^2_{\text{A}} < \sigma^2_0$ and calculate

  $$R = \frac{\sigma_0^2 - \delta_2}{\sigma_0^2},$$ then look up ϕ_2 in table 5.
- Choose the higher ϕ for determining the proper sample size.

It should be noted that if $\delta_1 = \delta_2$, the value for ϕ_1 will be much greater than that for ϕ_2. Alternatively, for a given sample size, δ_2 can be much less than δ_1 to achieve the same $\alpha/2$ risk for the two alternative hypotheses.

If $H_a: \sigma_A^2 \neq \sigma_B^2$, and if δ is the positive difference between σ_A^2 and σ_B^2 that is important, then R is calculated by either

$$R = \frac{\sigma_A^2 + \delta}{\sigma_A^2},$$

where σ_A^2 is assumed to be smaller than σ_B^2, or by

$$R = \frac{\sigma_B^2 + \delta}{\sigma_B^2},$$

where σ_B^2 is assumed to be smaller than σ_A^2.

The value of R calculated as above is exactly the same as $R = \sigma_B^2/\sigma_A^2$, where σ_B^2 is larger than σ_A^2, or σ_A^2/σ_B^2, where σ_A^2 is larger than σ_B^2.

The value of R is looked up, with appropriate values of α and β, in table 9 to obtain $\phi_A = \phi_B$. Note carefully: there is no way to use the above formulas to calculate R since you don't know σ_A^2 or σ_B^2. These formulas are shown only to assist the engineer in understanding and choosing R. Likewise, you have no idea which σ^2 is larger. Therefore, the best experiment design will have the same sample size for both populations.

Criterion Values

Sample variances are not distributed normally about the population variance; rather, they are distributed as chi square (χ^2). Thus, the criterion must be based on the χ^2 distribution.

If $H_a: \sigma_A^2 < \sigma_0^2$, then the proper criterion, $*S^2$, is obtained from the formula

$$*S^2 = \frac{\chi_\alpha^2 \sigma_0^2}{\phi}$$

where the χ^2 value is obtained from table 7 for a given α and ϕ. The χ^2 columns for the various alphas are labeled across the top of the table and the ϕ values along the left side. The column headed $\chi_{0.05}^2$ is the column to use when $\alpha = 0.05$. In this case, the alternative hypothesis is accepted if the calculated S^2 from the experiment is *less than* $*S^2$.

If $H_a: \sigma_A^2 > \sigma_0^2$, then the proper criterion is also obtained from this formula:

$$*S^2 = \frac{\chi_\alpha^2 \sigma_0^2}{\phi}.$$

However, in this case, the χ_α^2 value is obtained from table 8 for a given α and ϕ, and the alternative hypothesis is accepted if the calculated S^2 from the experiment is *greater than* $*S^2$.

If $H_a: \sigma_A^2 \neq \sigma_B^2$, then the proper criterion is F^*, which is looked up in table 10 with ϕ_1 and ϕ_2. ϕ_1 is the number of degrees of freedom ($N - 1$) of the population with the larger S^2, and ϕ_2 is the number of degrees of freedom for the smaller S^2. After S_A^2 and S_B^2 are calculated from the experiment, the ratio (F) of the larger S^2 value to the smaller S^2 value is computed. Thus, we speak of ϕ_1 being the degrees of freedom in the numerator and ϕ_2 being the degrees of freedom in the denominator. The set of three numbers at the intersection of the proper ϕ_1 column and ϕ_2 row is the value of F^* for $\alpha = 0.10$, $\alpha = 0.05$, and $\alpha = 0.01$. The alternative hypothesis is accepted if F is *greater than* F^*.

CASE 1

$$H_0{:}\sigma_1^2 = \sigma_0^2; \qquad H_a{:}\sigma_1^2 < \sigma_0^2$$

A milling machine is turning out parts with an average surface roughness of 100 microinches (μ_0). The variance of the roughness from part to part is 100 (σ_0^2). Some problems are caused by this variation in roughness; e.g., if any 10 of these parts are assembled together, and if each is about 105 microinches, or if each is about 95 microinches, the mechanism is satisfactory. However, if the 10 parts vary, for example, from 90 to 110 microinches, the mechanism is unsatisfactory. If all 10 parts are at 100 microinches, the mechanism is outstanding.

A potential vendor of a new type of milling machine states that his machine will produce parts with an average roughness of 100 microinches and that the variance will be substantially less than 100. Should this new machine be purchased?

Case 1 Experiment Design

A. State the object of the experiment.

$$H_0{:}\sigma_{\text{new}}^2 = \sigma_{\text{old}}^2 = 100$$
$$H_a{:}\sigma_{\text{new}}^2 < \sigma_{\text{old}}^2$$

B. Choose α, β, and δ, and calculate R.

$\alpha = 0.05$
$\beta = 0.01$
$\delta = 50$

$$R = \frac{\sigma_0^2 - \delta}{\sigma_0^2} \qquad R = \frac{100 - 50}{100} = 0.5$$

C. Find sample size in table 5.

$\phi = 66 \qquad N = 66 + 1 = 67$

D. Compute criterion.

From table 7:

χ^2 (for $\phi = 66$ and $\alpha = 0.05$) $= 48.44$

$$*S^2 = \frac{\chi^2\sigma_0^2}{\phi} \qquad *S^2 = \frac{(48.44)(100)}{66} = 73.4$$

E. Obtain 67 parts at random from those made by the new machine, test the parts, and calculate $\bar{X}$ and S^2.

$$\bar{X} = \frac{\sum^{i} X_i}{N} \qquad \bar{X} = 100$$

$$S^2 = \frac{\sum^{i} (X_i - \bar{X})^2}{N - 1} \qquad S_{\text{new}}^2 = 68$$

F. Compare S^2 with $*S^2$.

$(S^2 = 68) < (*S^2 = 73)$

G. Make the decision.

Accept $H_a: \sigma^2_{\text{new}} < \sigma^2_0$ with at least 95 percent confidence. Conclusion: purchase the new machine.

CASE 2

$$H_0: \sigma_1^2 = \sigma_0^2; \qquad H_a: \sigma_1^2 > \sigma_0^2$$

A production line is turning out parts with population mean tensile strength of 600 psi and a variance of 900. Speeding up the production line is proposed as a means of increasing the profit. If the speed should be increased, it is expected that the mean tensile strength will remain the same but the variance will probably increase. If the increase in variance is not too large, say 200, the increase in speed will be worthwhile; however, an increase in variance to a level of 2000 would be very undesirable. Note carefully we are willing to accept the new process even if $\sigma^2_{\text{new}} = 1100$, which is certainly a larger number than 900. The variance of the old process is $\sigma^2_{\text{old}} = 900$, but σ^2_0 for this experiment is $\sigma^2_{\text{old}} + 200 = 1100$. If σ^2_{new} is 1100, we want to accept the null hypothesis with only α risk of making the wrong decision.

Case 2 Experiment Design

A. State the object of the experiment.

$H_0: \sigma^2_{\text{high speed}} = \sigma^2_{\text{regular speed}} + 200$
$= (900 + 200) = 1100$

$H_a: \sigma^2_{\text{high speed}} > \sigma^2_{\text{regular speed}} + 200$
(Note: $\sigma^2_0 = 1100$.)

B. Choose α, β, and δ, and calculate R.

$\alpha = 0.05$
$\beta = 0.01$
$\delta = 900\ (2000 - 1100 = 900)$

$$R = \frac{\sigma_0^2 + \delta}{\sigma_0^2} \qquad R = \frac{1100 + 900}{1100} = 1.82$$

C. Find the sample size from table 6.

$\phi = 102,\ N = 102 + 1 = 103$

D. Compute criterion from table 8.

χ^2 (for $\phi = 102$ and $\alpha = 0.05$) $= 126.32$

$$*S^2 = \frac{\chi^2 \sigma_0^2}{\phi} \qquad *S^2 = \frac{(126.32)(1100)}{102} = 1362.3$$

E. Obtain 103 parts from the process with the conveyor speeded up, and measure the strength of the parts. Compute $\bar{X}$ and S^2.

$$\bar{X} = \frac{\sum^{i} X_i}{N} \qquad \bar{X} = 605 \text{ psi}$$

$$S^2 = \frac{\sum^{i} (X_i - \bar{X})^2}{N - 1} \qquad S^2_{\text{fast}} = 1624$$

F. Compare S^2 with $*S^2$. — $(S^2 = 1624) > (*S^2 = 1362)$

G. Make the decision. — Accept $H_a: \sigma^2_{\text{high speed}} > 1100$ with at least 95 percent confidence. Conclusion: do not increase the conveyor speed.

It should be noted at this point that the null and alternative hypotheses could have been designated with regard to the unacceptable level of variance (i.e., 2000) instead of as shown above. The two hypotheses would then read:

$$H_0: \sigma^2_{\text{high speed}} = 2000$$
$$H_a: \sigma^2_{\text{high speed}} < 2000$$

With the alternative hypothesis in this form, the experiment would be a case 1 experiment instead of a case 2 experiment. The values of α and β would have to be reversed from those shown in the case 2 experiment design, and the case 1 formula for determining R would have to be used. With $\alpha = 0.01$ and $\beta = 0.05$, R would be:

$$R = \frac{2000 - 900}{2000} = 0.55$$

The sample size and the criterion will be exactly the same as calculated in the case 2 example above, and, of course, the decision will be exactly the same.

CASE 3

$$H_0: \sigma_1^2 = \sigma_2^2; \qquad H_a: \sigma_1^2 \neq \sigma_2^2$$

Two manufacturers of welding equipment state that their machines will produce parts having a population mean tensile strength of 10,000 psi. Variance is also a major consideration, but the two manufacturers have no idea of the variance between the parts produced on their machines.

Case 3 Experiment Design

A. State the object of the experiment. — $H_0: \sigma^2_{\text{vendor A}} = \sigma^2_{\text{vendor B}}$; $H_a: \sigma^2_{\text{vendor A}} \neq \sigma^2_{\text{vendor B}}$

B. Choose α and β, and determine R. — $\alpha = 0.05$; $\beta = 0.10$

$$R = \frac{\sigma_A^2}{\sigma_B^2} \quad \text{or} \quad \frac{\sigma_B^2}{\sigma_A^2} \quad \text{such that } R > 1$$

Based on economic considerations, the experimenter calculated that if one machine has a variance three times that of the other machine, it must be rejected with $(1 - \alpha/2)(100)$ percent confidence. Therefore, $R = 3.0$.

C. Find the sample size in table 9. $\phi_A = \phi_B = 30 \quad N_A = N_B = 31$

D. Compute criterion. $F^*_{\phi_1,\phi_2}$ is obtained from table 10. $F^*_{(30,30)} = 1.84$ for $\alpha = 0.05$

E. Obtain 31 parts from machine A and 31 parts from machine B.
Compute $\bar{X}_A$, $\bar{X}_B$, S^2_A, and S^2_B.

$$\bar{X} = \frac{\sum^{i} X_i}{N}$$

$$S^2 = \frac{\sum(X_i - \bar{X})^2}{N - 1}$$

$\bar{X}_A = 10{,}000$ psi
$S^2_A = 160{,}000$
$\bar{X}_B = 10{,}050$ psi
$S^2_B = 250{,}000$

F. Compute $F = S^2_B/S^2_A$. $F = \dfrac{250{,}000}{160{,}000} = 1.56$

G. Compare F with F^*. $(F = 1.56) < (F^* = 1.84)$

H. Make the decision. Accept H_0. With at least 90 percent confidence, the right decision is to purchase either piece of welding equipment.

CASE 4 Combined Problem: $H_{a1}: \mu_A \neq \mu_B$; $\quad H_{a2}: \sigma^2_A \neq \sigma^2_B$

A magnetic tape is supplied in two versions. High signal output is important, but variation in signal output is equally important.

Case 4 Experiment Design

A. Define the object of the experiment.
$H_{01}: \mu_{\text{version A}} = \mu_{\text{version B}}$
$H_{a1}: \mu_{\text{version A}} \neq \mu_{\text{version B}}$
$H_{02}: \sigma^2_{\text{version A}} = \sigma^2_{\text{version B}}$
$H_{a2}: \sigma^2_{\text{version A}} \neq \sigma^2_{\text{version B}}$

B. Choose α, β, and δ for H_1. $\alpha_1 = 0.05 \quad \beta_1 = 0.01 \quad \delta = 1\sigma$
Choose α, β, and R for H_2. $\alpha_2 = 0.05 \quad \beta_2 = 0.10 \quad R = 3$

C. Compute N_1 for hypothesis 1 on assumption that $\sigma^2_1 = \sigma^2_2$.
Find N_2 for hypothesis 2 from table 9.
Choose the higher N for the experiment.

$N_{1A} = N_{1B} = 38$ is the proper sample size for making the decision about μ. $\phi_{2A} = \phi_{2B} = 30$; $N_{2A} = N_{2B} = 31$ is the proper sample size for making decisions about σ^2. Therefore, $N = 38$ is the proper sample size for the experiment.

D. Obtain 38 tapes of version A and 38 tapes of version B and test. Compute $\bar{X}_A$, $\bar{X}_B$, S_A^2, and S_B^2.

$$\bar{X} = \frac{\sum^{i} X_i}{N} \qquad S^2 = \frac{\sum(X_i - \bar{X})^2}{N - 1}$$

$\bar{X}_A = 255 \quad S_A^2 = 36$
$\bar{X}_B = 250 \quad S_B^2 = 25$
$\bar{X}_A - \bar{X}_B = 5.0$

Compute $\bar{X}_A$, $\bar{X}_B$
Compute $F = S_A^2/S_B^2$ or S_B^2/S_A^2.

$$F = \frac{S_A^2}{S_B^2} = 1.44$$

E. Obtain criterion for testing σ^2. Look up $F^*_{(\phi_1,\phi_2,\alpha)}$ in table 10.

$F^*_{(37,37,0.05)} = 1.74$

F. Compare F with F^*.

$(F = 1.44) < (F^* = 1.74)$

G. Make the decision about variances.

Accept H_{02}. With at least 90 percent confidence, the correct decision is that σ_A^2 is not different from σ_B^2.

H. Compute average variance.

$$S_{avg}^2 = \frac{\phi_A S_A^2 + \phi_B S_B^2}{\phi_A + \phi_B}$$

$S_{avg}^2 = 30.5 \quad \phi = 37 + 37 = 74$

$S_{avg} = 5.52$

I. Compute criterion for testing $(\mu_A - \mu_B)$.

$$(\bar{X}_A - \bar{X}_B)^* = t_{\alpha/2} S \sqrt{\frac{1}{N_A} + \frac{1}{N_B}}$$

$t_{0.025} = 2.00$ for $\phi = 74$.

$(\bar{X}_A - \bar{X}_B)^* = (2.00)(5.52)\sqrt{\frac{2}{38}} = 2.53$

J. Compare $(\bar{X}_A - \bar{X}_B)$ with $(\bar{X}_A - \bar{X}_B)^*$.

$[(\bar{X}_A - \bar{X}_B) = 5] > [(\bar{X}_A - \bar{X}_B)^* = 2.53]$

K. Make the decision about means.

Accept $H_{a1}: \mu_A > \mu_B$ with at least 97.5 percent confidence.

L. Make the engineering decision on the tape.

Magnetic tape A is better. It has a higher signal output, and there is no important increase in the variance.

CASE 5

$H_0: \mu_A = \mu_B$; Assumed $\sigma_A^2 = \sigma_B^2$

In a case 5 problem involving means (page 55), it was assumed that the variances were equal but unknown. After the experiment is conducted, the experimenter can calculate an estimate of S^2 for each population and conduct an F-test to determine whether the assumption of equal variance was correct. If the assumption is proven correct, proceed as in the example of case 5 (page 56). If the assumption is proven incorrect, then the criterion of case 6 must be used for the decision making.

Suppose a vendor proposes two different resins for a plastic part. The impact strength of the part is the only property of interest. It is expected that the variances will be equal. A difference of 1σ in impact strength is important to detect.

Case 5 Experiment Design

A. State the object of the experiment.

$H_0: \mu_A = \mu_B$
$H_a: \mu_A \neq \mu_B$

B. Choose the values of α, β, and δ.

$\alpha = 0.01$
$\beta = 0.01$
$\delta = 1\sigma$

C. Look up the value of U_α in table 2 (double-sided); look up the value of U_β in table 1;

$U_\alpha = 2.576$
$U_\beta = 1.282$

Compute $N_A = N_B$.

$$N_A = N_B = 2(U_\alpha + U_\beta)^2 \frac{\sigma^2}{\sigma^2}$$

$$N_A = N_B = 2(2.576 + 1.282)^2 \frac{\sigma^2}{\sigma^2} = 29.77$$

Compute $\phi = N_A + N_B - 2$.

$\phi = 30 + 30 - 2 = 58$

Compute $N_t = 2(t_\alpha + t_\beta)^2 \frac{\sigma^2}{\delta^2}$.

$$N_t = 2(2.66 + 1.30)^2 \frac{\sigma^2}{\sigma^2} = 31.36$$

D. Have the vendor supply 31 parts with resin A and 31 parts with resin B. Compute $\overline{X}_A$, $\overline{X}_B$, S_A^2 and S_B^2.

$\overline{X}_A = 185$ lbs
$\overline{X}_B = 200$ lbs
$S_A^2 = 80$
$S_B^2 = 20$

E. Calculate F and look up F^* (table 10).

$$F = \frac{80}{20} = 4$$

$F^*\ (30,30,.01) = 2.39$

F. Compare F with F^* and make decision on equality of variance.

$[F = 4] > [F^* = 2.39]$
Conclusion: $\sigma_A^2 \neq \sigma_B^2$
The data is case 6, not case 5 as assumed.

G. Complete criterion.

$$|\overline{X}_A - \overline{X}_B|^* = t_\alpha \sqrt{\frac{S_A^2}{N_A} + \frac{S_B^2}{N_B}}$$

$$|\overline{X}_A - \overline{X}_B|^* = 2.66 \sqrt{\frac{80}{31} + \frac{20}{31}} = 4.78$$

H. Compare $(\overline{X}_B - \overline{X}_A)$ with $|\overline{X}_A - \overline{X}_B|^*$

$[(\overline{X}_B - \overline{X}_A) = 15] > [|\overline{X}_A - \overline{X}_B|^* = 4.78]$

I. Make the decision. Accept H_a: resin B > resin A with at least 99.5 percent confidence.
Conclusion: Use resin B for the part.

It will be noted that, in this example, it makes no difference whether the criterion formula of case 5 or case 6 is used. If, however, some sample had been lost or $N_A = N_B$ were different for some reason, then the case 5 and case 6 criterion would be different.

Caution: All of the above tests on variances are based on the assumption that the underlying population is normally distributed. If this assumption cannot be made, the calculated sample size will be too small, in most cases, for the desired α and β risks. Consult a qualified statistician for help; the size of the error could be serious.

EXERCISES

1. A vendor is supplying 100,000 special lamps per year at a cost to the customer of $2,500,000 per year. The special lamp has a mean output (μ_0) of 295 lumens and a variance (σ^2) of 32. The specification for the lamp is 295 ± 5 lumens. Profits were predicted on the basis that 98 percent of the lamps produced would be within specification; however, the number of reject lamps has risen to an intolerable level. It is proposed that certain steps of the manufacturing process be changed in an attempt to reduce the variance. The proposed changes will increase the costs by $0.50 per lamp, but a decrease of 1 lumen in the standard deviation would increase profits by $.50 per lamp. It costs $150 to make an experimental lamp.

- Design the experiment. Justify your statistical decisions.
- If, with your sample size and α, β, and δ, you obtain a result of $S^2 = 18.4$, what is your engineering decision? Is your job now finished? If not, what would you do next?

2. A vendor supplies a sample of 11 parts made with material A and 11 parts made with material B. The vendor does not know which material is best in regards to mean abrasion but, based on the data given below, is sure that the variances of the two materials are the same. Do you agree?

Amount of material abraded, mg/1000 cycles

Material A	Material B
0.25	0.48
0.64	0.54
0.31	0.61
0.72	0.64
0.55	0.58
0.33	0.74
0.62	0.64
0.54	0.55
0.48	0.60
0.57	0.59
0.44	0.63

REFERENCES

Box, G. E. P., Hunter, W. G. and Hunter J. S. 1978. *Statistics for experimenters.* New York: John Wiley and Sons, Inc.

Ott, L. 1977. *An introduction to statistical methods and data analysis.* North Scituate, Mass.: Duxbury Press.

5 Sequential Experiments

As stated in Chapter 3, if the number of sample results available at the end of an experiment is less than the number calculated, the best procedure is to proceed with the analysis as if the available sample size is the correct sample size. Under such conditions, if H_a is accepted, the α risk is no worse than the specified α risk. Further, a glance at the operating characteristic curve in Chapter 2 (figure 2-11) shows that, if μ_1 is much greater than $\mu_0 + \delta$, the risk of accepting the null hypothesis is much less than the specified β risk. This implies that the computed number of trials is excessive in many cases where the alternative hypothesis is accepted. A better procedure, therefore, would be to make and test samples sequentially if it is possible to do so.

Sequential experimenting consists of taking one sample; measuring some property of it; applying a test of significance to the data; and then making a decision to accept H_0, to accept H_a, or to continue testing. A second sample is prepared and tested if, and only if, the experimenter accepts neither H_0 nor H_a.

Sequential analysis cannot usually be used with problems where the response of interest is, for example, the life of a machine. A machine might perform for 8 months before failing; at that point, a second machine would be started and might perform for 14 months, etc. The total test could go on for years. With such a problem, one must compute the proper N and start all N samples at approximately the same time. In this case, the number of samples might be excessive, but the total elapsed time will unquestionably be minimal.

Sequential analysis can and should be used if the cost of a sample is high but only a short time is required to prepare and test each sample. For example, the synthesis of a certain chemical is very costly, requiring \$1000 worth of raw material, two man-days of labor, and the use of expensive equipment. A run takes one day, and testing the quality of the output requires one-half day. With such a problem, the best procedure would be to make one batch on the first day, test it on the second day, and, if necessary, make a second batch on the third day, testing it on the fourth day, etc. The equipment could be used for other problems on the second and fourth days, and so on.

The philosophy for a good sequential experiment is exactly the same as for the simple comparative experiments described in Chapters 1 through 3. However, most of the procedures are different, or, at least, they look different.

The first step is exactly the same as described for previous experiment designs; the experimenter must choose α, β, and δ and must state the null and alternative hypotheses. Two criterion lines are calculated for all values of N:

$$U^* = \frac{\delta}{2} + \frac{a\sigma^2}{N\delta}, \qquad \text{and} \qquad L^* = \frac{\delta}{2} - \frac{b\sigma^2}{N\delta},$$

where $a = \ln[(1 - \beta)/\alpha]$, $b = \ln[(1 - \alpha)/\beta]$, and $N = 1, 2, 3, 4, \ldots$ If, after N trials, the mean of the trials minus μ_0 is greater than U^*, the alternative hypothesis is accepted with at least $(1 - \alpha)(100)$ percent confidence. If, after N trials, the mean of the trials minus μ_0 is less than L^*, the null hypothesis is accepted with at least $(1 - \beta)(100)$ percent confidence.

Notice that the above equations include μ_0, σ^2, α, β, and δ: the same terms that were used in calculating N and the criterion in simple comparative experiments. Though the equations look different, essentially the same sort of operations are being performed in a sequential experiment as are performed in simple comparative experiments.

Three experiment designs are presented in this chapter for sequential experiments:

- Case 1. σ^2 is known; $H_a: \mu_1 > \mu_0$
- Case 2. σ^2 is known; $H_a: \mu_1 \neq \mu_0$
- Case 3. σ^2 is unknown; $H_a: \mu_1 > \mu_0$

CASE 1 σ^2 Is Known; $H_a: \mu_1 > \mu_0$

At the present time, a cam-and-shaft assembly, which requires about one day of machining, is failing within the warranty period in about 20 percent of the field machines in which it is used. On an accelerated test, these production assemblies fail in six hours (μ_0). The variance (σ_0^2) of the accelerated test is 0.5. Certain changes in the machining operation are proposed as a means of improving the life of the part.

Case 1 Experiment Design

A. Define the object of the experiment. $H_0: \mu_1 = \mu_0 = 6$ hours
$H_a: \mu_1 > \mu_0$

B. Choose α, β, and δ. Specify σ^2. $\alpha = 0.05$
$\beta = 0.01$
$\delta = 0.40$ hours $\quad \delta^2 = 0.16$
$\sigma^2 = 0.50$

C. Compute $N = (U_\alpha + U_\beta)^2 \dfrac{\sigma^2}{\delta^2}$. $\quad N = (1.645 + 2.326)^2 \dfrac{0.50}{0.16} = 49.3$

D. Decide on a sequential plan. Make and test a part on alternate days.

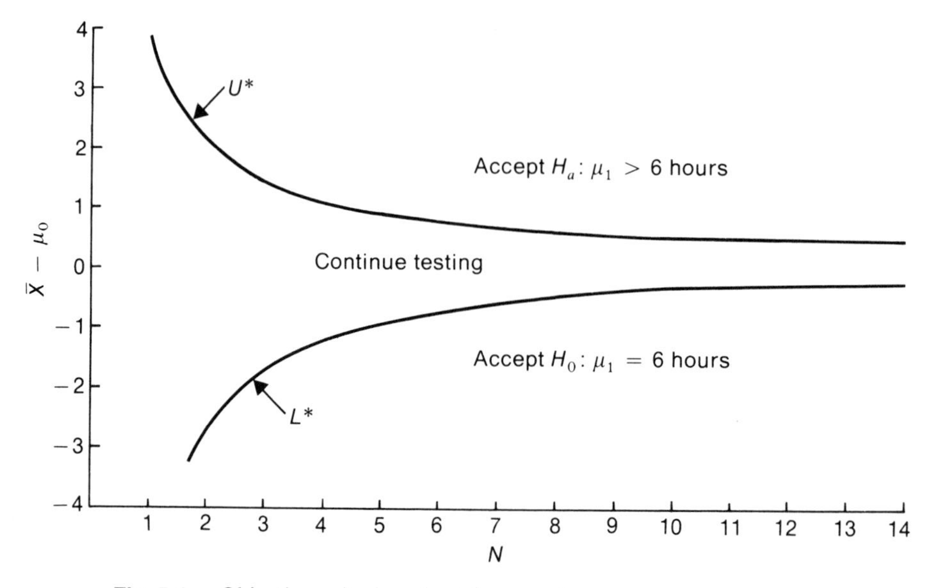

Fig. 5-1. Objective criterion chart for case 1.

E. Compute U^* and L^*.

$a = 2.986$
$b = 4.554$

$$U^* = \frac{\delta}{2} + \frac{a\sigma^2}{N\delta} \qquad U^* = \frac{0.40}{2} + \frac{2.986(0.5)}{(0.4)N} = 0.20 + \frac{3.73}{N}$$

$$L^* = \frac{\delta}{2} - \frac{b\sigma^2}{N\delta} \qquad L^* = \frac{0.40}{2} - \frac{4.554(0.5)}{(0.4)N} = 0.20 - \frac{5.69}{N}$$

F. Make up a chart (figure 5-1) to use as an objective criterion for making decisions.

G. Make and test the first assembly using the new procedure.

The first assembly has a life of 8.4 hours.
8.4 − 6.0 = 2.4 hours

H. Subtract μ_0 from the result. Plot on the chart and make a decision.

The value of 2.4 plotted at $N = 1$ is within the continue-testing space.

I. Make and test a second assembly. Add this result to the first one, divide by 2, and subtract 6.

The second sample result is 8.1. The cumulative result is 8.4 + 8.1 = 16.5.
$\bar{X} = 8.25$
$\bar{X} - 6 = 2.25$

J. Plot $\bar{X} - 6$ on the chart.

The value 2.25 plotted at $N = 2$ is in the accept-H_a space.

K. Make the decision.

Accept H_a:$\mu_1 > \mu_0$ with at least 95 percent confidence. Conclusion: change the process specification to incorporate the procedure used in this experiment.

Those who have diligently read the previous chapters know that, according to the calculation for N, a total of 49 trials was required; however, the experimenter has made a decision based on only 2 samples. Forty-nine was the proper sample size to detect a μ_1 of 6.4 hours ($\mu_0 + \delta$) with only β risk of accepting H_0. Many fewer trials are required if μ_1 is much better than 6.4, which is evidently so in this experiment. Why should the experimenter not make one more sample and test it just to be sure? One reason is that U^* is an objective criterion decided upon before the experiment was conducted for exactly $(1 - \alpha)(100)$ percent confidence. If it is tampered with, it will no longer be objective. Another reason is that the next result could be, say, 5.5, and the average of these three results would be 7.33 hours, which is back in the continue-testing area, though substantially greater than the mean (μ_0) of 6.0 hours.

The new user of statistical design frequently worries about remaining in the continue-testing area forever. The solution to this problem is simple: if still in the continue-testing area, the experiment should be stopped when the computed value of N has been reached. The experimenter would then apply the regular criterion, discussed in the previous chapters, and make a decision accordingly. Note, however, in figure 5-1, that as N increases, the two lines are converging; as N approaches 49, they will be very close together.

CASE 2

σ^2 Is Known; $H_a{:}\mu_1 \neq \mu_0$

The problem used in the case 1 experiment design can be used as the example for case 2 as well. The only difference is that, for case 2, two additional limit lines must be calculated. These additional lines are:

$$U_1^* = -\frac{\delta}{2} + \frac{b\sigma^2}{N\delta}$$

$$L_1^* = -\frac{\delta}{2} - \frac{a\sigma^2}{N\delta}$$

Also, for case 2, it is necessary to restate the alternative hypothesis as:

$$H_{a1}{:}\mu_1 > \mu_0$$
$$H_{a2}{:}\mu_1 < \mu_0.$$

Case 2 Experiment Design

The decision-making space will be divided into four areas:

- Accept $H_{a1}{:}\mu_1 > \mu_0$
- Accept $H_{a2}{:}\mu_1 < \mu_0$
- Accept $H_0{:}\mu_1 = \mu_0$
- Continue testing

From the previous problem, with $\alpha = 0.05$, $\beta = 0.01$, $\delta = 0.4$, and $\sigma^2 = 0.50$:

$$b = 4.554$$

$$a = 2.986$$

$$U^* = 0.20 + \frac{3.73}{N}$$

$$L^* = 0.20 - \frac{5.69}{N}$$

$$U_1^* = -0.20 + \frac{5.69}{N}$$

$$L_1^* = -0.20 - \frac{3.73}{N}$$

The chart for plotting the results and making a decision is drawn as shown in figure 5-2. (Note: if H_{a1} is to be accepted, the mean result minus 6 must only be above

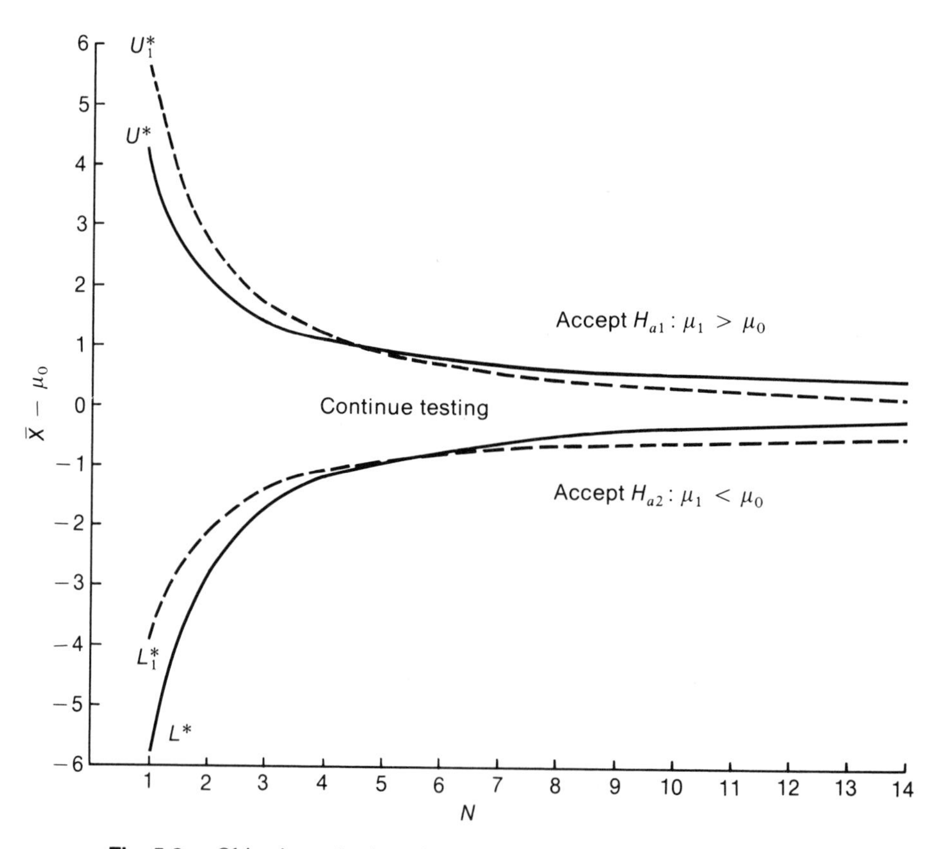

Fig. 5-2. Objective criterion chart for case 2.

U^*; if H_{a2} is to be accepted, the mean result minus 6 must only be below L_1^*. It can be computed that U_1^* will cross L^* at $N = 28$; therefore, at least 28 samples will be required to accept $H_0: \mu_1 = \mu_0$, which will occur if $\bar{X} - 6$ is between U_1^* and L^*.

CASE 3 σ^2 Is Unknown; $H_a: \mu_1 > \mu_0$

A typewriter ribbon presently being produced has a life of 50,000 characters. When tested on an accelerated testing device, the mean number of characters typed before failure is 5000 (μ_0). The variance is 250,000 (σ_0^2). It is thought that addition of a certain chemical to the ink formula will increase the life of the ribbon. One day is required to fabricate a ribbon, and it takes about an hour to perform the test. It is expected that the variance of the ribbons with the additive will differ from the variance for the regular production ribbons; i.e., σ_1^2 is unknown.

Case 3 Experiment Design

A. State the object of the experiment.

$H_0: \mu_{\text{new ribbon}} = \mu_{\text{old ribbon}} = 5000$
$H_a: \mu_{\text{new ribbon}} > \mu_{\text{old ribbon}}$

B. Choose α, β, and δ in terms of σ_{new}.

$\alpha = 0.05$
$\beta = 0.01$
$\delta = 0.50\sigma_{\text{new}}$

Compute N.

$N = 63$

$$N = (U_\alpha + U_\beta)^2 \frac{\sigma^2}{\delta^2}$$

Since N is large, decide upon a sequential experiment.

C. From table 13, obtain limits U^* and L^* for $\alpha = 0.05$, $\beta = 0.01$, and $\delta = 0.50\sigma$.

$$D = \frac{\delta}{\sigma} = \frac{0.5\sigma}{\sigma} = 0.5$$

N	L^*	U^*
6	[−3.91]	[2.75]
8	[−2.94]	2.59
10	−2.35	2.49
15	−1.47	2.36
20	−0.96	2.33
25	−0.59	2.31
30	−0.30	2.34
35	−0.06	2.37
40	0.14	2.41
45	0.32	2.45
50	0.48	2.50
60	0.75	2.59
70	0.99	2.69

D. Make up a chart from the above table (see figure 5-3).

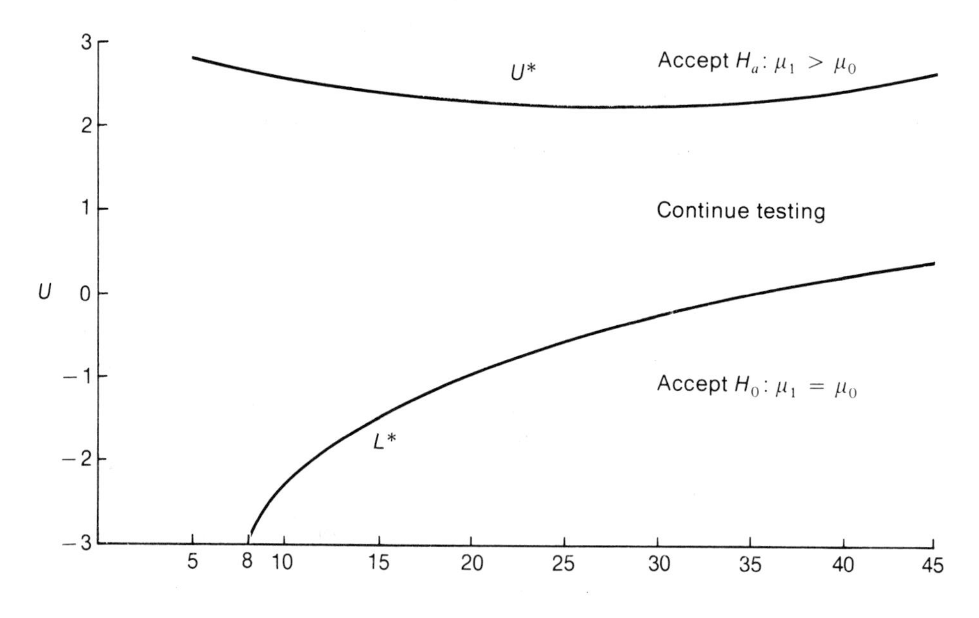

Fig. 5-3. Objective criterion chart for case 3.

E. Make and sequentially test samples. Calculate U_i after each sample result is obtained.

$$U_i = \frac{T_i}{\sqrt{SS_i}},$$

where $T_i = \sum^{i} (X_i - \mu_0)$ and $(SS)_i = \sum^{i} (X_i - \mu_0)^2$.

N	X_i	$X_i - \mu_0$	T_i	$\sqrt{SS_i}$	U_i
1	5500	500	500	500	1.00
2	5100	100	600	510	1.18
3	4900	−100	500	520	0.96
4	5900	900	1400	1039	1.35
5	5400	400	1800	1113	1.62
6	5100	100	1900	1118	1.70
7	5200	200	2100	1136	1.84
8	5000	0	2100	1136	1.84
9	6100	1100	3200	1581	2.02
10	5800	800	4000	1772	2.26
11	5700	700	4700	1905	2.47

F. Make the decision.

After 11 trials, $U_{11} = 2.47$ is greater than $U^* = 2.46$.
Therefore, accept $H_a: \mu_1 > 5000$. Conclusion: make all future formulas of the ink with the additive.

EXERCISES

1. A complicated platinum part is pressed at room temperature into a complicated titanium part. Both parts are expensive to manufacture, but the test is rather inexpensive, consisting only of measuring the force required to separate the two parts. (Note: the parts are scrap after the test.) At the present time, the mean force (μ_0) is 275 pounds, and the variance (σ^2) is

100. It is proposed that, if the platinum part is cooled to $-50°$ before being pressed into the titanium part, a higher force will be required to separate the parts at normal operating conditions.

- Design a sequential experiment, given that $\alpha = 0.01$, $\beta = 0.05$, and $\delta = 6$ pounds. Calculate N.
  Use the following numbers sequentially until a decision is reached on your chart, then calculate the 99 percent confidence interval on the new μ: 278, 284, 291, 286, 285, 284, 289, 297, 282, 275, 296, and 280.
- Use the same numbers as above, but start with the bottom number and go up the list.
- State your engineering decision.

2. Suppose that you are given the assignment in exercise 3 of Chapter 4 and that you have chosen $\alpha/2 = 0.25$, $\beta = 0.05$, $\delta = 1.0$, and $N_{40} = N_{30} = 78$. You then decide, however, to conduct a pair-comparison sequential experiment. Sequentially determine on which machine you should stop the experiment, and state your decision if the following results were obtained on the first 12 machines tested:

Machine	Cleanliness rating	
	30°	40°
1	5.5	5.1
2	4.8	4.5
3	6.2	6.2
4	6.9	6.8
5	6.4	5.9
6	3.7	3.6
7	5.7	5.8
8	5.9	5.2
9	5.1	4.9
10	6.0	5.7
11	5.1	4.8
12	7.8	7.1

REFERENCES

Davies, O. L. 1978. *Design and analysis of industrial experiments.* 2nd ed. New York: Longmans, Inc.

Wald, A. 1947. *Sequential analysis.* New York: John Wiley and Sons, Inc.

Part Two

Two-Level Multivariable Experiments

6 Two-Level Multivariable Experiments: Eight-Trial Hadamard Matrix Designs

If an experimenter is interested in determining the effects of more than one variable on a certain response, the usual procedure is to design a simple comparative experiment to determine the effect of the first variable; then a second comparative experiment to determine the effect of the second variable; and so on. It will be shown that this procedure is extremely inefficient. For example, suppose that for selected values of σ^2, α, β, and δ, $N_{\text{high}} = N_{\text{low}}$ and is calculated to be 4 and that three different variables (A, B, and C) are expected to influence μ. The term N_{high} refers to the sample size at what is called the high level of a variable; N_{low} is the sample size at the low level of the same variable.

Introduction to Matrix Experiments

With a one-factor-at-a-time procedure, a total of 24 samples would be required:

first,
4 samples at low A, low B, low C
4 samples at high A, low B, low C — Determines effect of A (but only at low B and low C).

then,
4 samples at best A, low B, low C
4 samples at best A, high B, low C — Determines effect of B (but only at the best level of A and low C).

finally,
4 samples at best A, best B, low C
4 samples at best A, best B, high C — Determines effect of C (but only at the best levels of A and B).

For the same experimental conditions, it will be shown that a single matrix-design experiment that included all the variables would permit a valid experiment to be made with a total of only eight samples: a savings of 67 percent in direct sample preparation and testing costs and time. Furthermore, the information obtained from those eight samples in a matrix-design experiment would be greater than would be obtained by a series of simple comparative experiments.

The experimental situation described above is referred to as a two-level, three-variable experiment; i.e., each of the three variables has a high level and a low level that are being investigated in the experiment. The eight trials required for a two-level, three-variable experiment may be represented in four different ways, in terms of:

- High and low levels of the variables.
- Treatment combinations.
- Vertices of a cube.
- Plus (+) and minus (−) signs.

To illustrate, suppose that eight samples were prepared as follows:

Trial	Level of A	Level of B	Level of C
1	low	low	low
2	high	low	low
3	low	high	low
4	high	high	low
5	low	low	high
6	high	low	high
7	low	high	high
8	high	high	high

Here, the term "high" indicates one level of a given variable (e.g., a 200° processing temperature), and "low" indicates a second level of the same variable (e.g., a 150° processing temperature). One term could also indicate samples from XYZ Company, and the other term could indicate samples from ABC Company. The terms low and high do not necessarily refer to the actual numerical values of quantitative variables; the usual convention is that the present value of a variable is the low level, and the value of the same variable that is expected to improve the product or process is the high level.

The eight trials listed above cover all combinations of the two levels and three variables. These treatment combinations may also be portrayed in a factorial block, as shown in figure 6-1.

Either the number 1 or the absence of a letter indicates the low level of a variable, while a letter represents the high level of the variable; e.g., "a" represents

		1	c
1	1	Trial 1 (1)	Trial 5 c
	a	Trial 2 a	Trial 6 ac
b	1	Trial 3 b	Trial 7 bc
	a	Trial 4 ab	Trial 8 abc

Fig. 6-1. Factorial block.

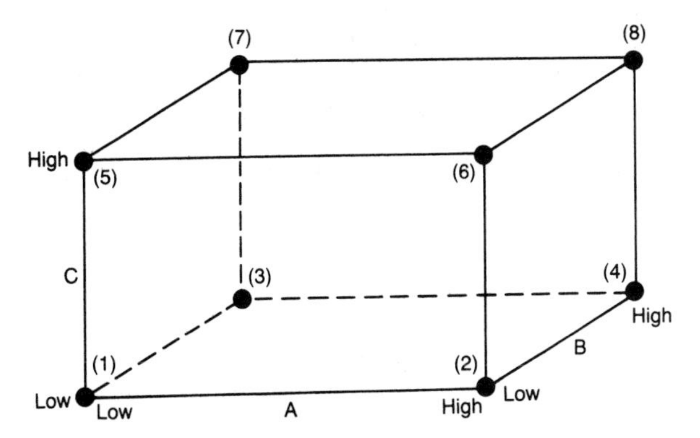

Fig. 6-2. Trial combinations in three-dimensional experimental space.

the high level of variable A, "b" represents the high level of variable B, etc. The block abc (also called the treatment combination abc) is in the "a" row, the "b" row, and the "c" column; therefore, this block designates the sample to be made at the high level of A, the high level of B, and the high level of C. This, of course, is trial number 8.

A third way of presenting the trials is shown in figure 6-2. Here, the trials are the vertices of the three-dimensional experimental space.

A fourth way of representing the trials of a matrix experiment is to let (−) signify the low level of a variable and (+) signify the high level of a variable. Thus, the eight trials described above may be represented as:

Trial	A	B	C
1	−	−	−
2	+	−	−
3	−	+	−
4	+	+	−
5	−	−	+
6	+	−	+
7	−	+	+
8	+	+	+

Assume that an experimenter has three variables of interest: temperature, time, and material supplier (vendor). Further assume that the experimenter wants to determine the effects on some response of interest of two levels of temperature (100° and 150°), two levels of time (5 minutes and 10 minutes), and two different vendors (vendor X and vendor Y). If the experimenter designates temperature as variable A, time as variable B, and vendor as variable C, the various designations of the three variables could be as follows:

- Variable A (Temperature)
  - 100° is designated as the low, or (−), or (1) level of variable A.
  - 150° is designated as the high, or (+), or (a) level of variable A.
- Variable B (Time)
  - 5 minutes is designated as the low, or (−), or (1) level of variable B.
  - 10 minutes is designated as the high, or (+), or (b) level of variable B.

- Variable C (Vendor)
  - Vendor X is designated as the low, or (−), or (1) level of variable C.
  - Vendor Y is designated as the high, or (+), or (c) level of variable C.

Thus, referring to the previous discussion, trial 8 could be represented by any of the following designations:

A	Variable B	C
high	high	high
+	+	+
a	b	c
150°	10 min	vendor Y

The 8-trial matrix experiment with three variables is also called a complete factorial experiment because all combinations of the variables are tested. It is also called a 2^3 experiment, where the 2 refers to the number of levels of the variables and the 3 refers to the number of variables. With this designation, the total number of treatment combinations can be immediately calculated: a 2-level, 3-variable experiment has $2^3 = 8$ treatment combinations; a 2^5 experiment has 32 treatment combinations; and so on.

If an experimenter makes a single sample at each of the eight treatment combinations specified above and obtains a test result on each sample, it is obvious that the difference between sample result 2 and sample result 1 is due, except for experimental error, to the high level of A vs the low level of A. The same can be said for samples 4 vs 3, 6 vs 5, and 8 vs 7. With these eight samples then, the effect of variable A has been measured four times. This is the exact value of N that was computed as being necessary for making a valid decision with α and β risks. The experimenter has four results at high A and four results at low A, and the difference between the mean of these four results at high A and the mean of the four results at low A is an estimate of $(\mu_{\text{high A}} - \mu_{\text{low A}})$:

$$(\bar{X}_{\text{high A}} - \bar{X}_{\text{low A}}) = \frac{(2-1)+(4-3)+(6-5)+(8-7)}{4}$$

$$= \frac{2+4+6+8}{4} - \frac{(1+3+5+7)}{4}.$$

The beauty of the matrix approach, moreover, is that the same eight results can be used to estimate $\mu_{\text{high B}} - \mu_{\text{low B}}$. Note that the difference in results between samples 3 and 1 is due, except for experimental error, to the high level of B vs the low level of B. The same is true for samples 4 vs 2, 7 vs 5, and 8 vs 6. Thus, the effect of $\mu_{\text{high B}} - \mu_{\text{low B}}$ is estimated four times, as required.

Finally, it is apparent that $\mu_{\text{high C}} - \mu_{\text{low C}}$ is estimated four times with the results of sample 5 vs 1, 6 vs 2, 7 vs 3, and 8 vs 4, as required.

Note carefully that in a problem such as that described above, where there are three variables and it is computed that $N_{\text{high}} = N_{\text{low}} = 4$ for a valid experiment, the experimenter makes only one sample of each treatment combination of the matrix. N refers to the number of comparisons that are to be obtained, not to the number of samples in each treatment combination. On the other hand, if $N_{high} = N_{low}$ is calculated to be 8 for a valid experiment, and there are three variables, then the experimenter must prepare and test two samples at each of eight treatment combinations; there are only four treatment combinations with high A, but N_{high} must be 8. It therefore follows that each treatment combination must be duplicated for N_{high} to be 8 for all variables. Two-level multivariable experiments are simple extensions of the single-variable cases described in Chapter 3.

This book will discuss only the two cases that are usually encountered in two-level multivariable experiments: case 2, where σ_i^2 are known and all σ_i^2 are equal; and case 5, where σ_i^2 are unknown and all σ_i^2 are equal. All the formulas, theory, procedure, and practices that were described for case 2 in Chapter 3 apply to multivariable matrix experiments if σ^2 is known. All the formulas, theory, procedures, and practices that were described for case 5 in Chapter 3 apply to multivariable experiments if σ^2 is unknown.

Other Advantages of Matrix Experiment Design

If, in a one-factor-at-a-time experiment, it is shown that high A is better than low A; and then, at high A, that low B is better than high B; and, finally, at high A and low B, that high C is better than low C, the experimental path of the one-factor-at-a-time experiment for three variables can be shown by the dotted lines in figure 6-3. In the total space of A, B, and C, the experimenter has explored only one of the four possible lines comparing the high level of A with the low level of A. The experimenter has no idea what the response difference is between (+1, +1, +1) and (−1, +1, +1). Furthermore, for example, the experimenter is able to interpolate between (−1, −1, −1) and (+1, −1, −1) but cannot extrapolate to (−1, +1, +1) or (+1, +1, +1).

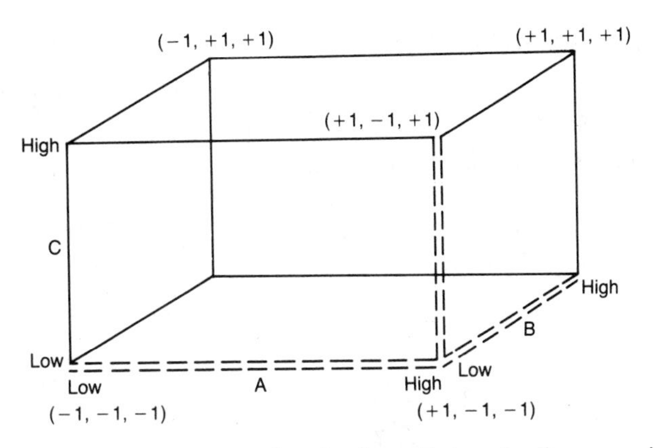

Fig. 6-3. Experimental path of one-factor-at-a-time experiment.

The experimenter is led to believe that the vertex $(+1, -1, +1)$ is the best combination of levels of A, B, and C. This is not necessarily true. It could very well be that, if interactions are present, any of the untested vertices could be far superior to $(+1, -1, +1)$.

By contrast, with a matrix (factorial) design, all the vertices of the space are tested, and the experimenter can interpolate anywhere within the three-dimensional space.

It will be shown later in this chapter that, with eight trials and three variables in a matrix design, the significance of interactions can be estimated as well as the main factors. This is an additional bonus of information provided by matrix designs with no additional cost. It will also be shown that, with factorial designs, if a mistake is made in the production of a sample or if a mistake is made in testing a properly made sample, the error will usually be apparent. This is not true with one-factor-at-a-time experiments without replication.

Most useful two-level multivariable factorial designs can be generated from Hadamard matrices, which were discovered by Jacques Hadamard, a French mathematician. Use of these matrices for experiment design was first described by Plackett and Burman, who limited their use to saturated designs. This author, however, has shown the applicability of these matrices to most two-level experiment designs.

In this chapter, most of the two-level experiment designs shown are suitable for "paper-and-pencil" calculations; Chapter 13 gives a computer program that will generate these designs and will also analyze the data from the experiments. This computer program will also be used to design and analyze the experiments discussed in Chapters 7, 8 and 9.

This computer program is written in APL, which is not readily available to many engineers. If not available, the matrix can be generated by hand with pencil and paper. The more difficult and time-consuming analysis of the data can be performed with any standard ANOVA program. An example of this analysis will be presented in Chapter 15.

Construction of Hadamard Matrices

The initial vectors used for cyclically generating Hadamard matrices of sizes 8×8, 16×16, 32×32, 64×64, and 128×128 are determined from group theory, an explanation of which is beyond the scope of this book. The size of the vector specifies the size of the experiment; e.g., an 8×8 Hadamard matrix specifies $T = 8$ trials and is useful for up to $T - 1 = 7$ variables at two levels for each variable. The primary vectors for cyclically generating the matrices are:

$T = 8$ $\quad + + + - + - -$

$T = 16$ $\quad + + + + - + - + + - - + - - -$

$T = 32$ $\quad + + + + + - - + + - + - - + - - - - + - + - + + + -$
$+ + - - -$

$T = 64$ + + + + + + − + − + − + + − − + + − + + + − + + − +
− − + − − + + + − − − + − + + + + − − + − + − − −
+ + − − − − + − − − − −

$T = 128$ + + + + + + + − + − + − + − − + + − − + + + − + + +
− + − − + − + + − − − + + − + + + + − + + − + − + +
− + + − − + − − + − − − + + + − − − − + − + + + + +
− − + − + − + + + − − + + − + − − − + − − + + + +
− − − + − + − − − − + + − − − − − + − − − − − −

Generation of the Hadamard matrix from the primary vector is shown in the following example for $T = 8$.

EXAMPLE

A. The proper vector is put in column form.

+
+
+
−
+
−
−

B. The vector is permuted six ($T - 2$) times, as shown.

+	−	−	+	−	+	+
+	+	−	−	+	−	+
+	+	+	−	−	+	−
−	+	+	+	−	−	+
+	−	+	+	+	−	−
−	+	−	+	+	+	−
−	−	+	−	+	+	+

C. A set of minus (−) signs is added at the bottom.

+	−	−	+	−	+	+
+	+	−	−	+	−	+
+	+	+	−	−	+	−
−	+	+	+	−	−	+
+	−	+	+	+	−	−
−	+	−	+	+	+	−
−	−	+	−	+	+	+
−	−	−	−	−	−	−

D. A column of plus (+) signs is added to the left, and the columns are numbered.

0	1	2	3	4	5	6	7
+	+	−	−	+	−	+	+
+	+	+	−	−	+	−	+
+	+	+	+	−	−	+	−

+	−	+	+	+	−	−	+
+	+	−	+	+	+	−	−
+	−	+	−	+	+	+	−
+	−	−	+	−	+	+	+
+	−	−	−	−	−	−	−

The column of pluses on the left serves no useful purpose for the experimenter. The column labeled 1 is called contrast column one, which will be used as described below, and all the other columns are given similar designations, with the last column called contrast column seven.

Use of the 8 × 8 Hadamard Matrix

The 8 × 8 Hadamard matrix is useful for all two-level matrix designs for testing up to seven hypotheses—e.g., $(H_o)_i: \mu_{\text{high}} = \mu_{\text{low}}$ with up to seven variables, where it is computed that the proper sampling size (N) is 4 or less for each level of the variables. $N_{\text{high}} = N_{\text{low}} = 2(U_\alpha + U_\beta)^2 \dfrac{\sigma^2}{\delta^2}$, or $N_{\text{high}} = N_{\text{low}} = 2(t_\alpha + t_\beta)^2 \dfrac{S^2}{\delta^2}$.

The following experiment designs can be obtained from the 8 × 8 matrix by proper labeling of the matrix:

- One variable, each treatment combination replicated four times. The effect of the variable and an estimate of the variance with 6 degrees of freedom is obtained.
- Two variables, each treatment combination replicated twice. The mean effects of both variables and their interaction are obtained. An estimate of the variance with 4 degrees of freedom is also obtained.
- Three variables. The mean effects of all three variables and all their interactions can be determined. No estimate of the variance can be obtained.
- Four variables. The mean effects of all variables can be determined if all three-factor interactions are assumed to be insignificant. Two-factor interactions are confounded with each other. (Confounding will be described later.) No estimate of the variance can be obtained.
- Five, six, or seven variables. The mean effects of the variables can be determined if all interactions are assumed to be insignificant. No interactions can be estimated, and no estimate of the variance can be obtained.

One Variable, Each Treatment Combination Replicated Four Times

The matrix is labeled as follows for this experiment design, and the treatment combinations required for the experiment are then derived:

Trial	0	A 1	2	3	4	5	6	7	Treatment combinations
1	+	+	−	−	+	−	+	+	a
2	+	+	+	−	−	+	−	+	a
3	+	+	+	+	−	−	+	−	a
4	+	−	+	+	+	−	−	+	(1)
5	+	+	−	+	+	+	−	−	a
6	+	−	+	−	+	+	+	−	(1)
7	+	−	−	+	−	+	+	+	(1)
8	+	−	−	−	−	−	−	−	(1)

Since there is only one variable, only one column (number 1) is used; this column specifies how the samples should be made with respect to that variable. The procedure for obtaining the treatment combinations is to read across the row for each trial. Where there is a plus (+) sign under a variable, specify the high, or upper, level of the variable; where there is a minus (−) sign, specify the low level of the variable. Thus, in the first row, trial 1, the sign under A is a plus; therefore, the treatment combination is (a). Since there are no other variables indicated in columns 2 through 7, the signs in these columns have no use in the design of this experiment.

It is apparent that this design is the same as either the case 2 or the case 5 simple comparative experiment of Chapter 3, where N, the number of comparisons, is 4. There are two distinct populations, and the null hypothesis is $H_0{:}\mu_{\text{high A}} = \mu_{\text{low A}}$.

The data from this experiment can be analyzed by reading down the columns. The mean effect of variable A is obtained by using the signs in column A with the results obtained on the eight trials and dividing by 4:

$$\begin{aligned}&(\text{result } 1 + \text{result } 2 + \text{result } 3 - \text{result } 4 + \text{result } 5 - \text{result } 6\\&\quad - \text{result } 7 - \text{result } 8)/4 = (\bar{X}_{\text{high A}} - \bar{X}_{\text{low A}}).\end{aligned}$$

The resultant value is compared with the proper criterion $|\bar{X}_{\text{high}} - \bar{X}_{\text{low}}|^* = U_\alpha \sigma \sqrt{\frac{1}{4} + \frac{1}{4}}$, if σ^2 is known, or $t_\alpha S \sqrt{\frac{1}{4} + \frac{1}{4}}$, if σ^2 is unknown, to make a decision about the difference in population means $(\mu_A - \mu_B)$.

All of the data can be condensed into a block diagram as shown in figure 6-4.

High A	Low A
Result 1	Result 4
2	6
3	7
5	8

Fig. 6-4. Block diagram of results.

It is now apparent that the four results at low A can be used to obtain an estimate of the variance (S^2) with three degrees of freedom ($N_{\text{low}} - 1$) by using the formula

$$S^2 = \frac{\sum(X_i - \bar{X})^2}{N_{\text{low}} - 1}.$$

Similarly, the four results at high A will also give an estimate of variance with three degrees of freedom. The average of these two estimates will be an estimate of the variance with six degrees of freedom.

Exactly the same estimate of variance can be obtained by using the six unlabeled columns of the Hadamard matrix. (An unlabeled column is defined as a column not used to determine a main effect or interaction.) Each unlabeled column gives an estimate of variance with one degree of freedom, and the average of the six estimates will have six degrees of freedom. The formula for determining an estimate of variance from an unlabeled column is:

$$S_i^2 = \frac{[\sum(\text{column sign})(\text{result})]^2}{T} \quad \text{with one degree of freedom,}$$

where i = column number and T = number of trials in the experiment. In this case, $T = 8$. For example, using column 2, an estimate of variance is obtained as follows:

$$(-\text{result } 1 + \text{result } 2 + \text{result } 3 + \text{result } 4 - \text{result } 5 + \text{result } 6 - \text{result } 7 - \text{result } 8)^2/8 = S_2^2 \text{ with one degree of freedom.}$$

Using column 3, another estimate of variance is obtained as follows:

$$(-\text{result } 1 - \text{result } 2 + \text{result } 3 + \text{result } 4 + \text{result } 5 - \text{result } 6 + \text{result } 7 - \text{result } 8)^2/8 = S_3^2 \text{ with one degree of freedom.}$$

When using unlabeled columns to determine an estimate of variance, it is important to remember that *all* unlabeled columns must be used. Even if one of the columns should give a very high estimate of variance, under no circumstances should it be presumed that this is a poor estimate that should be eliminated. All estimates from all unlabeled columns must be used to determine the average estimate of variance.

When the estimate of variance has been determined for each unlabeled column, the average estimate of variance is determined by the following formula:

$$S_{\text{avg}}^2 = \frac{\sum^{i} S_i^2}{j} \quad \text{with } j \text{ degrees of freedom,}$$

where i = contrast column numbers (in this case, 2, 3, 4, 5, 6, and 7), and j = the number of contrast columns used (in this case, $j = 6$).

Two Variables, Each Treatment Combination Replicated Twice

For this design, the Hadamard matrix is labeled as follows:

Trial	0	A 1	B 2	3	−AB 4	5	6	7	Treatment combinations
1	+	+	−	−	+	−	+	+	a
2	+	+	+	−	−	+	−	+	ab
3	+	+	+	+	−	−	+	−	ab
4	+	−	+	+	+	−	−	+	b

Trial	0	A 1	B 2	3	−AB 4	5	6	7	Treatment combinations
5	+	+	−	+	+	+	−	−	a
6	+	−	+	−	+	+	+	−	b
7	+	−	−	+	−	+	+	+	(1)
8	+	−	−	−	−	−	−	−	(1)

The reader is reminded of the convention for designating treatment combinations described earlier: the presence of the letter signifies the high level of the variable; the absence of the letter signifies the low level of the variable. Thus, treatment combination a means high A and low B.

The first contrast column is labeled A, and the second contrast column is labeled B. These columns are called contrast columns because they consist of both pluses and minuses. The first column (all pluses) is not a contrast column. Only contrast columns are useful for decision making.

Column 4, labeled −AB, is not an independent contrast; rather, this column measures the interaction of variables A and B. The reader will note that, for each trial, the sign in the −AB column is the reverse of the sign obtained when the sign in column 1(A) is multiplied by the sign in column 2(B). For example:

- In trial 1, A = (+), B = (−); (+)(−) = (−), which is the reverse of (+) in the −AB column.
- In trial 8, A = (−), B = (−); (−)(−) = (+), which is the reverse of (−) in the −AB column.

The treatment combinations are obtained by reading across the rows. Trial 1 has a (+) in the A column and a (−) in the B column; therefore, the first sample should be made at high A and low B, which is treatment combination a. Only the signs in the columns that designate variables are used in determining treatment combinations; the sign in the −AB column is not used to determine these treatment combinations.

This design is called a replicated complete factorial because all combinations of the two levels of A and B are included in the design and all combinations appear twice.

The data from an experiment are analyzed by reading down the columns in exactly the same manner as for the experiment (one variable replicated four times) described above:

$$(\bar{X}_{\text{high A}} - \bar{X}_{\text{low A}}) = (\text{trial 1} + \text{trial 2} + \text{trial 3} - \text{trial 4} + \text{trial 5} - \text{trial 6} - \text{trial 7} - \text{trial 8})/4.$$

$$(\bar{X}_{\text{high B}} - \bar{X}_{\text{low B}}) = (-\text{trial 1} + \text{trial 2} + \text{trial 3} + \text{trial 4} - \text{trial 5} + \text{trial 6} - \text{trial 7} - \text{trial 8})/4.$$

$$(\bar{X}_{\text{high AB}} - \bar{X}_{\text{low AB}}) = (\text{trial 1} - \text{trial 2} - \text{trial 3} + \text{trial 4} + \text{trial 5} + \text{trial 6} - \text{trial 7} - \text{trial 8})/4.$$

The four unlabeled columns (columns 3, 5, 6, and 7) are used to obtain an estimate of the variance with four degrees of freedom. For example, from column 3:

(−trial 1 − trial 2 + trial 3 + trial 4 + trial 5 − trial 6 + trial 7 − trial 8)^2/8 is an estimate of σ_0^2 with one degree of freedom.

EXAMPLE

A photocopier should operate under all temperature conditions from 100° to 200° and at humidities from 30 percent to 80 percent. The experimenter wants to determine the effects of temperature and humidity, and of any interaction between them, on the performance of the photocopier. The response of interest is the amount of toner buildup on the photoconductor. Variance is unknown.

A. State the object of the experiment. Note: the experimenter always states the alternative hypotheses for interactions as an absolute value. The sign is unimportant.

$(H_0)_1: \mu_{100°} = \mu_{200°}$
$(H_a)_1: \mu_{200°} > \mu_{100°}$
$(H_0)_2: \mu_{30\%} = \mu_{80\%}$
$(H_a)_2: \mu_{30\%} \neq \mu_{80\%}$
$(H_0)_3: \mu_{\text{AB interaction}} = 0$
$(H_a)_3: |\mu_{\text{AB interaction}}| > 0$

B. Make decisions on α, β, and δ.

$\alpha = 0.1$
$\beta = 0.1$
$\delta = 2.5\,\sigma$
$\sigma^2 = \text{unknown}$

C. Compute N.

$$N = 2(U_\alpha + U_\beta)^2 \frac{\sigma^2}{\delta^2} = 2(1.64 + 1.28)^2 \frac{\sigma^2}{(2.5\sigma)^2} = 2.72$$

Since one of the alternatives is double-sided, the size of N must be computed using table 2 (double-sided) for U_α and table 4 (double-sided) for t_α. Table 1 is used for U_β and table 3 is used for t_β.

If the 8 × 8 Hadamard matrix is used, $N_{\text{high}} = N_{\text{low}} = 4$. There will be four unlabeled columns available for estimating variance. ϕ will, therefore, be 4. $t_\alpha = 2.13$, $t_\beta = 1.53$ for $\phi = 4$. The required sample size, corrected for not knowing the variance, is, therefore,

$$N_{\text{high}} = N_{\text{low}} = 2(t_\alpha + t_\beta)^2 \frac{\sigma^2}{\delta^2}$$

$$= 2(2.13 + 1.53)^2 \frac{\sigma^2}{(2.5\sigma)^2} = 4.2$$

The required sample size, 4.2, is approximately equal to the sample size, 4, supplied by the 8 × 8 Hadamard matrix. Therefore, the 8 × 8 Hadamard matrix is the correct matrix. If the required sample size had been $N_{\text{high}} = N_{\text{low}} = 8$, for example, then this 8 × 8 Hadamard matrix would be too small and the next larger, 16 × 16, Hadamard matrix would be required. This 16 × 16 matrix will be described in the next chapter.

Specify the levels of the variables.

	Low level	High level
A = Temperature	100°	200°
B = Humidity	30%	80%

D. Label the 8 × 8 Hadamard matrix properly, and determine the eight treatment combinations.

Trial	0	A 1	B 2	3	−AB 4	5	6	7	Treatment combinations
1	+	+	−	−	+	−	+	+	a
2	+	+	+	−	−	+	−	+	ab
3	+	+	+	+	−	−	+	−	ab
4	+	−	+	+	+	−	−	+	b
5	+	+	−	+	+	+	−	−	a
6	+	−	+	−	+	+	+	−	b
7	+	−	−	+	−	+	+	+	(1)
8	+	−	−	−	−	−	−	−	(1)

E. Test the machine under the temperature and humidity conditions specified in each trial.

Trial	Temp, deg	Humidity, %	Toner buildup, mg/100 in.^{2}
1	200	30	16
2	200	80	32
3	200	80	28
4	100	80	15
5	200	30	14
6	100	80	17
7	100	30	9
8	100	30	12

F. Determine the effect of the variables and interaction by using the signs in the columns with the results obtained.

	A	B	−AB
	+16	−16	+16
	+32	+32	−32
	+28	+28	−28
	−15	+15	+15
	+14	−14	+14
	−17	+17	+17
	−9	−9	−9
	−12	−12	−12
$\sum X_{\text{high}} - \sum X_{\text{low}} =$	37	41	−19

$$\bar{X}_{200^\circ} - \bar{X}_{100^\circ} = \tfrac{37}{4} = 9.25$$
$$\bar{X}_{80\%} - \bar{X}_{30\%} = \tfrac{41}{4} = 10.25$$
$$\bar{X}_{+\text{AB}} - \bar{X}_{-\text{AB}} = -\tfrac{19}{4} = -4.75$$

G. Estimate the variance by using the signs in the unlabeled columns with the results obtained.

	Column number 3	5	6	7
	−16	−16	+16	+16
	−32	+32	−32	+32
	+28	−28	+28	−28
	+15	−15	−15	+15
	+14	+14	−14	−14
	−17	+17	+17	−17
	+9	+9	+9	+9
	−12	−12	−12	−12
$\sum X_{\text{high}} - \sum X_{\text{low}} =$	−11	+1	−3	+1

$$S_3^2 = \frac{(-11)^2}{8} = 15.125$$

$$S_5^2 = \frac{(1.0)^2}{8} = 0.125$$

$$S_6^2 = \frac{(-3)^2}{8} = 1.125$$

$$S_7^2 = \frac{(1)^2}{8} = 0.125$$

$$S_{avg}^2 = \frac{15.125 + 0.125 + 1.125 + 0.125}{4} = \frac{16.5}{4} = 4.125$$

with four degrees of freedom.

H. Determine test criterion.
$|\bar{X}_{high} - \bar{X}_{low}|^* = t_\alpha S\sqrt{\frac{1}{N_{high}} + \frac{1}{N_{low}}}$

For the single-sided alternative hypotheses,
$|\bar{X}_{high} - \bar{X}_{low}|^* = (1.53)(2.03)\sqrt{\frac{1}{4} + \frac{1}{4}} = 2.17$
For the double-sided alternative hypotheses,
$|\bar{X}_{high} - \bar{X}_{low}|^* = (2.13)(2.03)\sqrt{\frac{1}{4} + \frac{1}{4}} = 3.03$

I. Apply test criterion, and make the decision. Note: symbology for interaction causes some problems. $(\bar{X}_{+AB} - \bar{X}_{-AB})$ should be interpreted exactly the same as $(\bar{X}_{high\ AB} - \bar{X}_{low\ AB})$ or $(\bar{X}_{high\ A} - \bar{X}_{low\ A})$, etc.

$(|\bar{X}_{200} - \bar{X}_{100}| = 9.25) > (|\bar{X}_{high} - \bar{X}_{low}|^* = 2.17)$
Therefore, accept $(H_a)_1 : \mu_{200°} > \mu_{100°}$ with at least 90 percent confidence. (H_a was single-sided.)
$(|\bar{X}_{80\%} - \bar{X}_{30\%}| = 10.25) > (|\bar{X}_{high} - \bar{X}_{low}|^* = 3.03)$
Therefore, accept $(H_a)_2 : \mu_{80\%} > \mu_{30\%}$ with at least 95 percent confidence. (H_a was double-sided.)
$(|\bar{X}_{+AB} - \bar{X}_{-AB}| = 4.75) > (|\bar{X}_{high} - \bar{X}_{low}|^* = 2.17)$
Therefore, accept $(H_a)_3 : \mu_{AB\ interaction} > 0$ with at least 90 percent confidence. (H_a was single-sided.)

J. Determine the nature of interaction AB.

Make up a two-way chart (figure 6-5) and plot the AB interaction (figure 6-6).

	1	b
1	9 12 Avg = 10.5	15 17 Avg = 16
a	16 14 Avg = 15	32 28 Avg = 30

Fig. 6-5. Two-way chart of AB interaction.

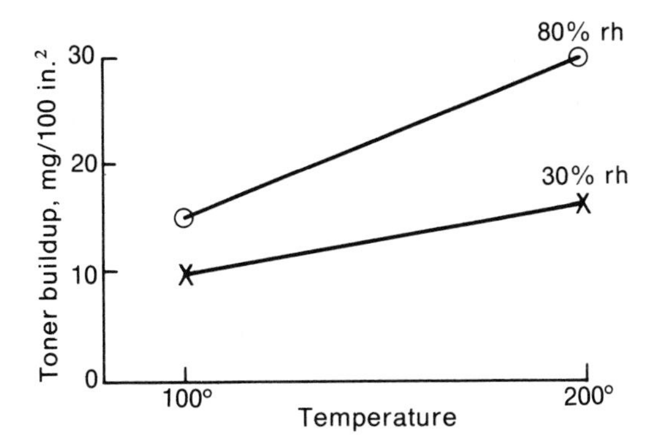

Fig. 6-6. Plot of AB interaction.

K. Make an engineering statement.

Increasing the temperature from 100° to 200° causes an increase in toner filming of 4.5 mg/100 in.^{2} if the relative humidity is 30 percent. The same increase in temperature causes an increase in toner filming of 14 mg/100 in.^{2} if the relative humidity is 80 percent. Increasing the relative humidity from 30 percent to 80 percent increases the toner filming by 5.5 mg/100 in.^{2} if the temperature is 100°. The same increase in relative humidity increases toner filming by 15 mg/100 in.^{2} if the temperature is 200°. Within the operating range of the machine, the toner buildup will vary from 9 to 32 mg/100 in.^{2} (see figure 6-5).

Many other hypotheses can be tested by using the data generated in the above experiment. For example, the question can be asked, "Is there a real effect of temperature at 30 percent relative humidity?" To answer this question, the experimenter must compare the applicable test data against a proper criterion. The alternative hypothesis being tested here is $H_a: \mu_{200°,30\%} > \mu_{100°,30\%}$, and for this question the relevant risk factor (α) is chosen to be 0.1. The general formula for the test criterion is that shown in item H, above; the value for t_α is taken from table 3 (single-sided) with $\phi = 4$, and the value for S is obtained from item G, above, by taking the square root of the value shown for S^2. The value for $N_{\text{high A}} = N_{\text{low A}} = 2$ because, of the eight trials, only four were conducted at 30 percent relative humidity and, of these four, two each were conducted at 100° and at 200°

Note carefully that, even though $N = 2$, t_α is looked up in table 3 for four degrees of freedom. This is because S^2 was estimated with $\phi = 4$, and t_α is a function of ϕ, not the specific number for N in any comparison.

From the above information, the experimenter calculates the test criterion as follows:

$$|\bar{X}_{\text{high}} - \bar{X}_{\text{low}}|^* = t_\alpha S \sqrt{\frac{1}{N_{\text{high}}} + \frac{1}{N_{\text{low}}}}$$

$$|\bar{X}_{\text{high}} - \bar{X}_{\text{low}}|^* = (1.53)(2.03)\sqrt{\tfrac{1}{2} + \tfrac{1}{2}} = 3.11$$

Using the test results (item E above) for the four trials at 30 percent relative humidity, the experimenter determines that $\bar{X}_{200°,30\%} = (16 + 14)/2 = 15$ and that $\bar{X}_{100°,30\%} = (9 + 12)/2 = 10.5$. Thus,

$$(\bar{X}_{200} - \bar{X}_{100})_{30\%} = (15 - 10.5) = 4.5.$$

Because the test-data value (4.5) is larger than the criterion value (3.11), it can be stated with greater than 90 percent confidence that:

$$\mu_{200°,30\%} > \mu_{100°,30\%}.$$

Another question that could be asked is, "Will the mean toner filming be less than 35 mg/100 in.^{2} at the most adverse conditions of temperature and humidity (200°, 80%)?" The hypothesis being tested is $H_a: \mu_{200°,80\%} < 35$ mg/100 in.^{2}. If the risk factor, α, is chosen to be 0.05 for this hypothesis, the formula used to determine the objective criterion is:

$$\bar{X}^* = \mu_0 - \frac{t_\alpha S}{\sqrt{N}}.$$

This formula is used because the experimenter now has the equivalent of a case 7 simple comparative experiment. From the information given:

$$\mu_0 = 35$$
$$S = 2.03$$
$$t_\alpha = 2.13 \text{ from table 3 (single-sided); } \phi = 4.$$
$$N = 2 \text{ (There are two sample results at 200°, 80 percent humidity.)}$$

From item E, above, $\bar{X}_{200°,80\%} = (32 + 28)/2 = 30$.

$$\bar{X}^* = 35 - \frac{(2.13)(2.03)}{\sqrt{2}} = 31.94$$

Because $\bar{X}$ (30) is less than the criterion (31.94), it can be stated with greater than 95 percent confidence that the mean toner filming will be less than 35 mg/100 in.^{2} at the most adverse combination of temperature and humidity.

Three Variables

An experiment design for three variables can be obtained from the 8×8 Hadamard matrix by the following labeling:

Trial	0	A 1	B 2	C 3	−AB 4	−BC 5	ABC 6	−AC 7	Treatment combinations
1	+	+	−	−	+	−	+	+	a
2	+	+	+	−	−	+	−	+	ab
3	+	+	+	+	−	−	+	−	abc
4	+	−	+	+	+	−	−	+	bc
5	+	+	−	+	+	+	−	−	ac
6	+	−	+	−	+	+	+	−	b
7	+	−	−	+	−	+	+	+	c
8	+	−	−	−	−	−	−	−	(1)

Several things can be observed from the above matrix:

- The treatment combinations are exactly the same (but in different order) as those in the 2^3 experiment discussed earlier.
- All the columns are used either for main factors or for interactions.
- There are no columns left over for estimating the variance.
- Every possible interaction appears at the top of some column.
- No interaction appears in two different columns.
- No column measures more than one interaction or main effects.

The first item above mentions that the treatment combinations in the Hadamard matrix are exactly the same as those in the 2^3 experiment described earlier. The reader should remember that the discussion earlier was intended only to introduce the topic of multivariable experiments (philosophy, economics, nomenclature, etc.). The Hadamard procedures are the only ones that the reader should use as examples for any experiments that they conduct.

For complete factorial designs, such as this three-variable design, the experimenter must know either σ or S before starting the experiment; otherwise, it would not be possible to compute an objective criterion for the experiment. As observed above, all columns are labeled in this design and none are available for estimating variance. Fortunately, there are several ways of getting around this dilemma:

- The experimenter can replicate the entire experiment. This will give two results at each treatment combination, from which an estimate of the variance can be obtained with one degree of freedom at each treatment combination. Since there are eight treatment combinations, the average estimate of σ^2 will have eight degrees of freedom. Under no circumstances should the experimenter reproduce only a few of the treatment combinations.
- If the variables are of the continuous type (temperature, time, pH, etc.), the experimenter can conduct any number of trials in the center of the system to obtain an estimate of the variance. This procedure will be described in detail in a later chapter.
- If the experimenter knows (or strongly suspects) before conducting the experiment that certain interactions cannot possibly exist, then the columns headed by these interactions can be used to estimate σ^2 by using the procedure previously discussed. For example, in the 8×8

matrix, if it is assumed that there is no possibility of −BC or ABC being real, then these two columns can be used to compute S_5^2 and S_6^2, and the average of these two will be S_{avg}^2 with two degrees of freedom.

The 8 × 8 Hadamard matrix with three variables is used exactly as described in the previous example for two variables. The specified treatment combinations are made and tested, and the results are analyzed by using the signs in the columns with the vector of results as previously described. Decisions are made on A,B,C, AB, AC, BC, and ABC with either $(1 - \alpha)$ or $(1 - \alpha/2)(100)$ percent confidence, depending upon whether the alternative hypothesis is single- or double-sided, if the alternative hypothesis is accepted, and with $(1 - \beta)(100)$ percent confidence if the null hypothesis is accepted. The criterion is the same:

$$|\bar{X}_{high} - \bar{X}_{low}|^* = t_\alpha S\sqrt{\tfrac{1}{4} + \tfrac{1}{4}} \text{ or } U_\alpha \sigma\sqrt{\tfrac{1}{4} + \tfrac{1}{4}}.$$

Two-way charts are constructed if any two-factor interaction effects are found to be greater than the criterion.

CASE 2

σ^2 Is Known

A process for welding two parts consists of controlling three variables: the temperature of the preheat, at 500°; the pressure between the two parts, at 400 psi; and the time of contact, at 30 seconds. The process is presently producing parts with a population mean tensile strength of 1420 psi (μ_0) and a population variance of 10,000 (σ_0^2). It is necessary that the quality of the parts be improved by increasing the tensile strength. It is thought that a temperature of 600°, a pressure of 450 psi, and a contact time of 40 seconds could, either singly or jointly, improve the tensile strength by as much as 300 psi (δ), which would be an important improvement.

Experiment Design

A. Define the object of the experiment. Let temperature = variable A, let pressure = variable B, and let time = variable C.

$(H_0)_A$: $\mu_{high\ A} = \mu_{low\ A}$, or
$(\mu_{high\ A} - \mu_{low\ A}) = 0$
$(H_a)_A$: $\mu_{high\ A} > \mu_{low\ A}$, or
$(\mu_{high\ A} - \mu_{low\ A}) > 0$
$(H_0)_B$: $\mu_{high\ B} = \mu_{low\ B}$
$(H_a)_B$: $\mu_{high\ B} > \mu_{low\ B}$
$(H_0)_C$: $\mu_{high\ C} = \mu_{low\ C}$
$(H_a)_C$: $\mu_{high\ C} > \mu_{low\ C}$

B. Choose the values of α, β, and δ. Specify σ^2.

$\alpha = 0.05$
$\beta = 0.01$
$\delta = 300$, $\delta^2 = 90{,}000$
$\sigma^2 = 10{,}000$, $\sigma = 100$

C. Compute $N = 2(U_\alpha + U_\beta)^2 \dfrac{\sigma^2}{\delta^2}$

$$N = 2(1.645 + 2.326)^2 \frac{10{,}000}{90{,}000} = 3.5$$

(Matrix experiments always use the two-population formula for N.)

A design with four samples at the high level and four samples at the low level of each variable is required. A $2^3 = 8$ trial experiment is adequate.

D. Define the treatment combinations.

Trial	A	B	C	Treatment combinations
1	−	−	−	(1)
2	+	−	−	a
3	−	+	−	b
4	+	+	−	ab
5	−	−	+	c
6	+	−	+	ac
7	−	+	+	bc
8	+	+	+	abc

E. Make one sample of each treatment combination, and determine tensile strength. (Simplify the arithmetic of analysis by subtracting a fixed value from each result.) The column of responses is called the response vector.

Trial	Result minus fixed value, psi	Response vector, psi
1	1460 − 1420 =	40
2	1680 − 1420 =	260
3	1820 − 1420 =	400
4	2010 − 1420 =	590
5	1420 − 1420 =	0
6	1750 − 1420 =	330
7	1890 − 1420 =	470
8	1900 − 1420 =	480

F. Compute criterion for decision

$$\frac{|\overline{X}_{\text{high B}} - \overline{X}_{\text{low}}|^*}{\sigma\sqrt{\dfrac{1}{N_{\text{high}}} + \dfrac{1}{N_{\text{low}}}}} = U_\alpha$$

$$|\overline{X}_{\text{high}} - \overline{X}_{\text{low}}|^* = (1.645)(100)\sqrt{\frac{1}{4} + \frac{1}{4}} = 116$$

G. Compute $(\overline{X}_{\text{high A}} - \overline{X}_{\text{low A}})$ and $(\overline{X}_{\text{high B}} - \overline{X}_{\text{low B}})$ and $(\overline{X}_{\text{high C}} - \overline{X}_{\text{low C}})$. Use the signs in the design matrix with the results in the response vector to obtain $\Sigma X_{\text{high}} - X_{\text{low}}$. Since there are four results at high and four at low, divide by 4 to obtain $\overline{X}_{\text{high}} - \overline{X}_{\text{low}}$.

$$(\overline{X}_{\text{high A}} - \overline{X}_{\text{low A}}) = \frac{(-40+260-240+590-0+330-470+480)}{4} = 187.5$$

$$(\overline{X}_{\text{high B}} - \overline{X}_{\text{low B}}) = \frac{(-40+260-400+590-0-330+470+480)}{4} = 328.5$$

$$(\overline{X}_{\text{high C}} - \overline{X}_{\text{low C}}) = \frac{(-40-260-400+590-0+0+330+470+480)}{4}$$

$$-2.5$$

H. Apply test criterion.

$(\overline{X}_{\text{high A}} - \overline{X}_{\text{low A}} = 187.5) > 116$, therefore, accept $(H_a)_A\!:\mu_{\text{high A}} > \mu_{\text{low A}}$.

$(\overline{X}_{\text{high B}} - \overline{X}_{\text{low B}} = 328.5) > 116$, therefore, accept $(H_a)_B: \mu_{\text{high B}} > \mu_{\text{low B}}$.
$(|\overline{X}_{\text{high C}} - \overline{X}_{\text{low C}}| = -2.5) < 116$, therefore, accept $(H_0)_C: \mu_{\text{high C}} = \mu_{\text{low C}}$

I. Make an engineering decision.

Change the process specification so that the temperature of the preheat is 600° and the pressure is 450 psi. This will increase the tensile strength by an estimated 187 + 328 = 515 psi. Increasing the time will not help.

At this point, the experimenter can calculate either single- or double-sided confidence intervals for the process with the temperature change only, the pressure change only, or both. The formulas used to calculate the confidence intervals are exactly the same as for the simple comparative experiments. The experimenter can also determine if there are any interactions between the variables.

CASE 5

σ^2 Is Unknown

A modified corona is being designed for charging a photoconductor. Three parameters of the design are of interest to the designer: width of the device; distance from the charging wire to the back wall of the device; and thickness of the wires. Some experimental evidence is available. It is known that the estimated variance of the output voltage (S_0^2) is 36 with $\phi = 10$; the present corona creates a voltage of about 525 volts; and the required voltage is greater than 600 volts. An improvement of 15 volts (δ) would be a significant improvement at this time.

Experiment Design

A. Define the object of the experiment. Let width = variable A, let distance = variable B, and let thickness = variable C.

$(H_0)_A: \mu_{6''} = \mu_{4''}$
$(H_a)_A: \mu_{6''} \neq \mu_{4''}$
$(H_0)_B: \mu_{1/2''} = \mu_{1''}$
$(H_a)_B: \mu_{1/2''} \neq \mu_{1''}$
$(H_0)_C: \mu_{100 \text{ mils}} = \mu_{150 \text{ mils}}$
$(H_a)_C: \mu_{100 \text{ mils}} \neq \mu_{150 \text{ mils}}$

Designate high and low levels of each variable. The value of the variable that is expected to be best is always designated as the high level.

	Low level	High level
A	6″	4″
B	½″	1″
C	100 mils	150 mils

B. Choose the values of α, β, and δ, and state S_0^2.

$\alpha = 0.02$
$\beta = 0.05$
$\delta = 15,\ \delta^2 = 225$
$S_0^2 = 36$ with $\phi_0 = 10$, $S_0 = 6$

C. Compute $N_{high} = N_{low}$ and define the treatment combinations.

$$N = 2(t_\alpha + t_\beta)^2 \frac{S^2}{\delta^2}$$

$t_\alpha = 2.76$
$t_\beta = 2.23$

$N_{high} = N_{low} = 7$. Since a 2^3 design has only 4 highs and 4 lows, and since at least 7 highs and 7 lows are required, a replicated 2-level, 3-variable design is required. A factorial block for the experiment is shown in the previous example. Note: there are 8 highs and 8 lows, as required for each variable.

D. Make two coronas at each treatment combination, and determine the voltage on the photoconductor with each corona.
Sum up the two results at each treatment combination.
Simplify the subsequent arithmetic by subtracting 1000 from each sum.

Corona design	Sample 1	Sample 2	Sum	Simplify by subtracting 1000
1	520	530	1050	50
2	530	535	1065	65
3	550	555	1105	105
4	550	560	1110	110
5	610	600	1210	210
6	615	615	1230	230
7	650	640	1290	290
8	650	665	1315	315

E. Compute variance estimates from replicate samples, and average them with the prior estimate if $S_1^2 = S_2^2$.

$$S^2 = \frac{\Sigma(X_i - \bar{X})^2}{N - 1}$$

$N = 2$ samples at each treatment combination.

Design
1 $S^2 = 50.0$, with $\phi = 1$
2 $S^2 = 12.5$, with $\phi = 1$
3 $S^2 = 12.5$, with $\phi = 1$
4 $S^2 = 50.0$, with $\phi = 1$
5 $S^2 = 50.0$, with $\phi = 1$
6 $S^2 = 0$, with $\phi = 1$
7 $S^2 = 50.0$, with $\phi = 1$
8 $S^2 = 112.5$, with $\phi = 1$
avg $= S_1^2 = 42.2$, with $\phi_1 = 8$

$$S_{avg}^2 = \frac{S_0^2\phi_0 + S_1^2\phi_1}{\phi_0 + \phi_1}$$

$$S_{avg}^2 \frac{(36)(10) + (42.2)(8)}{18} = 38.76$$

with $\phi_{avg} = 18$

F. Compute criterion.

$|\bar{X}_{high} = \bar{X}_{low}|$*

$$= t_\alpha S_{avg} \sqrt{\frac{1}{N_{high}} + \frac{1}{N_{low}}}$$

$t_{0.02}$ double-sided, with $\phi = 18$ can also be found in table 3 under $\alpha = 0.01$.

$t_\alpha = 2.55$

$|\bar{X}_{high} - \bar{X}_{low}|$*

$$= (2.55)(6.22) \sqrt{\frac{1}{8} + \frac{1}{8}} = 7.93$$

G. Compute $(\bar{X}_{4''} - \bar{X}_{6''})$, $(\bar{X}_{1''} - \bar{X}_{\frac{1}{2}''}$, and $(\bar{X}_{150\text{ mils}} - \bar{X}_{100\text{ mils}})$.

$$(\bar{X}_{4''} - \bar{X}_{6''}) = \frac{(-50 + 65 - 105 + 110 - 210 + 230 - 290 + 315)}{8} = 8.125$$

$$(\bar{X}_{1''} - \bar{X}_{\frac{1}{2}''}) = \frac{(-50 - 65 + 105 + 110 - 210 - 230 - 290 + 315)}{8} = 33.125$$

Use the signs in the design matrix (item D in previous example) with the vector of simplified sums in D above. This computed value must then be divided by 8 to obtain $(\overline{X}_{high} - \overline{X}_{low})$, because there are 8 high results and 8 low results in the experiment.

$$(X_{150\ mils} - X_{100\ mils}) = \frac{(-50-65-105-110+210+230+290+315)}{8} = 89.375$$

H. Apply test criterion.

$[(\overline{X}_{4''} - \overline{X}_{6'}) = 8.125] > [(|\overline{X}_{high} - \overline{X}_{low}|^* = 7.93)]$. Therefore, $\mu_{4'} > \mu_{6''}$ with at least 99 percent confidence. $[(\overline{X}_{1''} - \overline{X}_{\frac{1}{2}''}) = 33.125] > [|\overline{X}_{high} - \overline{X}_{low}|^* = 7.93]$. Therefore, $\mu_{1''} > \mu_{\frac{1}{2}''}$ with at least 99 percent confidence. $[(\overline{X}_{150\ mils} - \overline{X}_{100\ mils}) = 89.375] > [|\overline{X}_{high} - \overline{X}_{low}|^* = 7.93]$. Therefore, $\mu_{150\ mils} > \mu_{100\ mils}$ with at least 99 percent confidence.

I. Make an engineering decision.

If coronas are made with 150-mil wires, the resulting product will only be marginally good with 6-inch width and $\frac{1}{2}$-inch depth. If the product is made with 150-mil wires and 1-inch depth, the product will be exceptionally good (>600 volts) even with 6-inch width. If the product is made with 4-inch width and 150-mil wires, the product will probably be satisfactory even with $\frac{1}{2}$-inch depth. The best corona will be obtained if the product is made with 4-inch width, 1-inch depth, and 150-mil wires. The estimated voltage on the photoconductor will be 657 volts, which is the average of the two results at treatment combination 8.

Four Variables (Resolution IV Design)

If it may be assumed that the ABC interaction does not exist in a 4-variable problem, then a Resolution IV design for four variables can be obtained from the 8 × 8 Hadamard matrix by the following labeling:

							D		
		A	B	C	−AB	−BC	ABC	−AC	Treatment
Trial	I	1	2	3	4	5	6	7	combinations
1	+	+	−	−	+	−	+	+	ad
2	+	+	+	−	−	+	−	+	ab

Trial	I	A 1	B 2	C 3	−AB 4	−BC 5	D ABC 6	−AC 7	Treatment combinations
3	+	+	+	+	−	−	+	−	abcd
4	+	−	+	+	+	−	−	+	bc
5	+	+	−	+	+	+	−	−	ac
6	+	−	+	−	+	+	+	−	bd
7	+	−	−	+	−	+	+	+	cd
8	+	−	−	−	−	−	−	−	(1)

As may be seen, the column that was used for ABC is now used for the fourth variable, D. The treatment combinations are obtained in the usual fashion by reading across the rows. The first trial is made at high A, low B, low C, and high D.

As will be shown below, the above labeling of the matrix is only a preliminary labeling used to obtain the treatment combinations. Additional labeling will be required to use the matrix for analysis of the test results. After the additional labeling has been applied, analysis of the test data is accomplished by reading down the columns as previously described.

Looking at the above labeling of the Hadamard matrix, it may be seen that many possible interactions do not appear: AD, BD, CD, ABD, ACD, BCD, and ABCD. Using the multiplication rule, column 4 is found to be the inverse of the product of the signs in columns 3 and 6; i.e., the −CD interaction. Thus, column 4 measures both −AB and −CD. Similarly, column 5 is a measure of the −AD interaction as well as the −BC interaction, and column 7 is a measure of the −BD interaction as well as the −AC interaction. If, after obtaining the test data from this experiment, it is found that the effect measured by column 4 is highly significant, there is no way of determining from this experiment if this is due to AB or to CD. All the experiment shows is that either AB or CD, or both, are important interactions with $(1 - \alpha)$ confidence. Thus, AB is said to be confounded with CD.

If the multiplication rule is applied to the variables three at a time, it is seen that the product of the signs in columns B, C, and D is exactly the same as in column A. Thus, the variable A is confounded with the BCD interaction. It will also be observed that B is confounded with ACD, C is confounded with ABD, and D is confounded with ABC. For this design to be useful, the experimenter must be willing to assume that all three-factor interactions are insignificant and to have two-factor interactions confounded with each other.

Use of the multiplication procedure to determine the confounding in an experiment of this type is time consuming. An easier method is to use the following arithmetic rule:

> $A \times A = A^2 = I$ and $B \times B = B^2 = I$, etc.; i.e., any squared letter is equal to I. I signifies an identity element; i.e., $A \times I = A$, $B \times I = B$, etc.

In the above matrix, D was placed in the column headed by ABC. Therefore $ABC = D$. Multiply both sides of the equation by D and apply the above arithmetic rules:

$$\begin{array}{rl} ABC & = D \\ \times\ \ D & \times\ D \\ \hline ABCD & = D^2 = I \end{array}$$

The term ABCD is called the defining contrast of the experiment.

Multiply each column heading of the original matrix by the defining contrast and use the arithmetic rule defined above to obtain the confounding interactions. For example:

$$\begin{array}{l} \quad\ \mathrm{A} \\ \times\ \mathrm{ABCD} \\ \hline \mathrm{A^2BCD} \end{array} = \mathrm{IBCD} = \mathrm{BCD}$$

Therefore, A is confounded with BCD.

$$\begin{array}{l} \quad -\mathrm{AB} \\ \times\quad \mathrm{ABCD} \\ \hline -\mathrm{A^2B^2CD} \end{array} = -\mathrm{IICD} = -\mathrm{CD}$$

Therefore, $-$AB is confounded with $-$CD.

The total labeling of this Hadamard matrix, then, should be:

Trial	ABCD I	BCD A	ACD B	ABD C	−CD −AB	−AD −BC	D ABC	−BD −AC	Treatment combinations
1	+	+	−	−	+	−	+	+	ad
2	+	+	+	−	−	+	−	+	ab
3	+	+	+	+	−	−	+	−	abcd
4	+	−	+	+	+	−	−	+	bc
5	+	+	−	+	+	+	−	−	ac
6	+	−	+	−	+	+	+	−	bd
7	+	−	−	+	−	+	+	+	cd
8	+	−	−	−	−	−	−	−	(1)

Note that if the signs in the A, B, C, and D columns are multiplied for each trial, the result is always plus. Therefore the ABCD interaction is confounded with the first column of pluses.

In the heading for this section we called these designs Resolution IV without definition of the term or any statement of the significance of the IV. The term Resolution refers to the quality of the information that can be obtained from an experiment. The term is used in the same sense as it is used in describing the quality of an optical instrument; the greater the resolution, the better the quality. In experiment design, the higher the resolution number, the better the quality of the experiment. In this book, we will be concerned with designs of only three different resolutions:

- Resolution V, or better. All designs, regardless of numbers of trials or variables, in which all main factors and all two-factor interactions can be estimated if it is assumed that three-factor and higher interactions are zero. The design for three variables in the 8×8 Hadamard matrix was at least Resolution V.
- Resolution IV. All designs, regardless of numbers of trials or variables, in which all main effects and groups of interactions can be estimated if it is assumed that three-factor and higher interactions are zero. The design for four variables in eight trials is Resolution IV because all main factors are confounded with three-factor interactions, which are assumed

to be zero, and each two-factor interaction is confounded with another two-factor interaction.

- Resolution III. All designs, regardless of numbers of trials or variables, in which all main effects can be estimated if it is assumed that all interactions are zero.

Thus, Resolution V designs can be said to resolve, detect, or estimate main factors and two-factor interactions and are, therefore, better than Resolution IV designs, which can only resolve main factors and groups of two-factor interactions. Resolution III designs are the poorest designs, since they can only resolve main factors. With the 8 × 8 Hadamard matrix, it has been seen that, with one, two, or three variables, a design of at least Resolution V can be obtained. With four variables, however, it is impossible to obtain a Resolution V design in eight trials; it has been shown, however, that a Resolution IV design is obtainable if D is entered into the ABC column of the 8 × 8 Hadamard matrix. Furthermore, it will be shown in the next section that, with five, six, or seven variables and the 8 × 8 Hadamard matrix, only Resolution III designs can be obtained.

The experimenter must make a rational engineering judgment on what Resolution is required for any specific experiment. If all two-factor interactions *must* be estimated, and if all three-factor interactions can be reasonably assumed to be zero, or small, then a Resolution V design is adequate. Obviously, if the number of variables is four, then the 8 × 8 Hadamard matrix is too small. It will be shown later that, with a 16 × 16 Hadamard matrix (the next largest size after 8 × 8), a Resolution V design for five variables can be obtained.

One final hazard about the Resolution IV design should be mentioned. Suppose that God ordained in the experimental system being studied that there should be an AB interaction with a value of 36 and a CD interaction of −36. Except for some slight variation due to experimental error, column 4, when used with the experimental results, would give a value of zero as an estimate of the −AB − CD interaction pair. An experimenter should conclude only that there is no evidence of these interactions. It cannot be concluded that these interactions do not exist. There is nothing that can be done about this. If, on the other hand, it is found that (−AB − CD), A, and B are significant, and that C and D are insignificant, this is not justification for presuming that it is the −AB interaction that is significant rather than the −CD interaction.

Despite the limitations of the above design, it is a useful industrial tool. As the number of variables becomes very large, these designs become very desirable. Furthermore, if there are certain restrictions imposed on an experimenter, this might be the best possible design. For example, if an experimenter has four variables that must be studied, and there is only sufficient time or materials to conduct eight trials, the Resolution IV design is the best possible design. The effect of each variable and each interacting pair is measured four times. In addition, if there is an interaction in the system, its presence will be revealed even if it cannot be identified.

Suppose God ordained that only the combination of high A, high B, and low D would give a superior product; all other combinations would give average products. If an experimenter produces and tests the above eight treatment combinations, the second trial (treatment combination ab) will have a result substantially higher than the results of all the other trials. The experimenter will not know why that particular trial is good and will not know that there is an ABD interaction but will know how to produce good parts. Of course, if an experimenter is interested in why this one combination is superior, some follow-up trials can be made and tested.

EXAMPLE

A process was producing transistors to be used in a product that would be subjected to elevated temperatures. With the present process, the transistors had a turn-on voltage ($\bar{X}$) of 400 mV at room temperature. The designers had specified a turn-on voltage of at least 550 mV at room temperature. $S^2 = 200$, $\phi = 20$. Four variables of the process had been identified as probably contributing to the low turn-on voltage.

Variables	Present Level	Proposed Level
A Field Ox Pressure	.5 atm	15 atm
B P + Implanter	old	new
C 2nd Pad Ox dip	15 sec	5 sec
D P + Implanter Energy	90 KeV	120 KeV

A judgement was made that a Resolution IV design would be adequate.

A. State object of Experiment

$$H_0{:}\mu_{\text{high i}} = \mu_{\text{low i}}$$
$$H_a{:}\mu_{\text{high i}} > \mu_{\text{low i}}$$
$$i = \text{A,B,C,D}$$
$$H_0{:}|\mu_{ij}| = 0$$
$$H_a{:}|\mu_{ij}| > 0$$

B. Choose α, β, δ; state S^2 and ϕ

$$\alpha = .05$$
$$\beta = .05$$
$$\delta = 35$$
$$S^2 = 200$$
$$\phi = 20$$

C. Compute N

$$N_{\text{high}} = N_{\text{low}} = 2(t_\alpha + t_\beta)^2 \frac{S^2}{\delta^2}$$

$$N_{\text{H}} = N_{\text{L}} = 2(1.73 + 1.73)^2 \frac{200}{1225} = 3.91$$

Look up t_α and t_β in table 3

An 8-trial matrix is required for a valid experiment.
An 8-trial experiment with 4 variables will be Resolution IV as required.

D. Determine the 8 trials using the computer program (Chapter 13). Make and test each transistor for turn-on voltage.

Trial	Treatment Combination	Results
1	ad	445
2	ab	540
3	abcd	605
4	bc	560
5	ac	420
6	bd	620
7	cd	460
8	(1)	410

E. Determine the effects of the variables and interaction pairs using the signs in the Hadamard matrix with the test results. Note: The computer program of Chapter 13 will make these same calculations.

	A+BCD	B+ACD	C+ABD	D+ABC
	+445	−445	−445	+445
	+540	+540	−540	−540
	+605	+605	+605	+605
	−560	+560	+560	−560
	+420	−420	+420	−420
	−620	+620	−620	+620
	−460	−460	+460	+460
	−410	−410	−410	−410
	−40	+590	+30	+200
$\bar{X}_{high} - \bar{X}_{low} =$	−10	+147.5	+7.5	+50

	CD-AB	-AD-BC	-BC-AC
	+445	−445	+445
	−540	+540	+540
	−605	−605	−605
	+560	−560	+560
	+420	+420	−420
	+620	+620	−620
	−460	+460	+460
	−410	−410	−410
	+30	+20	−48
$\bar{X}_{high} - \bar{X}_{low} =$	7.5	+5.0	−12

F. Compute criterion.

$$|\bar{X}_{high} - \bar{X}_{low}|^* = t_\alpha S \sqrt{\frac{1}{N_{high}} + \frac{1}{N_{low}}}$$

$$|\bar{X}_{high} - \bar{X}_{low}|^* = 1.73\ (14.14)\sqrt{\frac{1}{4} + \frac{1}{4}} = 17.3$$

G. Compare $(\bar{X}_{high} - \bar{X}_{low})$ with $|\bar{X}_{high} - \bar{X}_{low}|^*$

$[(\bar{X}_{high\ B} - \bar{X}_{low\ B}) = 147.5] > [|\bar{X}_{high} - \bar{X}_{low}|^* = 17.3]$

$[|\bar{X}_{high\ D} - \bar{X}_{low\ D}) = 200] > [|\bar{X}_{high} - \bar{X}_{low}|^* = 17.3]$

H. Make the decision.

Accept $(H_a)_B$ and $(H_a)_C$.

Conclusion: The new implanter and 120 kev implanter energy are required for higher turn-on voltage. An increase of about 200 mV can be expected at the new process conditions. There is no evidence of any interactions.

Five, Six, or Seven Variables (Resolution III Designs)

Resolution III designs are also called main-factor designs. In these designs, only the main effects can be estimated; the experimenter must assume that all interactions are insignificant. The matrix labeling for a 6-variable experiment in eight trials is:

Trial	0	A 1	B 2	C 3	D −AB 4	E −BC 5	F ABC 6	−AC 7	Treatment combinations
1	+	+	−	−	+	−	+	+	adf
2	+	+	+	−	−	+	−	+	abef
3	+	+	+	+	−	−	+	−	abc
4	+	−	+	+	+	−	−	+	bcdf
5	+	+	−	+	+	+	−	−	acde
6	+	−	+	−	+	+	+	−	bde
7	+	−	−	+	−	+	+	+	cef
8	+	−	−	−	−	−	−	−	(1)

If seven variables are required, column 7 would then be used for the seventh variable (G). Thus, the proper relationship of columns and variables is:

Column	Variable
1	A
2	B
3	C
4	D
5	E
6	F
7	G

Normally, one does not calculate the confounding interactions in a Resolution III design. However, if the experimenter must know the confounding, the following general procedure is used to determine the identity elements when more than one variable is added to a design. For example, with a 5-variable, 8-trial design, where D is entered in the −AB column and E is entered into the −BC column:

$$\begin{array}{rl} D = & -AB \\ \times D & \times\ D \\ \hline I = & -ABD \end{array} \qquad \begin{array}{rl} E = & -BC \\ \times E & \times\ E \\ \hline I = & -BCE \end{array} \qquad \begin{array}{rl} I = & -ABD \\ \times I = & -BCE \\ \hline I = & (-ABD)(-BCE) = ACDE \end{array}$$

Thus, there are three identity elements: −ABD, −BCE, and ACDE in the defining contrast. Each original column heading must be multiplied by all three identities. In this design, then, A is confounded with −BD, −ABCE, and CDE. Every other contrast column will also measure four effects. If the experimenter can assume that BD, ABCE, and CDE are zero, then the first contrast column measures the effect of A.

With six variables, there are seven identity elements. Each main effect is confounded by the effect of seven interactions, but this creates no problem; the experimenter simply must assume that all these interactions are insignificant. With seven variables, there are 15 identity elements, and each main effect is confounded with 15 interactions, which must be assumed to be equal to zero.

It should be obvious at this point that, if it cannot be assumed that all interactions are zero, these are not the right designs.

EXAMPLE

The synthesis of a certain chemical involves five major variables. The quality of the chemical is determined by the percentage of a certain red impurity in the final product. The variables to be investigated are:

Variable	Parameter	Low level	High level
A	Reaction temperature	5°	15°
B	Catalyst amount	2.5 percent	3.5 percent
C	Mixing time	10 minutes	20 minutes
D	Solvent used in washing	acetone	toluene
E	Time of washing	24 hours	48 hours

It is expected that there are two-factor interactions and perhaps even a three-factor interaction. Each run is expensive ($2000) and takes three days. These runs must be made consecutively. An estimate of the variance (S^2) is known to be 1.0 with $\phi = 10$. Since time is short, as few trials as possible should be planned. The only interest is in obtaining a process that ultimately yields a product containing less than 1 percent of the red impurity. The present product contains about 13 percent (μ_0) of the red impurity. If any one of the variables improves the product by as much as 2.5% (δ), it would be a worthwhile improvement.

A. Define the object of the experiment.

$(H_0)_A: \mu_{5°} = \mu_{15°}$
$(H_a)_A: \mu_{5°} \neq \mu_{15°}$

$(H_0)_B: \mu_{2.5\%} = \mu_{3.5\%}$
$(H_a)_B: \mu_{2.5\%} \neq \mu_{3.5\%}$ etc.

B. Choose α, β, and δ.

$\alpha = 0.10$
$\beta = 0.05$
$\delta = 2.5$ percent of red contaminant
$S^2 = 1.0$ with $\phi = 10 \quad S = 1.0$

C. Compute N.

$$N_{\text{high}} = N_{\text{low}} = 2(t_\alpha + t_\beta)^2 \frac{S^2}{\delta^2}$$

$$N = 2(t_{0.10} + t_{0.05})^2 \frac{1}{(2.5)^2} = 4.19$$

$t_{0.10}$ is obtained from table 4.
$t_{0.05}$ is obtained from table 3.

D. Examine the choices of available experiment designs.

- A complete factorial requires $2^5 = 32$ trials and has $N = 16$ comparisons. This will enable the experimenter to estimate all two- and three-factor interactions and also to obtain a better estimate of the variance. The cost will be $64,000, the α and β risks will be much lower than required since N is greater than required, and 96 working days will be needed to obtain the results.

- A Resolution V design requires $2^{5-1} = 16$ trials, which enables the experimenter to estimate all two-factor interactions. This design will be discussed shortly. With this design $N = 8$, so α and β risks will be less than required. There is the slight possibility of obtaining erroneous information due to the confounding of two-factor interactions by a three-factor interaction or to the confounding of a main factor by a four-factor interaction. The cost will be \$32,000, and 48 days will be needed for the experiment.
- A Resolution III design requires $2^{5-2} = 8$ trials. The calculated α and β risks will be exactly as required. The experimenter will not be able to estimate two-factor interactions, but, if a two- or three-factor interaction is very significant, its effect will be demonstrated on one or two of the sample results. Furthermore, a couple of subsequent experiments should clarify the results, if necessary or desirable. The cost will be \$16,000, and only 24 days will be needed to obtain the sample results.

E. Make a decision on the design, and determine the treatment combinations.

Go with the Resolution III design.

Trial	0	A	B	C	D	E	6	7	Treatment combinations
1	+	+	−	−	+	−	+	+	ad
2	+	+	+	−	−	+	−	+	abe
3	+	+	+	+	−	−	+	−	abc
4	+	−	+	+	+	−	−	+	bcd
5	+	+	−	+	+	+	−	−	acde
6	+	−	+	−	+	+	+	−	bde
7	+	−	−	+	−	+	+	+	ce
8	+	−	−	−	−	−	−	−	(1)

F. Conduct the eight trials specified, and determine the percentage of red impurity in each lot of material produced.

Trial	Result, %
1	15.5
2	2.5
3	12.0
4	8.0
5	13.5
6	7.0
7	12.0
8	13.5

G. Compute criterion.

$$|\bar{X}_{\text{high}} - \bar{X}_{\text{low}}|^* = t_\alpha S\sqrt{\frac{1}{N_{\text{high}}} + \frac{1}{N_{\text{low}}}}$$

$t_\alpha = 1.81$ (Double-sided, table 4; $\phi = 10$.)

$N_{\text{high}} = N_{\text{low}} = 4$ (Always 4 in an 8×8 matrix.)

$|\bar{X}_{\text{high}} - \bar{X}_{\text{low}}|^* = (1.81)(1)\sqrt{\frac{1}{4} + \frac{1}{4}} = 1.27$

H. Determine average effects.

	A	B	C	D	E
	+15.5	−15.5	−15.5	+15.5	−15.5
	+2.5	+2.5	−2.5	−2.5	+2.5
	+12.0	+12.0	+12.0	−12.0	−12.0
	−8.0	+8.0	+8.0	+8.0	−8.0
	+13.5	−13.5	+13.5	+13.5	+13.5
	−7.0	+7.0	−7.0	+7.0	+7.0
	−12.0	−12.0	+12.0	−12.0	+12.0
	−13.5	−13.5	−13.5	−13.5	−13.5
$\sum X_{high} - \sum X_{low} =$	3.0	−25.0	7.0	4.0	−14.0
$\bar{X}_{high} - \bar{X}_{low} =$	0.75	−6.25	1.75	1.0	−3.50

I. Make *preliminary* decisions. A low value is desirable. If the mean effect is positive, the low level is better; if the mean effect is negative, the high level is better. If $(\bar{X}_{high} - \bar{X}_{low})$ is negative and $|\bar{X}_{high} - \bar{X}_{low}| > |\bar{X}_{high} - \bar{X}_{low}|^*$, accept H_a that μ_{high} is better than μ_{low}. If $(\bar{X}_{high} - \bar{X}_{low})$ is positive and greater than $|\bar{X}_{high} - \bar{X}_{low}|^*$, accept H_a that μ_{low} is better than μ_{high}.

Accept $(H_0)_A: \mu_{low\,A} = \mu_{high\,A}$.
Accept $(H_a)_B: \mu_{high\,B} < \mu_{low\,B}$ by 6.3 percent of red ingredient.
Accept $(H_a)_C: \mu_{low\,C} < \mu_{high\,C}$ by 1.8 percent of red ingredient.
Accept $(H_0)_D: \mu_{low\,D} = \mu_{high\,D}$.
Accept $(H_a)_E: \mu_{high\,E} < \mu_{low\,E}$ by 3.5 percent of red ingredient.
The best combination of the variable is, therefore, high B, low C, and high E. Trials 2 and 6 are both made with these levels.

J. Examine contrasts 6 and 7.

	Contrast 6	Contrast 7
	+15.5	+15.5
	−2.5	+2.5
	+12.0	−12.0
	−8.0	+8.0
	−13.5	−13.5
	+7.0	−7.0
	+12.0	+12.0
	−13.5	−13.5
$\sum X_{high} - \sum X_{low} =$	9.0	−8.0
$\bar{X}_{high} - \bar{X}_{low} =$	2.25	−2.0

Both of these contrasts are significant when compared to the criterion.

K. Examine all data and conclusions for discrepancies.

The first discrepancy is that trials 2 and 6 are both at the desired level of variables B, C, and E; yet, trial 2 has a response of 2.5, whereas trial 6 has a response of 7.0. This difference is not explainable by variance which has a value of only 1; therefore, an interaction must be significant. The second discrepancy is that contrasts 6 and 7, which are not associated with any variable but which do measure unidentified interactions, are significant. This also indicates that at least two interactions must be present.

Examine samples 2 and 6:

	A	B	C	D	E	Result
Sample 2	+	+	−	−	+	2.5
Sample 6	−	+	−	+	+	7.0

Except for the levels of A and D, they were made exactly the same, yet the results are very different. The difference must be due to A, to D, or to both. Remember that A was insignificant, but the low level of A was numerically better. D was also insignificant, but the low level of D was numerically better.

It is very possible that the combination − + − − + will be less than 2.5 percent; this treatment combination was not tested. It is also possible that the combination + + − + + could be less than 2.5 percent; this treatment combination also was not tested.

L. Make and test one sample of each of the untested treatment combinations discussed above. With trials 2 and 6 and these two additional trials, a complete factorial for variables A and D, with B high, C low, and E high, has been obtained, as shown in figure 6-7.

	A	B	C	D	E	Results, %
Trial 9	−	+	−	−	+	0.7
Trial 10	+	+	−	+	+	10.1
Trial 2	+	+	−	−	+	2.5
Trial 6	−	+	−	+	+	7.0

	1	d
1	Sample 9	Sample 6
a	.Sample 2	Sample 10

Fig. 6-7. Complete factorial of variables A and D.

M. Calculate criterion.

$$|\bar{X}_{\text{high}} - \bar{X}_{\text{low}}|^* = t_\alpha S \sqrt{\frac{1}{N_{\text{high}}} + \frac{1}{N_{\text{low}}}}$$

$N_{\text{high}} = N_{\text{low}} = 2$

This criterion is used for testing the hypotheses of this follow-up experiment,

$H_0: \mu_{\text{high A}} = \mu_{\text{low A}}$

$H_0: \mu_{\text{high D}} = \mu_{\text{low D}}$

when B is high, C is low and E is high.

Criterion $= (1.81)(1)\sqrt{\frac{1}{2} + \frac{1}{2}} = 1.81$

$(\bar{X}_{\text{high A}} - \bar{X}_{\text{low A}}) = 2.45$ percent.

$(\bar{X}_{\text{high D}} - \bar{X}_{\text{low D}}) = 6.95$ percent.

$(\bar{X}_{\text{high AD}} - \bar{X}_{\text{low AD}}) = 0.65$ percent.

Conclusion: contrary to the preliminary conclusions from the 8-trial experiment, variables A and D are better at the low level than at the high level when B is high, C is low, and E is high. (See figure 6-8.)

	1	d
1	0.7%	7.0%
a	2.5%	10.1%

Fig. 6-8. Two-way chart of A and D data when B is high, C is low, and E is high.

N. Make decisions.

Since D was not significant in the first eight trials, but was significant in the follow-up trials, D interacts with at least one of the other variables. Additional experiments could be conducted to further define the interaction, but a combination that appears satisfactory has been achieved and, on a sample of one, has a value less than the desired level.

The same reasoning as above applies to variable A. Recommend that the process be operated at low A, high B, low C, low D, and high E; i.e., 5°, 3.5 percent catalyst, 10 minutes mixing time, acetone solvent, and 48 hours washing. After several lots have been made at the new process level, a better estimate of the process average will be obtained. At that time, a decision would be made on the necessity for further experimentation.

The experimenter might wish to explore the possibility of a still lower reaction temperature (say, 0°), of more catalyst (say, 4 percent), of still lower mixing time (perhaps 8 minutes), and of longer washing time (perhaps 60 hours) to further decrease the amount of red impurity.

All the Hadamard matrices discussed in this chapter have been used to test hypotheses. Implicitly, the experimenter has also been evaluating a model of some process. With one variable in eight trials, the assumed model is

$$\text{Response} = b_0 + b_1\text{A}.$$

In the case of three variables in eight trials, the assumed model is

$$\text{Response} = b_0 + b_1\text{A} + b_2\text{B} + b_3\text{C} + b_{1,2}\text{AB} + b_{2,3}\text{BC} + b_{1,3}\text{AC} + b_{1,2,3}\text{ABC}.$$

For the case of seven variables in eight trials, the assumed model is

$$\text{Response} = b_0 + b_1\text{A} + b_2\text{B} + b_3\text{C} + b_4\text{D} + b_5\text{E} + b_6\text{F} + b_7\text{G}.$$

After analysis of the data, as described in this chapter, the data can be presented in model form as above. The procedure is as follows:

- b_0 is the mean of all the data.
- b_1 is $\frac{1}{2}(\bar{X}_{\text{high A}} - \bar{X}_{\text{low A}})$.
- $b_{1,2}$ is $-\frac{1}{2}(\bar{X}_{\text{high AB}} - \bar{X}_{\text{low AB}})$.

In the example on photocopiers, where the variables are temperature and humidity, the experimentally determined model would be

$$\text{toner buildup} = 17.875 + 4.625(\text{Temperature}) + 5.125(\text{Humidity}) + 2.375(\text{Temperature})(\text{Humidity}).$$

This model is, of course, in terms of coded values of temperature and humidity. Therefore, the model would predict that, at 200° and 30% humidity,

$$\text{toner buildup} = 17.875 + 4.625(+1) + 5.125(-1) + 2.375(+1)(-1) = 15.$$

If the experimenter wishes the model in terms of actual temperature and humidity, the data of the experiment can be entered into a computer regression analysis program. A model will be obtained with different coefficients, b_0, b_1, b_2, and $b_{1,2}$, because the values of A and B will be numbers like 100°, 220°, etc., and 30% or 60%, etc. The predicted value of toner buildup, however, will be exactly the same for both models.

It is noted at this point that, if the result on one or more of the trials in a Hadamard matrix is lost or cannot be obtained for some reason, the data cannot be analyzed by the methods described in this chapter. In such cases, a computer regression analysis program must be used to analyze the data. The results then will be given in terms of a regression equation. The experimenter should consult a qualified statistician regarding tests of significance for the coefficients, analysis of residuals, estimation of variance, etc.

The book by M. Natrella contains two excellent chapters (in layman's language) on regression analysis for those readers who want to become familiar with this method of analysis.

EXERCISES

1. God's law for the effect of three variables on the tension of a spring is as follows:
   - At the low level of all the variables, the value of tension is 25 pounds.
   - The effect of going from low A to high A is 5 pounds.
   - The effect of going from low B to high B is −2 pounds.

- The effect of going from low C to high C is 4 pounds.
- At high A and high C, there is an additive effect of 3 pounds.
- The variance is zero.

What results will be obtained on the eight trials of a Hadamard matrix?

2. The following results are obtained on an 8-trial Hadamard matrix with two variables:

Treatment combination	Results
a	32
ab	40
ab	38
b	37
a	34
b	36
(1)	26
(1)	24

- Do a complete analysis of the data. Use $\alpha = 0.05$ single-sided.
- Compute the estimate of the variance by both methods given in this text.
  - Using the proper columns of the Hadamard matrix.
  - Using replicate samples.

3. An 8-trial Hadamard matrix is used to investigate six variables in the early stages of a development project. The following results are obtained:

Treatment combination	Results
adf	48
abef	73
abc	76
bcdf	52
acde	50
bde	53
cef	36
(1)	38

What tentative engineering conclusions are reached by an analysis of these data? What are reasonable proposals for a follow-up experiment if the project objective is a response of 100?

4. An optical character recognition machine is being developed by an engineering team. The specification calls for a maximum error rate of 0.00002. How many characters must be passed through the machine if $\alpha = \beta = 0.05$ and $\rho_\alpha = 0.000015$ and $\rho_\beta = 0.000025$? How many characters must be examined at each treatment combination if an 8-trial Hadamard matrix is used with seven variables?

5. A textile manufacturer is breaking in a new type loom. There are seven possible variables that could influence the quality of the fabric produced. An 8-trial Hadamard matrix is proposed. $\alpha = \beta = 0.10$, $\lambda_\alpha = 1$ defect/100 sq. ft., $\lambda_\beta = 2$ defects/100 sq. ft. How many square feet of material must be produced at each treatment combination?

REFERENCES

John, P.W.M. 1971. *Statistical design and analysis of experiments.* New York: Macmillan Publishing Co.

Natrella, M. *National Bureau of Standards Handbook* 91. Washington, D.C.: U.S. Government Printing Office.

7 Two-Level Multivariable Experiments: Hadamard Matrices Greater Than Order 8

The previous chapter described the construction and use of the order-8 Hadamard matrix with up to seven variables when the proper sample size was calculated to be $(N_{\text{high}} = N_{\text{low}}) \leq 4$. If the number of variables is greater than 7, or if $N_{\text{high}} = N_{\text{low}}$ is calculated to be greater than 4, or if the desired resolution for the number of variables cannot be obtained with the order-8 matrix, then a larger Hadamard matrix must be used for the experiment. For example, if $N_{\text{high}} = N_{\text{low}}$ is calculated to be 3, and if there are only five variables but the experiment must be Resolution IV, then the order-8 matrix is not large enough. (In the previous chapter, it was shown that only four variables can be included in an 8×8 Hadamard matrix and still be Resolution IV.) Therefore, a larger matrix (16×16) is necessary to meet all the requirements of the experiment.

This chapter discusses all 2^n Hadamard matrices greater than order 8 (8×8) up to order 128; i.e., matrices of size 16×16, 32×32, 64×64, and 128×128. Of course, even larger matrices can be obtained, but they are not specifically discussed in this text.

All the principles, procedures, and uses described in Chapter 6 for the 8×8 Hadamard matrix apply to these larger matrices.

The primary vectors for cyclically generating these matrices are shown in Chapter 6. After describing the use of the higher-order 2^n Hadamard matrices, this chapter also presents primary vectors for generating four useful matrices that are not 2^n matrices: 12×12, 20×20, 24×24, and 44×44.

Labeling of the contrasts in complete factorial designs for all 2^n Hadamard matrices up to order 128 is given in figure 7-1. When the number of variables (A, B, C, D, etc.) is equal to or less than the number required for complete factorial designs, the variables are placed as shown in figure 7-1, and designs that are at least Resolution V are obtained.

If there are fewer variables than needed for a complete factorial design, the unlabeled contrasts must measure variance. For example, suppose there are only two variables (A and B) in an 8-trial experiment. Since there is no variable C,

Contrast	Number of trials					Contrast	Number of trials		Contrast	Number of trials	Contrast	Number of trials
	8	16	32	64	128		64	128		128		128
1	A	A	A	A	A	32	ACF	ABD	64	AD	96	ACF
2	B	B	B	B	B	33	AD	BCE	65	BE	97	BDG
3	C	C	C	C	C	34	BE	CDF	66	CF	98	ABCE
4	AB	D	D	D	D	35	CF	DEG	67	DG	99	BCDF
5	BC	AB	E	E	E	36	ABD	ABEF	68	ABE	100	CDEG
6	ABC	BC	AC	F	F	37	BCE	BCFG	69	BCF	101	ABDEF
7	AC	CD	BD	AB	G	38	CDF	ABCDG	70	CDG	102	BCEFG
8		ABD	CE	BC	AB	39	ABDE	ACDE	71	ABDE	103	ABCDFG
9		AC	ACD	CD	BC	40	BCEF	BDEF	72	BCEF	104	ACDEG
10		BD	BDE	DE	CD	41	ABCDF	CEFG	73	CDFG	105	ADEF
11		ABC	AE	EF	DE	42	ACDE	ABDFG	74	ABDEG	106	BEFG
12		BCD	ABC	ABF	EF	43	BDEF	ACEG	75	ACEF	107	ABCFG
13		ABCD	BCD	AC	FG	44	ABCEF	ADF	76	BDFG	108	ACDG
14		ACD	CDE	BD	ABG	45	ACDF	BEG	77	ABCEG	109	ADE
15		AD	ACDE	CE	AC	46	ADE	ABCF	78	ACDF	110	BEF
16			ABCDE	DF	BD	47	BEF	BCDG	79	BDEG	111	CFG
17			ABDE	ABE	CE	48	ABCF	ABCDE	80	ABCEF	112	ABDG
18			ABE	BCF	DF	49	ACD	BCDEF	81	BCDFG	113	ACE
19			AB	ABCD	EG	50	BDE	CDEFG	82	ABCDEG	114	BDF
20			BC	BCDE	ABF	51	CEF	ABDEFG	83	ACDEF	115	CEG
21			CD	CDEF	BCG	52	ABDF	ACEFG	84	BDEFG	116	ABDF
22			DE	ABDEF	ABCD	53	ACE	ADFG	85	ABCEFG	117	BCEG
23			ACE	ACEF	BCDE	54	BDF	AEG	86	ACDFG	118	ABCDF
24			ABCD	ADF	CDEF	55	ABCE	AF	87	ADEG	119	BCDEG
25			BCDE	AE	DEFG	56	BCDF	BG	88	AEF	120	ABCDEF
26			ADE	BF	ABEFG	57	ABCDE	ABC	89	BFG	121	BCDEFG
27			ABCE	ABC	ACFG	58	BCDEF	BCD	90	ABCG	122	ABCDEFG
28			ABD	BCD	ADG	59	ABCDEF	CDE	91	ACD	123	ACDEFG
29			BCE	CDE	AE	60	ACDEF	DEF	92	BDE	124	ADEFG
30			AD	DEF	BF	61	ADEF	EFG	93	CEF	125	AEFG
31			BE	ABEF	CG	62	AEF	ABFG	94	DFG	126	AFG
						63	AF	ACG	95	ABEG	127	AG

Note: All the even-order interactions above are really minus interactions.

Fig. 7-1. Assignment of contrasts in complete factorial experiments.

all of the contrasts that include C (i.e., contrasts 3, 5, 6, and 7) must measure variance.

If there are more variables than are required for a complete factorial design, figure 7-1 provides guidance on placement of the additional variables. For example, if there are four variables in eight trials, the best design (Resolution IV) is obtained by using contrast 6, the ABC interaction, for variable D (see Chapter 6). Then, the defining contrast is I = ABCD, and the confounding of the other contrasts can be obtained by multiplying each contrast by ABCD.

Note that at least Resolution IV designs can be obtained when the number of variables is greater than the number required for complete factorial designs but is equal to or less than one-half the number of trials in the design. For example, in a 16-trial design, a Resolution V design can be obtained with 5 variables, and Resolution IV designs can be obtained with 6, 7, or 8 variables; 5 variables or less produce at least Resolution V designs, and 9 variables or more (to the maximum of 15 variables with a 16-trial design) produce Resolution III designs.

16-Trial Designs

With the order-16 Hadamard matrix, four variables (A, B, C, and D) comprise a complete factorial design (see figure 7-1), which is at least a Resolution V design. Introduction of a fifth variable (E) will also produce a design which is at least Resolution V if the thirteenth contrast, the highest-order contrast (ABCD), is labeled as variable E.

The defining contrast is I = ABCDE. The product of this defining contrast and a two-factor interaction will always be a three-factor interaction, and the product of the defining contrast with a main effect will always be a four-factor interaction. To be at least Resolution V, the defining contrast must be at least five letters long.

The use of more than five but less than nine variables results in Resolution IV designs with the order-16 Hadamard matrix. (Remember, the upper limit on the number of variables for a Resolution IV design is one-half the number of trials in the design.) For these Resolution IV designs, the additional variables E through H are placed over the three-factor interaction contrasts as follows:

Contrast	Interaction		Variable
8	ABD	=	E
11	ABC	=	F
12	BCD	=	G
14	ACD	=	H

The confounding of two-factor interactions for this and all larger designs is given in tables in Chapter 13.

The procedure for determining the confounding of two-factor interactions in Resolution IV designs is briefly described here. Suppose, for example, that an experimenter wishes to conduct a 16-trial Resolution IV experiment with six variables. The first four contrast columns will be labeled A, B, C, and D. The eighth column (ABD) will be labeled E and the eleventh contrast column (ABC) will be labeled F. Then,

$$\begin{array}{rr} & \text{ABD} = \text{E} \\ \times & \text{E} = \text{E} \\ \hline & \text{ABDE} = \text{E}^2 = \text{I} \end{array} \qquad \begin{array}{rr} & \text{ABC} = \text{F} \\ \times & \text{F} = \text{F} \\ \hline & \text{ABCF} = \text{F}^2 = \text{I} \end{array}$$

and also,

$$\begin{array}{rl} & \text{I} = \text{ABDE} \\ \times & \text{I} = \text{ABCF} \\ \hline & \text{I} = \text{CDEF} \end{array}$$

Therefore the defining contrast for this experiment is

$$\text{I} = \text{ABDE} + \text{ABCF} + \text{CDEF}.$$

Note that if a two-factor interaction (e.g., $-\text{AB}$) is multiplied by the defining contrast, as shown below,

$$-\text{AB}(\text{I} = \text{ABDE} + \text{ABCF} + \text{CDEF}),$$

the result is

$$-\text{AB} = -\text{DE} - \text{CF} - \text{ABCDEF}.$$

Thus, the contrast column labeled $-\text{AB}$ really measures $-\text{AB} - \text{DE} - \text{CF} - \text{ABCDEF}$. Since we will generally assume in this text that third-order and higher interactions do not exist, then, if the $-\text{AB}$ column is significant, it is due to AB, DE, or CF or to a combination of these two-factor interactions.

If any element in the defining contrast is four letters long, and if there are no three-letter or smaller words, the design is Resolution IV.

It is obvious from this discussion that new variables must be introduced into columns that have at least three letters, e.g., ABC or BCD.

The use of from 9 through 15 variables produces Resolution III designs with the 16-trial Hadamard matrix. The first eight variables are assigned to contrasts as shown for Resolution IV designs; the remaining variables are assigned to the contrasts with an even number of letters (see figure 7-1).

When only three variables (A, B, and C) are used with the order-16 Hadamard matrix, all contrasts shown in figure 7-1 that include variable D (i.e., contrasts 4, 7, 8, 10, 12, 13, 14, and 15) are estimates of variance. The same principle would be used if there were only one or two variables.

32-Trial Designs

With the order-32 Hadamard matrix, a design that is at least Resolution V can be obtained with six variables by labeling the sixteenth contrast (ABCDE) with the sixth variable, F. The defining contrast is I = ABCDEF; each two-factor interaction is confounded with a four-factor interaction, and each main factor is confounded with a five-factor interaction. Of course, it is assumed that these four- and five-factor interactions are insignificant.

Resolution IV designs can be obtained with the order-32 Hadamard matrix for from 7 to 16 variables. The contrast columns containing an odd number of letters are used for these variables, and the columns are labeled as follows:

Contrast	Variable	Interaction	Identity element
1	A		
2	B		
3	C		
4	D		
5	E		
9	F	= ACD	I = ACDF
10	G	= BDE	I = BCDG
12	H	= ABC	I = ABCH
13	I*	= BCD	I = BCDI*
14	J	= CDE	I = CDEJ
16	K	= ABCDE	I = ABCDEK
18	L	= ABE	I = ABEL
23	M	= ACE	I = ACEM
26	N	= ADE	I = ADEN
28	0	= ABD	I = ABDO
29	P	= BCE	I = BCEP

(Note: I* = variable I, not identity element.)

Note that the identity elements listed above are only the first part of the total defining contrast for the design. Each of these identity elements must be multiplied two at a time, three at a time, four at a time, etc., to obtain the complete defining contrast. For example, with seven variables, the defining contrast is I = ACDF = BDEG = ABCEFG, and each contrast column is confounded with three interactions. On the other hand, with 16 variables in a 32-trial Hadamard matrix, there are 2047 identity elements in the defining contrast. Thus, each contrast column is confounded with 2047 other interactions. The two-factor interactions that are confounded with each other in Resolution IV designs are given in Chapter 13.

Resolution III designs can be obtained for from 17 to 31 variables by assigning the first 16 variables as shown above, and then assigning the remaining variables to contrast columns with an even number of letters.

64-Trial Designs

With the order-64 Hadamard matrix, designs that are at least Resolution V are obtainable for up to eight variables if the seventh variable (G) is placed over the nineteenth contrast and the eighth variable (H) is placed over the twenty-third contrast. Thus, the labeling of an order-64 Hadamard matrix for Resolution V designs with up to eight variables is as follows:

Contrast	Variable	Interaction	Identity element
1	A		
2	B		
3	C		
4	D		
5	E		
6	F		
19	G	= ABCD	I = ABCDG
23	H	= ACEF	I = ACEFH
			I = BDEFGH

Note that the defining contrast (I = ABCDG = ACEFH = BDEFGH) is composed of fifth-order or higher terms. Therefore, when two-factor interactions are multiplied by these terms, the results will always be third-order or higher, so second-order interactions will not be confounded with each other. With only seven variables, it is possible to obtain a Resolution VII design by placing the variable G in the sixteenth contrast column (ABCDEF); however, this design is not programmed in the computer program discussed in a later chapter and is not discussed further here.

Resolution IV designs can be obtained from the 64 × 64 Hadamard matrix for from 9 to 32 variables by labeling the contrasts with an odd number of letters in the interactions (i.e., 3 or 5). By referring to figure 7-1, the reader will see that the labeling is as follows:

Contrast	Variable
1	A
2	B
3	C
4	D
5	E
6	F
12	G = ABF
17	H = ABE
18	I* = BCF
22	J = ABDEF
24	K = ADF
27	L = ABC
etc.	

(Note: I* = variable I, not identity.)

Resolution III designs can be obtained with the order-64 Hadamard matrix for from 33 to 63 variables by labeling the contrast columns that have an even number of letters in the interaction in addition to those labeled above.

128-Trial Designs

With the order-128 Hadamard matrix, Resolution V designs can be obtained for up to 11 variables by using the following labeling:

Contrast	Variable	Interaction
1	A	
2	B	
3	C	
4	D	
5	E	
6	F	
7	G	
23	H	= BCDE
77	I*	= ABCEG
101	J	= ABDEF
125	K	= AEFG

(Note: I* = variable I, not identity.)

(Note: With only eight or nine variables, designs that are better than Resolution V can be obtained. These designs are not included in the computer program of Chapter 13 and so are not discussed here.)

Resolution IV designs can be obtained with the order-128 Hadamard matrix for from 12 to 64 variables by proper labeling of the contrast columns that contain odd numbers of letters in the interactions. Resolution III designs can be obtained with this order matrix for from 65 to 127 variables by placing the variables from 65 upwards over the contrast columns containing even numbers of letters in the interactions. (Remember, the first 64 variables would be placed as for Resolution IV designs.)

Smoak Modified Designs for Resolution IV Designs

These modified designs were discovered by Marvin Smoak when confronted with a development problem on a stepper motor. A stepper motor consists of two basic components; a rotor and a stator. There were eight variables and a 16-trial Resolution IV design had been selected. The treatment combinations for the 16 trials was obtained from the computer.

NT	A	B	C	D	E	F	G	H
1	+	−	−	−	+	+	−	+
2	+	+	−	−	−	−	+	+
3	+	+	+	−	−	+	−	−
4	+	+	+	+	+	+	+	+
5	−	+	+	+	−	−	+	−
6	+	−	+	+	−	−	−	+
7	−	+	−	+	−	+	−	+
8	+	−	+	−	+	−	+	−
9	+	+	−	+	+	−	−	−
10	−	+	+	−	+	−	−	+
11	−	−	+	+	+	+	−	−
12	+	−	−	+	−	+	+	−
13	−	+	−	−	+	+	+	−
14	−	−	+	−	−	+	+	+
15	−	−	−	+	+	−	+	+
16	−	−	−	−	−	−	−	−

Four of the variables were on the rotor and four on the stator. The response of interest was starting voltage. Normal procedure would be to designate the four rotor variables with the letters A, B, C, and D and the four stator variables with the letters E, F, G, and H. It can be seen from the matrix that there will be 16 different rotors and 16 different stators.

It was found that by using the following procedures, a Resolution IV design could be obtained that required only eight rotors and eight stators. Each rotor would be tested with two different stators and each stator would be tested with two different rotors to give the 16 distinct treatment combinations required for a valid Resolution IV Design.

- Write down the first treatment combination

	A	B	C	D	E	F	G	H
Trial 1	+	−	−	−	+	+	−	+

- Find a trial with the same levels of A and B of trial 1 and write down that treatment combination

	A	B	C	D	E	F	G	H
Trial 6	+	−	+	+	−	−	−	+

- Select A and B and the other two variables that have the same levels in trial 1 and 6. In this case they are G and H.
- Therefore, A, B, G, and H should be designated as rotor variables and C, D, E, and F should be designated as stator variables.
- Only eight different rotors and eight different stators are required.
- Each rotor will be assembled with a proper stator as designated by the Hadamard matrix; the combination will be tested and result recorded; the assembly will be disconnected and each component will be tested with a second proper component. Thus a total of 16 results will be obtained.

For a 32-trial, two-component system, each with eight variables, the letters A, B, C, D, F, H, I, and O for the first component will give a Resolution IV Smoak Design.

For a 32-trial, four-component system, each with four variables, letters are assigned to variables as follows:

Component 1	A, B, C, and H	8 Configurations
2	D, E, F, and M	8 Configurations
3	I, L, O, and P	8 Configurations
4	G, J, K, and N	8 Configurations

For a 32-trial, three-component system, one with eight variables and two with four variables each, letters are assigned as follows:

Component 1	A, B, C, D, F, H, I, and O	16 Configurations
2	E, L, M, and P	8 Configurations
3	G, J, K, and N	8 Configurations

For a 64-trial, two-component system, each with 16 variables, the letters A, B, C, D, J, L, M, O, Q, U, W, X, Z, D′, E′, and F′ for the first component will give a Resolution IV Smoak Design requiring only 32 unique configurations of each component.

For a 64-trial, four-component system, each with eight variables, letters are assigned to the variables as follows and will require only 16 of each configuration:

Component 1	A, B, C, D, L, M, Q, and X
2	J, O, U, W, Z, D′, E′, and F′
3	E, F, G, H, I, P, R, and A′
4	K, N, S, T, V, Y, B′, and C′

For a 64-trial, three-component system, one with 16 variables, two components each with eight variables, letters are assigned as follows:

Component 1	A, B, C, D, J, L, M, O, Q, U, W, X, Z, D′, E′, and F′	32 Configurations
2	E, F, G, H, I, P, R, and A′	16 Configurations
3	K, N, S, T, U, Y, B′, and C′	16 Configurations

There are several limitations to the use of the above designs.

- It must be possible to assemble and disassemble the components without damage.

- The test performed on the assembly must not damage or change the components.
- The results of each test must become available in a relatively short period of time.
- The components should be relatively expensive.

Other Hadamard Matrices Not of Order 2^n

Hadamard matrices other than the 2^n matrices discussed above also exist; however, they can only generate Resolution III designs. The following are some useful vectors:

$$
\begin{array}{ll}
N = 12 & + + - + + + - - - + - \\
N = 20 & + + - - + + + + - + - + - - - - + + - \\
N = 24 & + + + + + - + - + + - - + + - - + - + - - - - \\
N = 44 & + + - - + - + - - + + + - + + + + + - - - + - + + + \\
 & - - - - - + - - - + + - + - + + -
\end{array}
$$

The procedure for generating matrices from these vectors is the same as described in Chapter 6. The labeling of these matrices is exactly the same as for previously described Resolution III designs. All such designs up to 124 × 124 are programmed in the computer program described in Chapter 13, and the use of these matrices is exactly the same as for the other matrices described in this chapter.

EXAMPLE

A process department is responsible for production of an aluminum roll that is subsequently coated with a cast-plastic cover. Problems with the coated roll are being encountered in the field, with two failure modes being observed: (1) lack of adhesion of the plastic to the aluminum; and (2) tearing of the plastic. Both failure modes result in the plastic surface becoming pitted, which causes the roll to perform improperly.

Analysis of the problem by the original design engineer, the manufacturing engineer, a consultant from the plastic supplier, and the corporate statistician resulted in the following list of material and process variables that could influence the failure of the rolls:

A. Type of aluminum used in the rolls.
B. Surface roughness of the aluminum.
C. Cleaning procedure for the aluminum rolls.
D. Cleaning material for the aluminum rolls.
E. Molecular weight of the coating polymer.
F. Percent catalyst used with the polymer.
G. Initial temperature of the aluminum rolls.
H. Curing time in the mold.
I. Curing temperature of the mold.
J. Postcuring time.
K. Postcuring temperature.
L. Grinding pressure on the finished plastic surface.

A test for adhesion is available that consists of loosening an edge of the plastic and stripping the coating from the roll. If there is at least one area the size of a dime that has no plastic adhering to the aluminum, the part is a failure.

The test for tear strength (cohesion) consists of applying a blunt gouging device at a fixed angle to the plastic coating and measuring the force required to tear the material. It is expected that variables A through D, and possibly E and G, will influence adhesion, and it is expected that variables E through L will influence tear strength. The variance is known to be 8 for cohesion test.

A. Define the object of the experiment.

$(H_0)_1: \mu_{2024} = \mu_{6061}$
$(H_a)_1: \mu_{6061} > \mu_{2024}$

It is expected that aluminum alloy 6061 will perform better than aluminum alloy 2024, which is presently being used.

$(H_0)_2: \mu_{500\ \mu\text{in.}} = \mu_{1000\ \mu\text{in.}}$
$(H_a)_2: \mu_{1000\ \mu\text{in.}} > \mu_{500\ \mu\text{in.}}$

It is expected that the rougher surface on the aluminum (1000 μin.) will be better than the smooth surface (500 μin.).

(The other hypotheses are set up in a similar manner.)

B. Choose α, β, and δ for the cohesion test, and specify σ^2 and μ_0.

$\alpha = 0.10$
$\beta = 0.05$
$\delta = 2.0$ pounds
σ^2 is known to be 8.

The mean tear strength of the present product (μ_0) is 50 pounds.

Consider the adhesion test.

For the adhesion test, a decision is made that the results will be examined at the conclusion of the experiment and that the decisions will be made at $\alpha = 0.05$.

C. Compute $N_{\text{high}} = N_{\text{low}}$.

$$N_{\text{high}} = N_{\text{low}} = 2(U_\alpha + U_\beta)^2 \frac{\sigma^2}{\delta^2}$$

$N_{\text{high}} = N_{\text{low}} = 2(1.282 + 1.645)^2 \frac{8}{4} = 34.26$

A 64-trial Hadamard matrix is approximately the *minimum* correct size. Maybe a larger design will be required after further consideration of the problem.

D. Examine the choices of available experiment designs.

(a) A complete factorial is out of the question since, for 12 variables, it would require $2^{12} = 4096$ trials.

(b) A Resolution V design could be obtained with 256 trials (order-256 Hadamard matrix), but this too was considered excessive.

(c) If one of the variables were eliminated from the experiment, a Resolution V design could be obtained for 128 trials. However, it was considered that any one of the 12 variables could potentially be the prime solution to the problem, either individually or as part of an interaction.

(d) A Resolution IV design could be obtained with 64 trials (order-64 Hadamard matrix). Furthermore, it is expected that several of the variables will prove to be insignificant. Only variables A through D, and possibly E and G, are expected to influence adhesion. Only variables E through L are expected to influence tear strength. If these assumptions are true, then, as far as adhesion is concerned, there are only six variables in the experiment and, regarding tear strength, there are only eight variables at most. With only eight variables and 64 trials, the design is at least Resolution V for either response. Thus, this design should, in the final analysis, converge to a Resolution V design. Sixty-four trials were considered reasonable by the manufacturing people.

E. Use the computer program described in Chapter 13 to obtain the experiment design. Use the present state of the variable as the low level and the proposed better state of the variable as the high level.

(Note: The computer program outputs the design matrix in terms of 0 and 1 instead of the conventional designation of −1 and +1. Thus, in designing the experiment, the engineer should read 0 as the low level of the variable and 1 as the high level. In the computer program itself, the 0 is converted back to −1 for the calculations.)

```
NT  ABCDEFGHIJKL
 1  100000110111
 2  110000001010
 3  111000000011
 4  111100000101
 5  111110010001
 6  111111111111
 7  011111001000
 8  101111000010
 9  010111000001
10  101011000100
11  010101010101
12  101010101010
13  110101100010
14  011010100000
15  001101100001
16  100110100101
17  110011110000
18  011001011010
19  101100111000
20  110110011000
21  111011111001
22  011101011100
23  101110101100
24  110111110110
25  011011001110
26  101101010110
```

```
27  0 1 0 1 1 0 1 0 1 1 1 1
28  0 0 1 0 1 1 1 1 0 0 1 1
29  1 0 0 1 0 1 0 1 1 1 1 1
30  0 1 0 0 1 0 1 0 1 0 0 1
31  0 0 1 0 0 1 1 0 0 1 1 1
32  1 0 0 1 0 0 1 1 0 0 0 1
33  1 1 0 0 1 0 0 1 1 1 1 0
34  1 1 1 0 0 1 1 0 1 1 0 1
35  0 1 1 1 0 0 1 1 0 0 1 0
36  0 0 1 1 1 0 0 1 1 0 1 1
37  0 0 0 1 1 1 1 1 1 1 0 0
38  1 0 0 0 1 1 0 0 1 1 0 1
39  0 1 0 0 0 1 0 1 0 0 1 1
40  1 0 1 0 0 0 1 1 1 1 1 0
41  1 1 0 1 0 0 0 0 1 1 0 0
42  1 1 1 0 1 0 0 1 0 1 1 1
43  1 1 1 1 0 1 1 0 1 0 1 1
44  0 1 1 1 1 0 1 0 0 1 1 0
45  0 0 1 1 1 1 1 1 0 1 0 1
46  1 0 0 1 1 1 0 0 1 0 1 1
47  0 1 0 0 1 1 0 0 0 1 1 1
48  1 0 1 0 0 1 0 1 0 0 0 0
49  0 1 0 1 0 0 1 1 1 0 1 1
50  0 0 1 0 1 0 0 1 1 1 0 1
51  0 0 0 1 0 1 1 0 1 0 0 0
52  1 0 0 0 1 0 1 0 0 0 1 1
53  1 1 0 0 0 1 1 0 0 1 0 0
54  0 1 1 0 0 0 1 1 0 1 0 0
55  0 0 1 1 0 0 0 0 1 1 1 1
56  0 0 0 1 1 0 0 1 0 0 1 0
57  0 0 0 0 1 1 1 1 1 0 1 0
58  1 0 0 0 0 1 0 1 1 0 0 1
59  0 1 0 0 0 0 1 1 1 1 0 1
60  0 0 1 0 0 0 0 0 1 0 0 1
61  0 0 0 1 0 0 0 0 0 1 1 0
62  0 0 0 0 1 0 0 1 0 1 0 0
63  0 0 0 0 0 1 1 0 1 1 1 0
64  0 0 0 0 0 0 0 0 0 0 0 0
```

F. Obtain raw material, and set up experimental production procedures.

(a) Order sufficient tubular material of alloy 2024 and alloy 6061 for at least 32 rolls of each alloy. Some extra material of each should be purchased for possible follow-up experiments.

(b) Order one drum of the present high-molecular-weight polymer and one drum of the proposed low-molecular-weight polymer.

(c) Order one drum of the present cleaning material and one drum of the proposed cleaning material.
(d) When the above materials are received, make up one roll at each of the 64 treatment combinations specified by the computer program.

G. Compute criterion for tear-strength response.

$$|\bar{X}_{\text{high}} - \bar{X}_{\text{low}}|^* = U_\alpha \sigma \sqrt{\tfrac{1}{32} + \tfrac{1}{32}}$$
$$= (1.282)(2.828)(0.25)$$
$$= 0.91 \text{ pound}$$

H. Test samples and record results in the same order as given by the experiment design.
The adhesion test provides only a qualitative response; however, it can be partially quantified by assigning numbers as follows:

Very bad = −1
Bad = 0
Very good = +1

Sample No.	Adhesion	Tear strength
1	V. bad −1	65
2	V. bad −1	53
3	Bad 0	51
4	Bad 0	53
5	Bad 0	54
6	V. good 1	68
7	Bad 0	49
8	Bad 0	51
9	Bad 0	55
10	Bad 0	49
11	Bad 0	56
12	Bad 0	53
13	Bad 0	57
14	Bad 0	52
15	V. good 1	62
16	Bad 0	57
17	V. bad −1	58
18	Bad 0	56
19	V. good 1	54
20	Bad 0	49
21	Bad 0	61
22	Bad 0	52
23	V. good 1	52
24	Bad 0	62
25	Bad 0	58
26	Bad 0	65
27	Bad 0	65
28	Bad 0	63
29	Bad 0	67
30	V. bad −1	53
31	Bad 0	69
32	Bad 0	56
33	V. bad −1	56
34	Bad 0	62
35	V. good 1	53
36	Bad 0	53
37	Bad 0	55
38	V. bad −1	57
39	V. bad −1	56
40	Bad 0	63
41	Bad 0	50
42	Bad 0	59
43	V. good 1	60
44	V. good 1	60
45	Bad 0	59

Sample No.	Adhesion	Tear strength
46	V. good 1	52
47	V. bad −1	61
48	Bad 0	53
49	Bad 0	56
50	Bad 0	50
51	Bad 0	59
52	V. bad −1	53
53	V. bad −1	61
54	Bad 0	57
55	Bad 0	60
56	Bad 0	48
57	V. bad −1	58
58	V. bad −1	54
59	V. bad −1	58
60	Bad 0	51
61	Bad 0	57
62	V. bad −1	50
63	V. bad −1	67
64	V. bad −1	50

I. Analyze tear-strength data using the computer program of Chapter 13.

The following variables and interactions are significant for tear strength:

$$\bar{X}_{\text{high E}} - \bar{X}_{\text{low E}} = -1.97$$
$$\bar{X}_{\text{high F}} - \bar{X}_{\text{low F}} = 3.78$$
$$\bar{X}_{\text{high G}} - \bar{X}_{\text{low G}} = 4.78$$
$$\bar{X}_{\text{high J}} - \bar{X}_{\text{low J}} = 4.28$$
$$\bar{X}_{\text{high K}} - \bar{X}_{\text{low K}} = 3.97$$
$$\bar{X}_{\text{high L}} - \bar{X}_{\text{low L}} = 2.78$$
$$BE + AH + JK = 3.78$$

As expected, variables A, B, C, and D had no effect on tear strength. Surprisingly, however, the change due to variable E is minus, indicating that the change in molecular weight had a detrimental effect. There is an interaction of BE and/or AH and/or JK; from theoretical considerations of this system, it is probable that the true interaction is JK. Note, also, that B is not expected to have any effect; therefore, BE is also not expected to exist. Likewise for AH.

J. Make up a 6-way chart to present the data (figure 7-2).

(Note: The chart is laid out such that tear-strength results increase from the upper left corner down to the lower right corner. In the upper left block, the response is 49 and the trial number is 20, etc.) It is apparent that the block of 16 results at high J and high K is by far the superior block. Within that block of results, low E, high F, high G, and

			1				k			
			1		g		1		g	
			1	*l*	1	*l*	1	*l*	1	*l*
1	1	e	20 49	5 54	14 52	30 53	56 48	36 53	12 53	52 53
		1	64 50	60 51	19 54	32 56	2 53	3 51	35 53	49 56
	f	e	7 49	9 55	17 58	21 61	8 51	46 52	57 58	28 63
		1	48 53	58 54	51 59	15 62	18 56	39 56	13 57	43 60
j	1	e	62 50	50 50	23 52	16 57	33 56	42 59	44 60	27 65
		1	4 53	41 50	54 57	59 58	61 57	55 60	40 63	1 65
	f	e	10 49	38 57	37 55	45 59	25 58	47 61	24 62	6 68
		1	22 52	11 56	53 61	34 62	26 65	29 67	63 67	31 69

Fig. 7-2. Six-way chart of test data.

		1		g	
		1	*l*	1	*l*
1	e	49 50 48 } $\bar{X} = 49$ (56)	54 50 53 } $\bar{X} = 52$ (59)	52 52 53 } $\bar{X} = 52$ (60)	53 57 53 } $\bar{X} = 54$ (65)
	1	50 53 53 } $\bar{X} = 52$ (57)	51 50 51 } $\bar{X} = 51$ (60)	54 57 53 } $\bar{X} = 55$ (63)	56 58 56 } $\bar{X} = 57$ (65)
f	e	49 49 51 } $\bar{X} = 50$ (58)	55 57 52 } $\bar{X} = 55$ (61)	58 55 58 } $\bar{X} = 57$ (62)	61 59 63 } $\bar{X} = 61$ (68)
	1	53 52 56 } $\bar{X} = 54$ (65)	54 56 56 } $\bar{X} = 55$ (67)	59 61 57 } $\bar{X} = 59$ (67)	62 62 60 } $\bar{X} = 61$ (69)

(Note: The circled sample results are those obtained at high J and high K.)

Fig. 7-3. Alternative method of presenting test data.

high L have small but significant effects on improving the product.

An alternative method of presenting the data is to make up the chart without the interacting variables and then to indicate the interaction by circling the results that are unusually good (or bad) due to the combination of the interacting variables. Figure 7-3 illustrates this method.

K. Analyze the adhesion data using the computer program of Chapter 13.

The adhesion response is actually a qualitative response in that the roll is adjudged as either good, bad, or very bad. However, for use in the computer program, the data can be made quantitative by designating, for example:

$$\text{Good} = +1$$
$$\text{Bad} = 0$$
$$\text{Very bad} = -1$$

Using this response scale, only variables C, D, and G show a large average positive response:

$$\bar{X}_{\text{high C}} - \bar{X}_{\text{low C}} = 0.69$$
$$\bar{X}_{\text{high D}} - \bar{X}_{\text{low D}} = 0.75$$
$$\bar{X}_{\text{high G}} - \bar{X}_{\text{low G}} = 0.19$$

These are significant effects when compared with a criterion calculated by using the 20 degrees of freedom of third-order and higher interaction as an estimate of the variance:

$$S = 0.28 \text{ with } \phi = 20$$

$$|\bar{X}_{\text{high}} - \bar{X}_{\text{low}}|^* = (1.73)(0.28)\sqrt{\tfrac{1}{32} + \tfrac{1}{32}} = 0.12$$

for 95 percent confidence.

Three interaction contrast columns are also significant:
contrast 9, which is CD; contrast 48, which is CG + AI + FL; and contrast 52, which is DG + EJ + BK.
From theoretical considerations, it is believed that only the CD interaction can exist. Therefore, the other interactions will not be considered at this time.
A two-way chart of C and D could be made up to show all the results; however, a quick inspection of the treatment combination for the eight samples with very good adhesion data reveals the following:

	Sample No.	A	B	C	D	E	F	G	H	I	J	K	L
V. good	6	1	1	1	1	1	1	1	1	1	1	1	1
V. good	15	0	0	1	1	0	1	1	0	0	0	0	1
V. good	19	1	0	1	1	0	0	1	1	1	0	0	0
V. good	23	1	0	1	1	1	0	1	0	1	1	0	0
V. good	35	0	1	1	1	0	0	1	1	0	0	1	0
V. good	43	1	1	1	1	0	1	1	0	1	0	1	1
V. good	44	0	1	1	1	1	0	1	0	0	1	1	0
V. good	46	1	0	0	1	1	1	0	0	1	0	0	1

Obviously, the majority of good results have high C, high D, and high G in common. Sample 46 doesn't seem to belong, however, since C is low and G is low.

From the design matrix, it is seen that there is one additional sample, 45, which is made with high C, high D, and high G, that should be good but in reality was bad. A check of the samples and recorded data revealed that, in fact, a switch had been made: sample 46 was, in fact, bad, and sample 45 was very good. Thus, the entire experiment became consistent.

L. Make engineering decision.

To eliminate the adhesion problem, use the new cleaning material and the new cleaning procedure for the aluminum rolls. Also, use the high initial temperature for the rolls in the coating process.

To substantially improve the tear strength, change the catalyst concentration to the high level, change to the high initial temperature of the molds, change postcuring time to the high level, change the postcuring temperature to the high level, and change the grinding procedure to the high level. By making these changes, an increase of approximately 40 percent in tear strength will be achieved.

Summary

All of the regular Hadamard matrix designs up to this point can be summarized in the chart on the following page.

The experimenter must first determine how many variables are to be included in the experiment. Then a variety of choices are available depending upon the time available, and the cost and information desired or required.

For example, suppose there are seven Variables. The choices are:

- A 128-trial design. All interactions can be estimated. The cost might be $10,000 for 128 trials and the time required four weeks.

- A 64-trial design. All main effects and two-factor interactions can be estimated. It must be assumed that all 3-factor and high interactions are zero. The cost will be $5,000 and two weeks will be required.
- A 16-trial design. All main effects and groups of 2-factor interactions can be estimated. It must be assumed that all 3-factor and higher interactions are zero. The cost will be $1,250 and $\frac{1}{4}$ week will be required.
- An 8-trial design. Only main effects can be estimated. It must be assumed that all interactions are zero. Cost will be $625 and approximately one day will be required.

MATRIX SIZE

	8	16	32	64	128
Complex Matrix	up to 3 variables	4	5	6	7
Resolution V	3	5	6	7–8	8–11
Resolution IV	4	6–8	7–16	9–32	12–64
Resolution III	5–7	9–15	17–31	31–36	65–127

In the above example, cost can range from $625 to $10,000. The experimenter must decide, for example, whether the information to be obtained by the $10,000 experiment is worth the cost. Likewise the time can vary between one day and four weeks. Again, the experimenter must decide, "Is the information of the complete matrix worth four weeks of work?"

It will be apparent that, as the number of variables gets larger, the number of choices decrease. For example, with 64 variables the only choices are a 128-trial Resolution IV design or a 64-trial Resolution III design.

EXERCISES

1. What is the complete labeling of a 32-trial Hadamard matrix with seven variables if F is put in column ABCDE and G is put in column ACDE?
2. What is the complete labeling of a 32-trial Hadamard matrix with seven variables if F is put in column ABCD and G is put in column BCDE?
   - Is this a better design than that given in 1?
   - Why?
3. Design a 16-trial Hadamard matrix with six variables. List all the confoundings. What resolution is this design? What interactions must be assumed to be zero? Suppose contrast column 10 is highly significant. What is your engineering statement?
4. A manufacturing process with five major variables was studied with a 16-trial Hadamard matrix. $\sigma = 2$, $\alpha = \beta = 0.05$. The following results were obtained:

Treatment Combination	Results (mW)
ae	20
ab	28
abce	27
abcd	29
bcde	18
acde	19
bd	19
ac	18
abde	28
bc	19
cd	21
ad	17
be	19
ce	19
de	20
(1)	18

Completely analyze the data and suggest further changes in the variables to obtain a product with still higher power output (mW).

REFERENCES

Box, G. E. P., Hunter, W. G., and Hunter, J. S. 1978. *Statistics for experimenters.* New York: John Wiley and Sons, Inc.

Daniel, C. 1976. *Applications of statistics to industrial experimentation.* New York: John Wiley and Sons, Inc.

John, P.W.M. 1971. *Statistical design and analysis of experiments.* New York: Macmillan Publishing Co.

8 John's Three-Quarter Fractional Factorials

There is one major limitation on the 2^n Hadamard matrix designs described in Chapters 6 and 7: for the Resolution IV and V designs, the gap between the number of trials required for each available matrix becomes larger and larger as the order of the matrix increases (i.e., 8, 16, 32, 64, 128, etc.). Dr. Peter John, of the University of Texas, alleviated this problem by developing the concept of a three-quarter fractional factorial. His designs provide Resolution IV and, in some cases, Resolution V designs with order 12, 24, 48, 96, etc., matrices.

Figure 8-1 shows how John's designs fill in some of the gaps between the Hadamard matrix designs discussed in Chapter 6 and 7:

	Number of Trials									
	6 John	8 Hadamard	12 John	16 Hadamard	24 John	32 Hadamard	48 John	64 Hadamard	96 John	128 Hadamard
Resolution V+	1*–2	1–3	1–4	1–5	1–5	1–6	1–8	1–8	1–11	1–11
Resolution IV	3	4	5–6	6–8	6–12	7–16	9–24	9–32	12–48	12–64

* Number of variables that can be included in the design matrix for the given resolution.

Fig. 8-1. Comparison of John's designs with Hadamard matrix designs.

The generation and use of the John's designs shown in figure 8-1 are described in this chapter.

Resolution V Designs

The maximum number of variables that can be included in a given matrix and still be Resolution V with John's designs are shown in figure 8–1:

- Four variables in 12 trials.
- Five variables in 24 trials.

- Seven and eight variables in 48 trials.
- Nine to eleven variables in 96 trials.

Each of these designs, except the 24-trial design, is described in the following subsections. The 24-trial John's design requires more trials for five variables than the 16-trial Hadamard matrix, but there is no increase in precision. This 24-trial matrix is only valuable for Resolution IV designs.

Four Variables in 12 Trials

John's Resolution V designs start out with a Hadamard matrix; i.e., the 16 × 16 Hadamard matrix is used to obtain the 12-trial John's design, etc. In each case, one-fourth of the Hadamard trials are eliminated by the procedure described below.

For this case (four variables in 12 trials), the starting point is the 16 × 16 Hadamard matrix, which is shown below:

Trial	0	A 1	B 2	C 3	D 4	−AB 5	−BC 6	−CD 7	ABD 8	−AC 9	−BD 10	ABC 11	BCD 12	−ABCD 13	ACD 14	−AD 15	Treatment combination
1	+	+	−	−	−	+	−	−	+	+	−	+	−	+	+	+	a
2	+	+	+	−	−	−	+	−	−	+	+	−	+	−	+	+	ab
3	+	+	+	+	−	−	−	+	−	−	+	+	−	+	−	+	abc
4	+	+	+	+	+	−	−	−	+	−	−	+	+	−	+	−	abcd
5	+	−	+	+	+	+	−	−	−	+	−	−	+	+	−	+	bcd
6	+	+	−	+	+	+	+	−	−	−	+	−	−	+	+	−	acd
7	+	−	+	−	+	+	+	+	−	−	−	+	−	−	+	+	bd
8	+	+	−	+	−	+	+	+	+	−	−	−	+	−	−	+	ac
9	+	+	+	−	+	−	+	+	+	+	−	−	−	+	−	−	abd
10	+	−	+	+	−	+	−	+	+	+	+	−	−	−	+	−	bc
11	+	−	−	+	+	−	+	−	+	+	+	+	−	−	−	+	cd
12	+	+	−	−	+	+	−	+	−	+	+	+	+	−	−	−	ad
13	+	−	+	−	−	+	+	−	+	−	+	+	+	+	−	−	b
14	+	−	−	+	−	−	+	+	−	+	−	+	+	+	+	−	c
15	+	−	−	−	+	−	−	+	+	−	+	−	+	+	+	+	d
16	+	−	−	−	−	−	−	−	−	−	−	−	−	−	−	−	(1)

The sixteen trials of the Hadamard matrix are separated into four blocks of four trials each by using the signs in the (−AB) column and the (BCD) column:

Block	−AB	BCD	Treatment combinations
i	+	+	bcd, ac, ad, b
ii	−	+	ab, abcd, c, d
iii	+	−	a, acd, bd, bc
iv	−	−	abc, abd, cd, (1)

Block i is eliminated from the matrix, and the remaining 12 trials, consisting of blocks ii, iii, and iv, make up a Resolution V design for the four variables A, B, C, and D.

The reader should not presume that this design with four variables and with only 12 trials is "better" than the full Hadamard matrix design with four variables in 16 trials. They are both Resolution V, but the precision of the estimates will now be shown to be better with the 16-trial design.

When a series of trials is blocked as above and some trials are eliminated, the interaction columns used in the blocking become identities for the remaining trials, and there will be confounding of interaction. Therefore,

$$I = (-AB) + BCD + (-ACD).$$

The $(-ACD)$ is also an identity because $(-AB)(BCD) = (-ACD)$.

If all 12 trials were used for estimating variable A, for example, then variable A would be confounded with variable B, with the interaction ABCD, and with the interaction $-CD$:

$$\begin{array}{r} I = -AB + BCD + (-ACD) \\ \times\ A \quad\quad\quad\quad\quad\quad\quad\quad\quad\quad \\ \hline A = -B + ABCD + (-CD) \end{array}$$

Therefore, all 12 trials cannot be used for estimating A, B, AB, etc.

To estimate main effect and two-factor interactions, the blocks are paired—(ii and iii), (ii and iv), and (iii and iv)—to provide three different sets of eight treatment combinations. One of these sets of eight treatment combinations is used to estimate C, D, and CD. Since only eight results are used to estimate, for example, C in the John's three-quarter fractional factorial design, the precision of the estimate is not as good as the precision in the full factorial Hadamard matrix, where 16 trials are used to estimate every factor or interaction. It will be seen later that the objective criterion will also be larger with these John's designs than with the full Hadamard matrix design.

The pairing of blocks ii, iii, and iv results in the following defining contrasts for the eight trials in each pair of blocks:

- Blocks ii and iii

$$\begin{array}{rrl} \text{ii} & I = & -(-AB) + (BCD) + (-ACD) \\ +\ \text{iii} & I = & +(-AB) - (BCD) + (-ACD) \\ \hline & 2I = & 0 \quad + \quad 0 \quad + 2(-ACD) \\ & I = & -(ACD) \end{array}$$

- Blocks ii and iv

$$\begin{array}{rrl} \text{ii} & I = & -(-AB) + (BCD) + (-ACD) \\ +\ \text{iv} & I = & -(-AB) - (BCD) - (-ACD) \\ \hline & 2I = & -2(-AB) + \quad 0 \quad + \quad 0 \\ & I = & (AB) \end{array}$$

- Blocks iii and iv

$$\begin{array}{rrl} \text{iii} & I = & +(-AB) - (BCD) + (-ACD) \\ +\ \text{iv} & I = & -(-AB) - (BCD) - (-ACD) \\ \hline & 2I = & 0 \quad - 2(BCD) + \quad 0 \\ & I = & -(BCD) \end{array}$$

The eight results that comprise blocks ii and iii can only estimate a few of the main effects and two-factor interactions that are desired. To determine which interactions or main factors can be estimated, all the column headings of the full Hadamard matrix must be multiplied by the defining contrast $I = -ACD$, using the rule discussed in the previous chapter (A^2, B^2, etc. $= I$, and $I \times A = A$, etc.), to determine the confounding:

$$
\begin{aligned}
A \times (-ACD) &= -CD \\
B &= -ABCD \\
C &= -AD \\
D &= -AC \\
-AB &= BCD \\
-BC &= ABD \\
-CD &= A \\
ABD &= -BC \\
-AC &= D \\
-BD &= ABC \\
ABC &= -BD \\
BCD &= -AB \\
-ABCD &= B \\
ACD &= -I \\
-AD &= C
\end{aligned}
$$

These eight trials cannot be used to estimate A, since A is confounded with −CD; however, these eight trials can be used to estimate B, which is confounded only with a four-factor interaction that is assumed to have a value of zero. If a variable is confounded with a two-factor interaction, or if a two-factor interaction is confounded by another two-factor interaction, neither can be estimated by the given eight results in that pair of blocks. All other effects and interactions can be estimated. From blocks ii and iii, B, −AB, −BC, and −BD can be estimated.

Blocks ii and iv have the defining contrast I = AB.

$$
\begin{aligned}
A \times AB &= B \\
B &= A \\
C &= ABC \\
D &= ABD \\
-AB &= -I \\
-BC &= -AC \\
-CD &= -ABCD \\
ABD &= D \\
-AC &= -BC \\
-BD &= -AD \\
ABC &= C \\
BCD &= ACD \\
-ABCD &= -CD \\
ACD &= BCD \\
-AD &= -BD
\end{aligned}
$$

Therefore, the eight trials of blocks ii and iv can be used to estimate C, D, and CD.

Blocks iii and iv have the defining contrast I = −(BCD), and it can be shown with the procedure above that these blocks can estimate A, AB, AC, and AD.

To summarize:

- Blocks ii and iii estimate B, AB, BC, and BD.
- Blocks ii and iv estimate C, D, and CD.
- Blocks iii and iv estimate A, AB, AC, and AD.

Note that all possible main factor and two-factor interactions are included. Note also that AB is estimated twice; the average of these two estimates is compared to the criterion, which will be, for AB only,

$$|\bar{X}_{\text{high}} - \bar{X}_{\text{low}}|^* = U_\alpha\sigma(0.61) \quad \text{or} \quad t_\alpha S(0.61).$$

For all other main effects and interactions, the criterion will be

$$|\bar{X}_{\text{high}} - \bar{X}_{\text{low}}|^* = U_\alpha\sigma\sqrt{\tfrac{1}{4} + \tfrac{1}{4}} \quad \text{or} \quad t_\alpha S\sqrt{\tfrac{1}{4} + \tfrac{1}{4}}.$$

EXAMPLE

An experiment is needed to evaluate the effect of changing four variables (spring length, shaft diameter, slot diameter, and type of lubricant) on the life of a spring clutch. There are only 12 machines available for testing the clutch. Previous experience with clutches indicates that the mean life is 10 months, normally distributed, with an estimated variance of 1 month with 10 degrees of freedom. An improvement of two months would be very important. There is definitely a possibility of two-factor interactions, and it is necessary to estimate the value of them if they do exist. It is desirable to have $\beta = 0.1$; any reasonable α is satisfactory. A Resolution V design is required.

The variables and their high and low levels are defined as follows:

Variable	Low level	High level
A = Spring length	2 in.	3 in.
B = Shaft diameter	1/4 in.	3/8 in.
C = Slot diameter	1/2 in.	3/4 in.
D = Lubricant	Low viscosity	High viscosity

A. Define the object of the experiment.

$(H_0)_1: \mu_{2''} = \mu_{3''}$
$(H_a)_1: \mu_{2''} \neq \mu_{3''}$

$(H_0)_2: \mu_{1/4''} = \mu_{3/8''}$
$(H_a)_2: \mu_{1/4''} \neq \mu_{3/8''}$

$(H_0)_3: \mu_{1/2''} = \mu_{3/4''}$
$(H_a)_3: \mu_{1/2''} \neq \mu_{3/4''}$

$(H_0)_4: \mu_{\text{low vis}} = \mu_{\text{high vis}}$
$(H_a)_4: \mu_{\text{low vis}} \neq \mu_{\text{high vis}}$

$(H_0)_5: \mu_{ij} = 0; \quad i \neq j = \text{A, B, C, D}$
$(H_a)_5: \mu_{ij} \neq 0$

B. The John's 12-trial, three-quarter fractional factorial is the only possible design. For this design, $N_{high} = N_{low} = 4$. Since S^2, δ, and β are known, α can be computed:

$$2(t_\alpha + t_\beta)^2 \frac{S^2}{\delta^2} = N$$

$$(t_\alpha + t_\beta)^2 = \frac{N\delta^2}{2S^2}$$

$$t_\alpha = \sqrt{\frac{N\delta^2}{2S^2}} - t_\beta$$

Note: t_β is single-sided.

$N = 4$
$\delta = 2$ months
$S^2 = 1$ with 10 degrees of freedom
$\beta = 0.1 \quad t_\beta = 1.37$ from table 3

$$t_\alpha = \sqrt{\frac{(4)(4)}{(2)(1)}} - 1.37$$

$t_\alpha = 2.83 - 1.37 = 1.46$
From table 4, α(double-sided) $\doteq 0.18$

C. Determine the treatment combinations to be tested.

As shown in the preceding subsection, the 12 trials are:

ii ab, abcd, c, d
iii a, acd, bd, bc
iv abc, abd, cd, (1)

D. Make up 12 clutches according to item C, above, and install them on the 12 test machines.
Record all failures of the clutch to engage. When the rate of failure is greater than five per day over a period of one week, a clutch will be considered a failure.

Block	Results (months to failure)	
ii	ab	10.7
	abcd	16.6
	c	11.4
	d	13.2
iii	a	10.8
	acd	14.2
	bd	11.4
	bc	12.5
iv	abc	12.9
	abd	11.7
	cd	13.1
	(1)	10.6

E. Analyze the results.
The results would normally be put in standard order for analysis by the computer program. For those readers not having a computer, the pen and paper method of analysis is shown.

Go to the Hadamard matrix and record the signs under the main factor and interaction columns beside the treatment combinations as listed above (see figure 8-2). List AB twice and, for identification purposes, list one as AB_1, the other AB_2.

	A	B	C	D	−AB₁	−BC	−CD	−AC	−BD	−AD	−AB₂
ii ab	+	+	−	−	−	+	−	+	+	+	−
abcd	+	+	+	+	−	−	−	−	−	−	−
c	−	−	+	−	−	+	+	+	−	−	−
d	−	−	−	+	−	−	+	−	+	+	−
iii a	+	−	−	−	+	−	−	+	−	+	+
acd	+	−	+	+	+	+	−	−	+	−	+
bd	−	+	−	+	+	+	+	−	−	+	+
bc	−	+	+	−	+	−	+	+	+	−	+
iv abc	+	+	+	−	−	−	+	−	+	+	−
abd	+	+	−	+	−	+	+	+	−	−	−
cd	−	−	+	+	−	+	−	+	+	+	−
(1)	−	−	−	−	−	−	−	−	−	−	−

Fig. 8-2. Blocks used for estimating main effects and interactions.

The blocks used for estimating each main effect and interaction are indicated by outline in figure 8-2. Thus, the effect of A is measured by blocks iii and iv by reading the outlined signs down column A and using these signs with the column of results, as shown below:

$$(\bar{X}_{\text{high A}} - \bar{X}_{\text{low A}}) = (+10.8 + 14.2 - 11.4 - 12.5 + 12.9 + 11.7 - 13.1 - 10.6)/4 = 0.5.$$

Similarly, the −BC interaction is measured from blocks ii and iii:

$$(\bar{X}_{\text{high BC}} - \bar{X}_{\text{low BC}}) = (+10.7 - 16.6 + 11.4 - 13.2 - 10.8 + 14.2 + 11.4 - 12.5)/4 = -1.35$$

F. Summarize the results.

$$\text{A } (\bar{X}_{3''} - \bar{X}_{2''}) = 0.50$$
$$\text{B } (\bar{X}_{3/8''} - \bar{X}_{1/4''}) = 0.40$$
$$\text{C } (\bar{X}_{3/4''} - \bar{X}_{1/2''}) = 1.95$$
$$\text{D } (\bar{X}_{\text{high vis}} - \bar{X}_{\text{low vis}}) = 2.25$$
$$-AB_1 = -.75$$
$$-AC = -0.25$$
$$-AD = -0.20$$
$$-BC = -1.35$$
$$-BD = 0.10$$
$$-CD = -0.45$$
$$-AB_2 = .15$$
$$\text{Average } -AB = 0.75$$

G. Compute the criterion.

$$|\bar{X}_{\text{high}} - \bar{X}_{\text{low}}|^* = t_\alpha S \sqrt{\frac{1}{N_{\text{high}}} + \frac{1}{N_{\text{low}}}}$$

for all estimates except AB.

$$|\bar{X}_{\text{high}} - \bar{X}_{\text{low}}|^* = (1.46)(1)\sqrt{\tfrac{1}{4} + \tfrac{1}{4}} = 1.022$$

Note: The value of $\bar{X}_{\text{high}}$ is based on only four results. Likewise for $\bar{X}_{\text{low}}$. Therefore, $N_{\text{high}} = N_{\text{low}} = 4$.

$|\bar{X}_{\text{high AB}} - \bar{X}_{\text{low AB}}|^* = t_\alpha S(0.61)$

Note: $t_\alpha = 1.46$ from item B, above.

$$|\bar{X}_{\text{high}} - \bar{X}_{\text{low}}|^*_{AB} = (1.46)(1)(0.61) = 0.89$$

H. Compare each $(\bar{X}_{\text{high}} - \bar{X}_{\text{low}})$ with the criterion, and make decisions.

$\mu_{\text{high C}}$ is better than $\mu_{\text{low C}}$, $\mu_{\text{high D}}$ is better than $\mu_{\text{low D}}$, and μ_{BC} is greater than zero with at least 91% confidence.

I. Make up a two-way chart of the BC interaction (figure 8-3). Use only the data from blocks ii and iii.

	1	c
1	d 13.2 a 10.8 avg = 12.00	c 11.4 acd 14.2 avg = 12.80
b	ab 10.7 bd 11.4 avg = 11.05	abcd 16.6 bc 12.5 avg = 14.55

Fig. 8-3. Two-way chart of BC interaction.

Plot the BC interaction (figure 8-4).

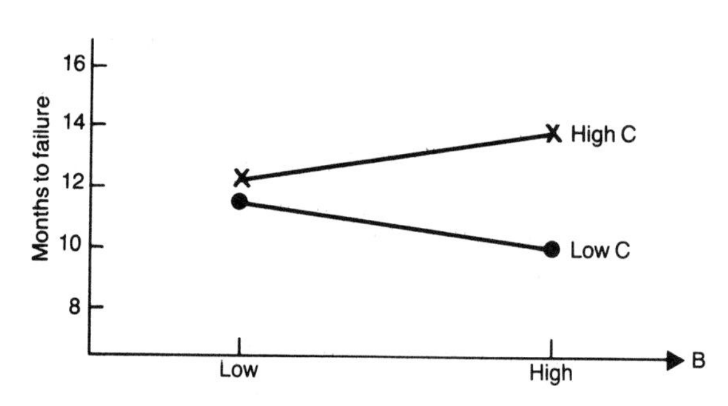

Fig. 8-4. Plot of BC interaction.

J. Make engineering decision.

Make the clutch with 3/8-in. shaft diameter and 3/4-in. slot diameter. Use high-viscosity lubricant. Spring length has no influence on the life of the clutch; therefore, a specification of 2 in. to 3 in. on the spring length is satisfactory.

Seven and Eight Variables in 48 Trials

The starting point for a 48-trial John's design is the 64 × 64 Hadamard matrix. This matrix is divided into four blocks by the defining contrast I = ABE = CDF = ABCDEF, which is the same whether seven or eight variables are used with the 48-trial design.

Seven Variables: This defining contrast is used to divide the 64 × 64 Hadamard matrix into four blocks, one of which will be discarded:

	I = ABE	+ CDF	+ ABCDEF
Block i	+	+	+
Block ii	−	+	−
Block iii	+	−	−
Block iv	−	−	+

Eliminate block i; then,

ii + iii; I = −ABCDEF

ii + iv; I = −ABE

iii + iv; I = −CDF

The first six contrast columns in the 64-trial Hadamard matrix will be labeled as usual with the letters A, B, C, D, E, and F. Since a design for seven variables is desired, one additional contrast column must be labeled G; the best design is obtained if column ABCD is labeled G. This results in the additional identity element I = ABCDG in all pairs of blocks. Therefore, the total defining contrasts are as follows:

Blocks ii + iii; I = ABCDEF + ABCDG + EFG

Blocks ii + iv; I = ABE + ABCDG + CDEG

Blocks iii + iv; I = CDF + ABCDG + ABFG

Note in blocks ii and iii (ABCDEF)(ABCDG) = EFG. The CDEG and ABFG are obtained in similar fashion.

To determine the main effects and two-factor interactions that can be evaluated in each block, proceed as follows:

1. List all the possible factors and interactions of variables A, B, C, D, and E, and multiply each of them by the three identity elements in each pair of blocks. In blocks ii + iii, for example:

I	= (ABCDEF)	= (ABCDG)	= (EFG)
A*	= BCDEF	= BCDG	= AEFG
B*	= ACDEF	= ACDG	= BEFG
AB*	= CDEF	= CDG	= ABEFG
C*	= ABDEF	= ABDG	= CEFG
AC*	= BDEF	= BDG	= ACEFG
BC*	= ADEF	= ADG	= BCEFG
ABC	= DEF	= DG*	= ABCEFG
D*	= ABCEF	= ABCG	= DEFG
AD*	= BCEF	= BCG	= ADEFG
BD*	= ACEF	= ACG	= BDEFG
ABD	= CEF	= CG*	= ABDEFG
CD*	= ABEF	= ABG	= CDEFG

I	= (ABCDEF)	= (ABCDG)	= (EFG)
ACD	= BEF	= BG*	= ACDEFG
BCD	= AEF	= AG*	= BCDEFG
ABCD	= EF	= G	= ABCDEFG
E	= ABCDF	= ABCDEG	= FG
AE*	= BCDF	= BCDEG	= AFG
BE*	= ACDF	= ACDEG	= BFG
ABE	= CDF	= CDEG	= ABFG
CE*	= ABDF	= ABDEG	= CFG
ACE	= BDF	= BDEG	= ACFG
BCE	= ADF	= ADEG	= BCFG
ABCE	= DF*	= DEG	= ABCFG
DE*	= ABCF	= ABCEG	= DFG
ADE	= BCF	= BCEG	= ADFG
BDE	= ACF	= ACEG	= BDFG
ABDE	= CF*	= CEG	= ABDFG
CDE	= ABF	= ABEG	= CDFG
ACDE	= BF*	= BEG	= ACDFG
BCDE	= AF*	= AEG	= BCDFG
ABCDE	= F	= EG	= ABCDFG

* Indicates main factors and two-factor interactions that are estimated by blocks ii and iii.

2. Wherever a main factor or two-factor interaction occurs on a row and there are no other main factors or two-factor interactions on that row, that main factor or two-factor interaction can be evaluated. For example, A, which is starred, can be evaluated because A is confounded with BCDEF, BCDG, and AEFG in blocks ii and iii. E cannot be evaluated using ii and iii because it is confounded with FG.
3. Notice that the three-factor contrast ABE is confounded with CDF, CDEG, and ABFG. If this contrast is found to be highly significant, and if it can be assumed that four-factor interactions are nonexistent, it can be stated that ABE and/or CDF is significant.
4. Notice that, in the block defined by ii + iii, it is only necessary to use the interaction of A, B, C, D, and E in the leftmost column under I. All the other possible interactions are generated with the identity elements by multiplication.
5. An alternative procedure is to multiply all the five-, six- and seven-factor interactions by the identity elements. For example, with blocks ii and iv:

I	= (ABE)	= (ABCDG)	= (CDEG)
ABCDEFG	= CDFG	= EF*	= ABF
ABCDEF	= CDF	= EFG	= ABFG
ABCDEG	= CDG	= E	= AB
ABCDFG	= CDEFG	= F*	= ABEF
ABCEFG	= CFG	= DEF	= ABDF
ABDEFG	= DFG	= CEF	= ABCF
ACDEFG	= BCDFG	= BEF	= AF*
BCDEFG	= ACDFG	= AEF	= BF*
ABCDE	= CD	= EG	= ABG
ABCDF	= CDEF	= FG*	= ABEFG
ABCEF	= CF*	= DEFG	= ABDFG
ABDEF	= DF*	= CEFG	= ABCFG
ACDEF	= BCDF	= BEFG	= AFG
BCDEF	= ACDF	= AEFG	= BFG

etc.

To summarize, in the 48-trial John's three-quarter fractional factorial with seven variables:

- The 32 trials defined by I = ABCDEF + ABCDG + EFG, which are blocks ii + iii, are used to estimate A, B, C, D, AB, AC, AD, AE, AF, AG, BC, BD, BE, BF, BG, CD, CE, CF, CG, DE, DF, and DG.
- The 32 trials defined by I = ABE + ABCDG + CDEG, which are blocks ii + iv, are used to estimate C, D, F, G, AC, AD, AF, AG, BC, BD, BF, BG, CF, DF, EF, and FG.
- The 32 trials defined by I = CDF + ABCDG + ABFG, which are blocks iii + iv, are used to estimate A, B, E, G, AC, AD, AE, BC, BD, BE, CE, CG, DE, DG, EF, and EG.

It will be noted that some of the main factors and interactions are estimated by two or even three of the pairs of blocks. The effect of the variable in these cases is averaged. It should also be noted that, for all variables estimated by only one pair of blocks, the criterion is given by

$$|\bar{X}_{\text{high}} - \bar{X}_{\text{low}}|^* = t_\alpha S\sqrt{\tfrac{1}{16} + \tfrac{1}{16}} \quad \text{or} \quad U_\alpha \sigma\sqrt{\tfrac{1}{16} + \tfrac{1}{16}}.$$

For those variables estimated in two or three pairs of blocks, the criterion is $t_\alpha(0.43)S$ or $U_\alpha(0.43)\sigma$.

Eight Variables: The defining contrast for selecting the blocks for an experiment with eight variables is exactly the same as for seven variables:

$$I = ABE + CDF + ABCDEF.$$

In addition to the first six contrast columns, which are designated A through F, two additional columns must be designated as G and H. Choosing G = ABCD, as before, and, in addition, H = ACEF will create the identities

$$I = ABCDG \quad \text{and} \quad I = ACEFH.$$

Blocks ii + iii, defined by I = ABCDEF, for example, then have seven identities:

$$\begin{aligned}
I &= ABCDEF &&= ABCDEF\\
I &= ABCDG &&= ABCDG\\
I &= ACEFH &&= ACEFH\\
I &= (ABCDEF)(ABCDG) &&= EFG\\
I &= (ABCDEF)(ACEFH) &&= BDH\\
I &= (ABCDG)(ACEFH) &&= BDEFGH\\
I &= (ABCDEF)(ABCDG)(ACEFH) &&= ACGH
\end{aligned}$$

The defining contrast is therefore, I = ABCDEF + ABCDG + ACEFH + EFG + BDH + BDEFGH + ACGH. Blocks ii and iii are used to estimate A, C, AB, AD, AE, AF, BC, BE, BF, BG, CD, CE, CF, DE, DF, DG, EH, and FH.

Blocks ii + iv have the defining contrast

$$I = ABE + ACEFH + ABCDG + CDEG + BCFH + BDEFGH + ADFGH$$

and measure C, D, F, G, H, AC, AD, AF, AG, AH, BD, BG, DF, DH, EF, EH, FG, and GH.

Blocks iii + iv have the defining contrast

$$I = CDF + ACEFH + ABCDG + ABFG + ADEH + BDEFGH + BCEGH$$

and measure A, B, E, G, H, AC, BC, BD, BE, BH, CE, CG, CH, DG, EF, EG, FH, and GH.

The criteria for this design are exactly the same as for the previous case of seven variables in 48 trials.

Nine to Eleven Variables in 96 Trials

The defining contrast for selection of the blocks is

$$I = AB + CDEFG + ABCDEFG.$$

The first seven contrast columns in the Hadamard matrix are labeled A through G. In addition, the following interactions are labeled as main factors:

$$BCDE = H$$
$$ABCEG = I^* \quad \text{(Variable I, not identity.)}$$
$$ABDEF = J$$
$$AEFG = K$$

The defining contrast for each pair of blocks will have 15 elements in it when 11 variables are used. If the reader encounters such a problem it is presumed that either the computer program will be used or that the defining contrasts can be calculated by the reader. The same procedure as explained for the smaller matrices is used to determine which factors and interactions are measured by each pair of blocks. Needless to say, this is a tedious task, so it is certainly recommended that a computer be used for an experiment of this or larger size.

The criterion for these cases is given by

$$|\bar{X}_{\text{high}} - \bar{X}_{\text{low}}|^* = U_\alpha \sigma \sqrt{\tfrac{1}{32} + \tfrac{1}{32}} \quad \text{or} \quad t_\alpha S \sqrt{\tfrac{1}{32} + \tfrac{1}{32}}$$

for all variables and interactions that are measured by only one pair of blocks. Where the main factor or interaction is measured by two or three pairs of blocks, the criterion is:

$$|\bar{X}_{\text{high}} - \bar{X}_{\text{low}}|^* = U_\alpha \sigma (0.31) \quad \text{or} \quad t_\alpha S(0.31).$$

Resolution IV Designs

Three Variables in 6 Trials

All Resolution IV designs are obtained by using the defining contrast

$$I = (-AC) + (-BC) + (-AB).$$

For the 6-trial designs, the 8-trial Hadamard matrix is divided into blocks as follows:

−AC	−BC	−AB	Block	Treatment combinations
+	+	−	i	ab, c
+	−	+	ii	a, bc
−	+	+	iii	ac, b
−	−	−	iv	abc, (1)

The six trials to be used are, therefore, a, bc, ac, b, abc, and (1), and the defining contrast for blocks ii + iii is AB.

To determine which contrasts can be evaluated by blocks ii and iii, multiply each main factor and interaction by AB to obtain the confounding:

$$\begin{aligned} A \times AB &= B \\ B \times AB &= A \\ AB \times AB &= I \\ {}^{*}C \times AB &= ABC \\ {}^{*}AC \times AB &= BC^{*} \\ BC \times AB &= AC \\ ABC \times AB &= C \end{aligned}$$

A is confounded with −B, so this contrast cannot be evaluated with blocks ii and iii. C is confounded with −ABC, so, on the assumption that ABC is insignificant, C can be evaluated. Likewise, AC is confounded with −BC; therefore, we can evaluate (AC − BC) using blocks ii and iii.

The confounding is readily apparent if the matrix of blocks ii and iii is drawn as follows:

C = −ABC

Block	Treatment combination	Total	A	B	C	−AB	−BC	ABC	−AC
ii	a	+	+	−	−	+	−	+	+
ii	bc	+	−	+	+	+	−	−	+
iii	ac	+	+	−	+	+	+	−	−
iii	b	+	−	+	−	+	+	+	−

A = −B

−BC = −(−AC)

−AB = Total

The same procedure is followed with blocks ii and iv to show that A and (AC − AB) can be evaluated. In blocks iii and iv, B and (BC − AB) can be evaluated.

Five or Six Variables in 12 Trials

With 12 trials, a Resolution IV design with up to six variables can be obtained. The initial defining contrast used to divide the 16 trials into four blocks is I = (−AC) + (−BC) + (−AB). What is called an alias set can be generated by multiplying the defining contrast by main factors or interactions as follows:

$$\begin{aligned} I &= AC = BC = AB \\ C &= A = B = ABC \\ D &= ACD = BCD = ABD \\ CD &= AD = BD = ABCD \end{aligned}$$

The procedure for introducing additional variables to obtain a Resolution IV design is to equate additional variables with the three-factor interactions in the alias sets, being sure that one higher-order interaction in each row is left over. In the saturated Resolution IV case (i.e., six variables in 12 trials), set E = ACD and F = BCD to obtain the additional identities. (Under no circumstances can the ABC or the ABCD interactions be used if a Resolution IV design is required.) The additional identities, therefore, are:

I = ACDE

I = BCDF

I = ABEF, by multiplication of ACDE by BCDF.

The four blocks are obtained from the 16-trial Hadamard matrix using the defining contrast I = (−AC) + (−BC) + (−AB)

−AC	−BC	−AB	Block	Treatment combinations
+	+	−	i	abef, abd, cd, cef
+	−	+	ii	ae, bcdf, bce, adf
−	+	+	iii	acde, bde, acf, bf
−	−	−	iv	abc, abcdef, def, (1)

For blocks ii and iii, the initial defining contrast is I = −AB; then, in addition, I = ACDE and I = BCDF because of the two added variables. Finally, I = ABEF by multiplication of ACDE and BCDF. Note that −BCDE is obtained by multiplying −AB by ACDE; −ACDF is obtained by multiplying −AB by BCDF; and −EF is obtained by multiplying −AB and ABEF. Then, the defining contrast is

$$I = -AB + ACDE + BCDF + ABEF - BCDE - ACDF - EF.$$

The contrasts that are measurable in blocks ii and iii can be determined by multiplying each original contrast column heading by the defining contrast:

A = B; therefore, not measurable.

B = A; therefore, not measurable.

AB = I; therefore, not measurable.

*C = ABC = ADE = BDF = ABCEF = BDE = ADF = CEF; therefore, C can be estimated.

*AC = −BC = DE = ABDF = BCEF = ABDE = −DF = ACEF; therefore, (AC − BC + DE − DF) can be estimated.

BC = −AC, etc.

ABC = C, etc.

*D = ABD = ACE = BCF = ABDEF = BCE = ACF = DEF; therefore, D can be estimated.

*AD = −BD = CE = ABCF = BDEF = ABCE = −CF = ADEF; therefore, (AD − BD + CE − CF) can be estimated.

*BD = −AD, etc.

ABD = D, etc.

*CD = ABCD = AE = BF = ABCDEF = −BE = −AF = CDEF; therefore, (CD + AE + BF − BE − AF) can be estimated.

ACD = BCD = E = ABF = BCDEF = ABE = F = ACDEF; therefore, E and F are not measurable.

BCD = ACD, etc.

ABCD = CD, etc.

With blocks ii and iii, then, C and D and the following groups of interactions can be estimated:

(AC − BC + DE − DF)
(AD − BD + CE − CF)
(CD + AE + BF − BE − AF)

The computation of the value of the contrasts is performed by reading down the proper column in the 16 × 16 Hadamard matrix, using only the eight rows that constitute blocks ii and iii. For example:

$$(\bar{X}_{\text{high C}} - \bar{X}_{\text{low C}}) = (-\text{ae} + \text{bcdf} + \text{bce} - \text{adf} + \text{acde} - \text{bde} + \text{acf} - \text{bf})/4.$$

The value of (AC − BC + DE − DF) can be obtained from either the AC or BC column in the Hadamard matrix using only the eight rows of blocks ii and iii.

For blocks ii and iv, the defining contrast is

I = BC + ACDE + BCDF + ABEF + ABDE + DF + ACEF,

and A and E are measurable along with the following groups of interactions:

(AB + AC + DE + EF)
(AD + CE + BE + AF)
(CD + BD + AE + BF + CF)

For blocks iii and iv, the defining contrast is

$$I = AC + ACDE + BCDF + ABEF + DE + ABDF + BCEF,$$

and B and F are measurable along with the following groups of interactions:

(AB + BC + EF + DF)
(BD + CF + BE + AF)
(CD + AE + AD + BF + CE)

Nine to Twelve Variables in 24 Trials

The 24-trial, three-quarter fractional factorial of Resolution IV is derived from the Hadamard matrix of order 32 using the following alias set:

I	= AC	= BC	= AB
C	= A	= B	= ABC
D	= ACD	= BCD	= ABD
E	= ACE	= BCE	= ABE
CDE	= ADE	= BDE	= ABCDE
CD	= AD	= BD	= ABCD
CE	= AE	= BE	= ABCE
DE	= ACDE	= BCDE	= ABDE

Up to a total of 12 variables are introduced into the design by setting additional variables in place of three-factor interactions in the first three columns only:

F = CDE J = BCD
G = ACD K = BCE
H = ACE L = BDE
I* = ADE (I* = variable I, not identity.)

The main effects C, D, E, and F and the interactions CD, CE, and DE are measured by blocks ii and iii, defined by I = AB. The main effects A, G, H, and I* and the interactions AD, AE, and AB are measured by blocks ii and iv, defined by I = BC. The main effects B, J, K, and L and the interactions BD, AB, and BE are measured by blocks iii and iv, defined by I = AC. Of course, the interactions given above are confounded by an array of other two-factor interactions, which can be determined by the previously described procedure of multiplying identities two at a time, three at a time, etc. (Note: the element of the groups that are confounded together are given in Chapter 13.)

Seventeen to Twenty-four Variables in 48 Trials

The 48-trial, three-quarter fractional factorial of Resolution IV is derived from the Hadamard matrix of order 64 using the following alias set:

I	= AC	= BC	= AB
C	= A	= B	= ABC
D	= ACD	= BCD	= ABD
E	= ACE	= BCE	= ABE
F	= ACF	= BCF	= ABF
CDE	= ADE	= BDE	= ABCDE
CDF	= ADF	= BDF	= ABCDF
CEF	= AEF	= BEF	= ABCEF
DEF	= ACDEF	= BCDEF	= ABDEF
CD	= AD	= BD	= ABCD
CE	= AE	= BE	= ABCE
CF	= AF	= BF	= ABCF
DE	= ACDE	= BCDE	= ABDE
DF	= ACDF	= BCDF	= ABDF
EF	= ACEF	= BCEF	= ABEF
CDEF	= ADEF	= BDEF	= ABCDEF

Up to a total of 24 variables may be introduced into the design by equating additional variables to odd interactions in the first three columns:

G = CDE	K = ACD	R = BCD	
H = CDF	L = ACE	S = BCE	
I* = CEF	M = ACF	T = BCF	(I* = variable I,
J = DEF	N = ADE	U = BDE	not identity.)
	O = ADF	V = BDF	
	P = AEF	W = BEF	
	Q = ACDEF	X = BCDEF	

Main effects C, D, E, F, G, H, I*, and J are measured with blocks ii and iii, where I = AB; main effects A, K, L, M, N, O, P, and Q are measured with blocks ii and iv, where I = BC; and main effects B, R, S, T, U, V, W, and X are measured with blocks iii and iv, where I = AC. Note that when there are fewer than 24 variables, some of the main factors are measured in more than one block.

Groups of interactions can be estimated, and are listed in Chapter 13.

Thirty-three to Forty-eight Variables in 96 Trials

The alias set for the 96-trial, three-quarter fractional factorial of Resolution IV is:

I	= AC	= BC	= AB
C	= A	= B	= ABC
D	= ACD	= BCD	= ABD
E	= ACE	= BCE	= ABE
F	= ACF	= BCF	= ABF
G	= ACG	= BCG	= ABG
CDE	= ADE	= BDE	= ABCDE
CDF	= ADF	= BDF	= ABCDF
CDG	= ADG	= BDG	= ABCDG
CEF	= AEF	= BEF	= ABCEF
CEG	= AEG	= BEG	= ABCEG
CFG	= AFG	= BFG	= ABCFG
DEF	= ACDEF	= BCDEF	= ABDEF
DEG	= ACDEG	= BCDEG	= ABDEG
DFG	= ACDFG	= BCDFG	= ABDFG
EFG	= ACEFG	= BCEFG	= ABEFG
CDEFG	= ADEFG	= BDEFG	= ABCDEFG
CD	= AD	= BD	= ABCD
CE	= AE	= BE	= ABCE
CF	= AF	= BF	= ABCF
CG	= AG	= BG	= ABCG
DE	= ACDE	= BCDE	= ABDE
DF	= ACDF	= BCDF	= ABDF
DG	= ACDG	= BCDG	= ABDG
EF	= ACEF	= BCEF	= ABEF
EG	= ACEG	= BCEG	= ABEG
FG	= ACFG	= BCFG	= ABFG
CDEF	= ADEF	= BDEF	= ABCDEF
CDEG	= ADEG	= BDEG	= ABCDEG
CDFG	= ADFG	= BDFG	= ABCDFG
CEFG	= AEFG	= BEFG	= ABCEFG
DEFG	= ACDEFG	= BCDEFG	= ABDEFG

Resolution IV designs with up to 48 variables are obtained by equating the additional variables to the following odd interactions:

G = CDE	A* = AFG
H = CDF	B* = ACDEF
I = CDG	C* = ACDEF
J = CEF	D* = ACDEG
K = CEG	E* = ACDFG
L = CFG	F* = ACEFG
M = DEF	G* = ADEFG
N = DEG	H* = BCD
O = DFG	I* = BCE
P = EFG	J* = BCF
Q = CDEFG	K* = BCG
R = ACD	L* = BDE
S = ACE	M* = BDF
T = ACF	N* = BDG
U = ACG	O* = BEF
V = ADE	P* = BEG
W = ADF	Q* = BFG
X = ADG	R* = BCDEF
Y = AEF	S* = BCDEG
Z = AEG	T* = BCDFG
	U* = BCEFG
	V* = BDEFG

EXAMPLE

A chemical process for producing typewriter ribbons is to be studied with the intent of improving the adhesion of the ink to the substrate. The variables that could influence this property of the ribbon are:

	High level	Low level
A. Speed of the web through the oven	10 ft/min	13 ft/min
B. Temperature of the oven	300°F	270°F
C. Mixing time of the ink	1 hour	2 hours
D. Amount of solvent in the ink	10%	15%
E. Presence of an additive in the formulation	Yes	No
F. Airflow in the oven	200 cu ft/min	175 cu ft/min

Because producing the samples on semi-production equipment is time consuming and expensive, a minimum number of samples is desired. The possibility of two-factor interactions exists, but it is not important to determine them at this time. A high α and β can be tolerated, and σ^2 is known to be 5.

A. Decide upon the design of the experiment.

A Resolution IV John's three-quarter fractional factorial of 12 trials is indicated. The main factors will not be confounded with two-factor interactions. If there are two-factor interactions, they will probably be detected within some group of two-factor interactions for later study.

B. State the null hypothesis. State the alternative hypothesis.

$(H_0)_i: (\mu_i)_{\text{low}} = (\mu_i)_{\text{high}}$
$(H_a)_i: (\mu_i)_{\text{low}} \neq (\mu_i)_{\text{high}}$
$i = $ A, B, C, D, E, F

C. Choose α and δ.
$N_{\text{high}} = N_{\text{low}}$ is fixed at 4 for this design.
β can be calculated by rearranging the formula:

$$N = 2(U_\alpha + U_\beta)^2 \frac{\sigma^2}{\delta^2}$$

$$U_\beta = \sqrt{\frac{N\delta^2}{2\sigma^2}} - U_\alpha$$

U_α is obtained from table 2.

$\alpha = 0.10 \qquad \delta = 4$
$N_{\text{high}} = N_{\text{low}} = 4$ in a 12-trial three-quarter fractional factorial experiment.

$$U_\beta = \sqrt{\frac{(4)(16)}{(2)(5)}} - 1.645$$
$$= \sqrt{\tfrac{64}{10}} - 1.645 = 0.88$$

$\beta = 0.2$ from table 1

D. Determine the treatment combinations to be tested. See page 156.

ae, bcdf, bce, adf (block ii)
acde, bde, acf, bf (block iii)
abc, abcdef, def, (1) (block iv)

Compute criterion for determining significance.

$$|\bar{X}_{\text{high}} - \bar{X}_{\text{low}}|^* = U_\alpha \sigma \sqrt{\frac{1}{N_{\text{high}}} + \frac{1}{N_{\text{low}}}}$$

$N_{\text{high}} = N_{\text{low}} = 4$

$$|\bar{X}_{\text{high}} - \bar{X}_{\text{low}}|^* = (1.645)(2.24)\sqrt{\tfrac{1}{4} + \tfrac{1}{4}} = 2.57$$

E. Make the indicated 12 runs and measure the adhesion of the product.

Treatment combination	Block	Result
ae	ii	21 in./lb
bcdf		24
bce		18
adf		24
acde	iii	24
bde		18
acf		24
bf		21
abc	iv	19
abcdef		25
def		23
(1)		20

F. Compute the values of all contrasts using the signs in the proper column and the responses.

From blocks ii and iii:

	C	D	AC −BC DE −DF	AD −BD CE −CF	CD AE BF −BE −AF
	−21	−21	+21	+21	−21
	+24	+24	+24	+24	−24
	+18	−18	+18	−18	+18
	−24	+24	+24	−24	+24
	+24	+24	−24	−24	−24
	−18	+18	−18	+18	+18
	+24	−24	−24	+24	+24
	−21	−21	−21	−21	−21
$\sum X_{\text{high}} - \sum X_{\text{low}} =$	6	6	0	0	−6
$\bar{X}_{\text{high}} - \bar{X}_{\text{low}} =$	1.5	1.5	0	0	−1.5

From blocks ii and iv:

	A	E	AB AC DE EF	AD CE BE AF	CD BD AE BF CF
	+21	+21	+21	+21	−21
	−24	−24	+24	+24	−24
	−18	+18	+18	−18	+18
	+24	−24	+24	−24	+24
	+19	−19	−19	+19	+19
	+25	+25	−25	−25	−25
	−23	+23	−23	+23	+23
	−20	−20	−20	−20	−20
$\sum X_{\text{high}} - \sum X_{\text{low}} =$	4	0	0	0	−6
$\bar{X}_{\text{high}} - \bar{X}_{\text{low}} =$	1	0	0	0	−1.5

From blocks iii and iv:

	B	F	AB BC EF DF	BD CF BE AF	CD AE AD BF CE
	−24	−24	+24	+24	—24
	+18	−18	+18	−18	+18
	−24	+24	+24	−24	+24
	+21	+21	+21	+21	−21
	+19	−19	−19	+19	+19
	+25	+25	−25	−25	−25
	−23	+23	−23	+23	+23
	−20	−20	−20	−20	−20
$\sum X_{\text{high}} - \sum X_{\text{low}} =$	−8	12	0	0	−6
$\bar{X}_{\text{high}} - \bar{X}_{\text{low}} =$	−2	3	0	0	−1.5

G. Make preliminary decision.

1. $[(\bar{X}_{\text{high F}} - \bar{X}_{\text{low F}}) = 3] > [|\bar{X}_{\text{high}} - \bar{X}_{\text{low}}|^* = 2.57]$

Therefore, with 95% confidence, high airflow (200 cu ft/min) is better than low airflow (175 cu ft/min).

2. CD + AE + AD + BF + CE = −1.5
   CD + AE + BF − BE − AF = −1.5
   CD + BD + AE + BF + CF = −1.5

These effects are not significant at the 95% level; however, the interactions that appear in all three contrasts are CD, AE, and BF. Therefore, it can be concluded with less than 95% confidence that one or more of these interactions probably has at least a small influence on the system.

3. List the trials in order of quality to obtain some clue to the CD, AE, or BF interactions:

abcdef	25
bcdf	24
adf	24
acde	24
acf	24
def	23
ae	21
bf	21
(1)	20
abc	19
bce	18
bde	18

Note that, in the top half, only acde, with a response of 24, does not have F in the treatment combination. This implies that either high A and E and/or high C and D have a significant positive interaction effect. However, the treatment combination ae (21) does not have unusually good quality. Therefore, the implication is that the true interaction is CD. Note this is only an implication, not proof.

The effect of B is negative 2, C is positive 1.5, D is positive 1.5. These effects are not significant at the 95% level. However, the direction of these effects should be considered in the final process specification.

H. Make decisions on the process.

Make future production ribbons under the following process conditions:

Decision	Process specification
A is not an important variable.	10 ft/min and 13 ft/min are equally good.
Probably low B is better.	270°F.
Probably high C is better.	One hour mixing.
Probably high D is better.	10% solvent.
E is not an important variable.	Additive is not effective.
Definitely, high F is better.	200 cu ft/min airflow.

If it is possible to conduct some additional trials, the experimenter should design trials so that the CD interaction can be confirmed or rejected.

EXERCISES

1. An experimenter has planned a complete factorial for four variables in 16 trials. The objective of the project is a paper feeding mechanism that has fewer than five paper jams per 10,000 sheets fed into the mechanism. Variance (σ^2) is known to be 2, and α is chosen to be 0.5, double-sided.

	(−)	(+)
A = Hardness of feed rollers	$35R_m$	$45R_m$
B = Spring tension on guide	10 lb	15 lb
C = Number of rollers	2	3
D = Angle of feed	30°	40°

The experimental trials are divided into four blocks with the defining contrast I = (−AB) = (−ACD) = (BCD).

a. What treatment combinations are in each of the four blocks?

Block	(−AB)	(−ACD)	(BCD)
i	+	+	+
ii	−	−	+
iii	−	+	−
iv	+	−	−

b. Suppose the following block of trials is completed first, with the following results:

Treatment combinations	Results, jams/10,000 sheets
ab	19
d	6
abcd	20
c	11

What conclusion can be drawn from these four data points?

c. Suppose the following block of trials is completed next, with the indicated results:

Treatment combinations	Results, jams/10,000 sheets
(1)	12
abd	16
cd	10
abc	22

What conclusions can be drawn from the eight data points?

d. The third block of trials is then completed, with the following results:

Treatment combinations	Results, jams/10,000 sheets
bd	17
a	15
bc	19
acd	14

What conclusion can be drawn from the 12 data points?

Is it necessary to test the remaining treatment combinations? What is your predicted values for paper jams at the four untested treatment conditions?

2. Show that blocks iii and iv in the 12-trial John's three-quarter fractional factorial can only estimate A, AB, AC, and AD.
3. Show that, in the 48-trial matrix, blocks ii and iii have the identity (−ABCDEF), that blocks ii and iv have the identity (−ABE), and that blocks iii and iv have the identity (−CDF).

REFERENCE

John, P. W. M. 1971. *Statistical design and analysis of experiments.* New York: Macmillan Publishing Co.

9 Unbalanced Resolution V Designs

The Resolution V designs described in previous chapters are of limited value to the engineer or scientist in industry because of the large number of trials required for a relatively small number of variables. In most situations, the experimenter is limited to a choice between Resolution III or Resolution IV designs. The designs described in this chapter make Resolution V designs a feasible alternative in many industrial experiments because of the substantially fewer number of trials required, as compared to the Resolution V designs described in Chapters 7 and 8.

The following efficient Resolution V designs for up to 12 variables were available until the discovery of these unbalanced Resolution V designs. (Note again, as in the case of the John's designs, these special designs are "better" only in the sense that they require fewer trials; there is a loss in the precision of the estimates in some cases.)

No. of variables	No. of contrasts to be estimated	No. of trials required	Design type
5	15	16	2^{5-1} fractional factorial
6	21	32	2^{6-1} fractional factorial
7	28	48	$3(2^{7-3})$ John's 3/4 fractional factorial
8	36	48	$3(2^{8-4})$ John's 3/4 fractional factorial
9	45	96	$3(2^{9-4})$ John's 3/4 fractional factorial
10	55	96	$3(2^{10-5})$ John's 3/4 fractional factorial
11	66	96	$3(2^{11-6})$ John's 3/4 fractional factorial
12	78	192	$3(2^{13-7})$ John's 3/4 fractional factorial

Three types of Resolution V designs are available: Class A, Class B, and Class AB.

Class A Designs

The following better (i.e., fewer trials required) class A special designs are available:

No. of variables	No. of contrasts	No. of trials
6	21	28
7	28	36
8	36	44
9	45	69
10	55	81
11	66	93

Six-Variable, 28-Trial Design

The six variables are designated as A, B, C, D, E, and F. The design consists of two parts: (1) a Resolution V design in 16 trials for the five variables A, B, C, D, and E, with variable F set at the low level for all trials and with confounding I = ABCDE; and (2) a Resolution IV design for the six variables A, B, C, D, E, and F (namely, the 16 trials with the defining contrast I = ABDE = ABCF = CDEF).

The treatment combinations for the first part of the design are:

ae	bcde	abde	be
ab	acde	bc	ce
abce	bd	cd	de
abcd	ac	ad	(1)

The treatment combinations for the second part of the design are:

*aef	*bcd	abde	*bef
ab	*acd	*bce	*cf
*abcf	*bdf	*cdef	de
*abcdef	*ace	*adf	(1)

It is apparent that the four treatment combinations not marked with an asterisk are also treatment combinations of the first part of the design and, therefore, need not be repeated. The total design, then, consists of 16 trials + 12 trials = 28 trials.

The computer program described in Chapter 13 can be used to analyze the data. The procedure is as follows:

(a) The 16 results for the treatment combinations of the first part of the design are used to estimate the mean effect of the following:

A	AB	BC	CD	DE
B	AC	BD	CE	
C	AD	BE		
D	AE			
E				

The computer command is 16 FAN 5 for this first part of the design.

(b) The 16 results for the treatment combinations of the second part of the design are used to estimate the mean effect of the following:

A (AB + DE + CF)
B (BC + AF)
C (CD + EF)
D (AC + BF)
E (CE + DF)
F

The computer command for this part of the design is 16 FAN 6.

(c) The estimate of the interactions underlined above can be unconfounded by subtracting the value of the confounding interactions that were determined in part (a). If the estimate of BC was 12 from the first set and the estimate of (BC + AF) is 32 from the second set, then the estimate of AF is 32 − 12 = 20.

(d) The effects that are estimated by both parts of the design are averaged.

(e) The criterion for determining the significance of variable F and the interactions AB, AC, AD, AE, BC, BD, BE, CD, CE, and DE is

$$|\bar{X}_{\text{high}} - \bar{X}_{\text{low}}|^* = U_\alpha \sigma \sqrt{\tfrac{1}{8} + \tfrac{1}{8}}; \quad \text{that is, } U_\alpha \sigma(0.5) \text{ if } \sigma^2 \text{ is known;}$$

or

$$|\bar{X}_{\text{high}} - \bar{X}_{\text{low}}|^* = t_\alpha S \sqrt{\tfrac{1}{8} + \tfrac{1}{8}} = t_\alpha S(0.5) \quad \text{if } S^2 \text{ is known.}$$

(f) The criterion for determining the significance of the average of the variables that are measured in both parts of the design (A, B, C, D and E) is

$$|\bar{X}_{\text{high}} - \bar{X}_{\text{low}}|^* = U_\alpha \sigma(0.40) \quad \text{if } \sigma^2 \text{ is known,}$$

or

$$|\bar{X}_{\text{high}} - \bar{X}_{\text{low}}|^* = t_\alpha S(0.40) \quad \text{if } S^2 \text{ is known.}$$

(g) The criterion for determining the significance of the interaction groups (AB + DE + CF), (BC + AF), (CD + EF), (AC + BF), and (CE + DF) is

$$|\bar{X}_{\text{high}} - \bar{X}_{\text{low}}|^* = U_\alpha \sigma \sqrt{\tfrac{1}{8} + \tfrac{1}{8}} \quad \text{if } \sigma^2 \text{ is known,}$$

or

$$|\bar{X}_{\text{high}} - \bar{X}_{\text{low}}|^* = t_\alpha S \sqrt{\tfrac{1}{8} + \tfrac{1}{8}} \quad \text{if } S^2 \text{ is known.}$$

EXAMPLE

A batch process is being used in development engineering to coat a plastic material onto small clay beads. The present coating thickness (μ_0) is 1 mil, but the product should have a thickness of 1.5 mils. The variance (σ_0^2) is known to be 0.01. Six variables have been identified as possible causes of low thickness; there is also a high probability that two-factor interactions are critical to obtaining the proper thickness. Each batch takes one-half day to process and evaluate. About three weeks is all the time available for the experiment; thus, the sample size is limited to about 30 trials. A conventional 32-trial Resolution V design is possible, but the special Resolution V design in 28 trials is desirable because of the substantial savings in both cost and time.

A. Define variables and their levels and the hypotheses.

		Old level	New level
A	Flow rate of beads	1 lb/min	0.9 lb/min
B	Viscosity of coating	200 cps	250 cps
C	Temperature (top)	275°F	300°F
D	Temperature (bottom)	150°F	125°F
E	Length of chamber	12 feet	15 feet
F	Nozzle diameter	3 inches	4 inches

$$\left.\begin{aligned} H_0&:\mu_{\text{low}} = \mu_{\text{high}} \\ H_a&:\mu_{\text{high}} > \mu_{\text{low}} \end{aligned}\right\} \text{For all main effects and interactions.}$$

B. Choose α. Since the number of trials is prechosen, β and δ need not be calculated.

$\alpha = 0.05$

C. Compute criteria. Since all alternative hypotheses are single-sided, use $U_{0.05}$ single-sided from table 1. The criteria for this design are given in (e), (f), and (g) of the previous section.

For A, B, C, D and E,

$$\begin{aligned} |\bar{X}_{\text{high}} - \bar{X}_{\text{low}}|^* &= U_\alpha \sigma(0.40) \\ &= (1.645)(0.1)(0.40) \\ &= 0.0658 \end{aligned}$$

For all other factors and interactions,

$$\begin{aligned} |\bar{X}_{\text{high}} - \bar{X}_{\text{low}}|^* &= U_\alpha \sigma(0.50) \\ &= (1.645)(0.1)(0.50) \\ &= 0.082 \end{aligned}$$

D. Make the 28 runs defined by the treatment combinations described earlier, and measure the thickness on a proper sample of the coated beads.

Treatment combination	Response
ae	0.94
*ab	1.31
abce	1.27
abcd	1.23
bcde	1.54
acde	1.06
bd	1.35
ac	1.00
*abde	1.36
bc	1.39
cd	1.28
ad	1.07
be	1.32
ce	1.08
*de	1.12
*(1)	0.98
aef	0.97
abcf	1.03
abcdef	1.05
bcd	1.55
acd	1.02
bdf	1.18
ace	0.99
bce	1.41
cdef	1.32
adf	1.13
bef	1.07
cf	1.08

* Indicates treatment combinations of the first part of the design that must be used a second time in analysis of the second part of the design.

E. Analyze the data using the computer program described in Chapter 13.

(a) The first 16 trials with five variables give the following significant variables and interactions:

Effect of A = −0.104
B = +0.279
D = +0.089
AC = +0.079

(Note: AC is not quite significant but is much larger than the other variables and interactions.)

(b) The second 12 trials plus the proper 4 trials from (a) above give the following significant variables and interactions:

Effect of A = −0.106
B = −0.169
D = +0.111
F = −0.114
AC + BF = +0.211

(c) Compute effect of BF.

BF = (BF + AC) − AC
BF = 0.211 − 0.079 = 0.132

(d) Therefore, D is significant
AC is significant
BF is significant

F. Make up a two-way chart of AC from the data of the 16-trial Resolution V part of the experiment (figure 9–1).

	1	c
1	1.35 1.32 1.12 0.98 $\bar{X}$ = 1.19	1.54 1.39 1.28 1.08 $\bar{X}$ = 1.32
a	0.94 1.31 1.36 1.07 $\bar{X}$ = 1.17	1.27 1.23 1.06 1.00 $\bar{X}$ = 1.14

Fig. 9-1. Two-way chart of AC from Resolution V part of experiment.

Make up a two-way chart of BF from the data of the 16-trial Resolution IV part of the experiment (figure 9-2).

	1	f
1	1.12 0.98 1.02 0.99 $\bar{X}$ = 1.03	0.97 1.32 1.13 1.08 $\bar{X}$ = 1.13
b	1.31 1.36 1.55 1.41 $\bar{X}$ = 1.41	1.03 1.05 1.18 1.07 $\bar{X}$ = 1.08

Fig. 9-2. Two-way chart of BF from Resolution IV part of experiment.

G. Make final decision on the future operating condition of the process.

Flow rate	1 lb/min	low A
Viscosity	250 cps	high B
Temperature (top)	300°F	high C
Temperature (bottom)	125°F	high D
Nozzle diameter	3 inches	low F
Length of chamber	12 or 15 feet	

The expected coating thickness at this treatment combination is 1.55 mils, which meets the specification.

The data can now be presented in a factorial block using only the variables A, B, C, D, and F with labeling, so that the results are better as one goes from the upper left to lower right (figure 9-3).

Notice that four of the blocks have two results and that the replicate results are very close together. In particular, note that the best combination, low A, high B, high C, high D, and low F, is duplicated, which confirms that one of these is not an accidental high result.

Eight of the total treatment combinations in the complete 2^5 factorial were not made, yet one can be almost absolutely certain that none of these treatments would come close to having a result of 1.54.

Note that, with the above information, the experimenter can in the future predict how to obtain any thickness of coating. For example, one would expect to get coatings even thicker than 1.55 mils with higher D, higher B, higher C, lower F, and lower A.

		1				d			
		1		c		1		c	
		1	b	1	b	1	b	1	b
a	f	aef 0.97			abcf 1.03	adf 1.13			abcdf 1.05
a	1	ae 0.94	ab 1.31	ac 1.00 ace 0.99	abce 1.27	ad 1.07	abde 1.36	acde 1.06 acd 1.02	abcd 1.23
1	f		bef 1.07	cf 1.08			bdf 1.18	cdef 1.32	
1	1	(1) 0.98	be 1.32	ce 1.08	bc 1.39 bce 1.41	de 1.12	bd 1.35	cd 1.28	bcde 1.54 bcd 1.55

Fig. 9-3. Factorial block of experiment data.

Seven-Variable, 36-Trial Design

The Resolution V design for seven variables in 36 trials can be obtained by adding a 16-trial Resolution IV design to the 6-variable, 28-trial experiment described in the previous section. The total number of different treatment combinations is then 16 + 12 + 8 = 36.

Resolution V design for 5 variables in 16 trials	+	Resolution IV design for 6 variables in 16 trials	+	Resolution IV design for 7 variables in 16 trials
ae		*aef		aef
ab		ab		*abg
abce		*abcf		abcf
abcd		*abcdef		*abcdefg
bcde		*bcd		*bcdg
acde		*acd		acd
bd		*bdf		bdf
ac		*ace		*aceg
abde		abde		abde
bc		*bce		bce
cd		*cdef		cdef
ad		*adf		*adfg
be		*bef		*befg
ce		*cf		*cfg
de		de		*deg
(1)		(1)		(1)

* Indicates trials that must be made for the second and third parts of the experiment.

The first part of the design measures

A	AB	BC	CD	DE
B	AC	BD	CE	
C	AD	BE		
D	AE			
E				

The second part of the design measures

A (AB + DE + CF)
B (BC + AF)
C (CD + EF)
D (AC + BF)
E (CE + DF)
F

The third part of the design measures

A (AB + DE + CF)
B (BC + AF + DG)
C (CD + EF + BG)
D (AC + BF + EG)
E (BD + AE + CG)
F (CE + DF + AG)
G (AD + BE + FG)

Thus, the two-factor interactions of A, B, C, D, and E are estimated from the first part of the design. The two-factor interactions of F with A, B, C, D, and E are estimated (by subtracting) from the second part of the design, and the two-factor interactions of G are estimated (by subtracting) from the third part of the design.

The mean effect of variable F is estimated from the second part of the design and the third part of the design, but the main effect of variable G is measured only by the third part of the design. The mean effects of variables A, B, C, D, and E are measured by all three parts of the design.

The procedure to be used for this design is described below.

(a) The mean effects of interactions AB, AC, AD, AE, BC, BD, CD, BE, CE, and DE from part 1 of the design are compared to the criterion

$$|\bar{X}_{\text{high}} - \bar{X}_{\text{low}}|^* = U_\alpha \sigma(0.50) \quad \text{or} \quad t_\alpha S(0.50).$$

(b) The mean effects of the interaction groups (BC + AF), (CD + EF), (AC + BF), and (CE + DF) from part 2 of the design are compared to the criterion

$$|\bar{X}_{\text{high}} - \bar{X}_{\text{low}}|^* = U_\alpha \sigma(0.50) \quad \text{or} \quad t_\alpha S(0.50).$$

(c) The mean effects of the interaction groups (BC + AF + DG), (CD + EF + BG), (AC + BF + EG), (BD + AE + CG), (CE + DF + AG), and

(AD + BE + FG) from part 3 of the design are compared to the criterion

$$|\bar{X}_{\text{high}} - \bar{X}_{\text{low}}|^* = U_\alpha\sigma(0.50) \quad \text{or} \quad t_\alpha S(0.50).$$

(d) Variable G is measured only by the third part of the experiment, and the average effect is compared to the criterion

$$|\bar{X}_{\text{high}} - \bar{X}_{\text{low}}|^* = U_\alpha\sigma(0.50) \quad \text{or} \quad t_\alpha S(0.50).$$

(e) The average of the mean effects of (AB + DE + CF) from both the second and third parts of the experiment is compared to the criterion

$$|\bar{X}_{\text{high}} - \bar{X}_{\text{low}}|^* = U_\alpha\sigma(0.43) \quad \text{or} \quad t_\alpha S(0.43).$$

(f) The average of the mean effects of F from the second and third parts of the experiment is compared to the criterion

$$|\bar{X}_{\text{high}} - \bar{X}_{\text{low}}|^* = U_\alpha\sigma(0.43) \quad \text{or} \quad t_\alpha S(0.43).$$

(g) The average of the mean effects of A, B, C, D, and E from all three parts of the experiment is compared to the criterion

$$|\bar{X}_{\text{high}} - \bar{X}_{\text{low}}|^* = U_\alpha\sigma(0.36) \quad \text{or} \quad t_\alpha S(0.36).$$

Eight-Variable, 44-Trial Design

This design is obtained from the 7-variable design by adding the following Resolution IV design for eight variables in 16 trials. The total number of treatment combinations is then 16 + 12 + 8 + 8 = 44.

*aefh
*abgh
abcf
*abcdefgh
bcdg
*acdh
*bdfh
aceg
abde
*bceh
cdef
adfg
befg
*cfgh
*degh
(1)

Analysis of the data from this fourth part of the total design will give an estimate of A, B, C, D, E, F, G, and H and groups of interactions. The confounding of the interactions in this fourth part of the design is:

$$\begin{aligned} &AB + DE + CF + \underline{GH} \\ &BC + AF + DG + \underline{EH} \\ &CD + EF + BG + \underline{AH} \\ &AC + BF + EG + \underline{DH} \\ &BD + AE + CG + \underline{FH} \\ &CE + DF + AG + \underline{BH} \\ &AD + BE + FG + \underline{CH} \end{aligned}$$

Thus, all the two-factor interactions of H can be estimated by subtraction (from the third part) of the known values of the confounding interactions. All the two-factor interactions of G (except GH) will be estimated from the third part of the design, and all the two-factor interactions of F (except FG and FH) will be estimated from the second part of the design.

The criteria for AB, AC, AD, AE, BC, BD, BE, CD, CE, DE, (AB + DE + CF), (BC + AF), (CD + EF), (AC + BF), (CE + DF), (BC + AF + DG), (CD + EF + BG), (AC + BF + EG), (BD + AE + CG), (CE + DF + AG), and (AD + BE + FG) are exactly the same as for the 36-trial design with seven variables.

The criterion for variable H is

$$|\bar{X}_{\text{high}} - \bar{X}_{\text{low}}|^* = U_\alpha\sigma(0.50) \quad \text{or} \quad t_\alpha S(0.50).$$

The criterion for the average of the two estimates of variable G is

$$|\bar{X}_{\text{high}} - \bar{X}_{\text{low}}|^* = U_\alpha\sigma(0.43) \quad \text{or} \quad t_\alpha S(0.43).$$

The criterion for the average of the three estimates of variable F is

$$|\bar{X}_{\text{high}} - \bar{X}_{\text{low}}|^* = U_\alpha\sigma(0.39) \quad \text{or} \quad t_\alpha S(0.39).$$

The criterion for the average of the four estimates of variables A, B, C, D, and E is

$$|\bar{X}_{\text{high}} - \bar{X}_{\text{low}}|^* = U_\alpha\sigma(0.34) \quad \text{or} \quad t_\alpha S(0.34).$$

Nine- to Eleven-Variable Designs

The starting design for these special Class A designs is the conventional Resolution V $3(2^{n-j})$ John's three-quarter fractional factorial for eight variables in 48 trials. If this design is combined with the 24-trial $3(2^{n-j})$ Resolution IV design with nine variables, which will have three treatment combinations in common with the first part, then all main factors and two-factor interactions can be estimated with 48 + 21 = 69 trials.

The Resolution V design for 10 variables consists of the 69 trials given above plus the 12 new treatment combinations that are required for the John's 24-trial design with 10 variables. Twelve of these trials will already have been made. The total number of trials is, therefore, 48 + 21 + 12 = 81.

Finally, the Resolution V design for 11 variables is obtained by adding the Resolution IV design for 11 variables in 24 trials, which will add 12 more new trials for a total of 93.

It should be noted that the computation of criteria for these designs is a difficult task, and a statistician should be consulted for analyses of these designs.

Class B Designs

A second class of Resolution V designs can be obtained by combining certain Resolution III designs with small complete factorials. The following designs have been so obtained:

No. of variables	No. of contrasts to be estimated	No. of trials required
7	28	38
8	36	46
9	45	71
12	78	143

Seven-Variable, 38-Trial Design

The Resolution III part of the design is the 32-trial, 7-variable design with the defining contrast I = ABCDEF = ABCDG = EFG. The treatment combinations are:

aeg	abef	dfg	abceg
ab	bc	ef	bcdfg
abcfg	acdfg	afg	acdeg
abcd	bdef	bfg	abdeg
abcdef	ceg	ac	bcef
bcdeg	ad	bd	cd
cdef	beg	acef	deg
adef	cfg	abdfg	(1)

All main factors and two-factor interactions can be estimated by these 32 trials except the following:

E = FG

F = EG

G = EF

These confoundings, however, can be easily resolved by an 8-trial complete factorial with the following treatment combinations:

e
ef
efg
fg
eg
f
g
(1)

where A, B, C, and D are all at the low level. It will be noticed, however, that two of the treatment combinations occur in both parts of the design, namely ef and (1). Therefore, the total number of trials is 32 + 6 = 38.

In many cases it will be better to set A, B, C, and D at their high level while varying E, F, and G in the second part of the experiment. The treatment combinations for the eight trials of the second part of the experiment will then be:

abcd e
abcd ef
abcd efg
abcd fg
abcd eg
abcd f
abcd g
abcd

and the two parts of the experiment will have abcd and abcdef in common.

The computer program described in Chapter 13 can be used to analyze these experiments. Interpretation of the computer output, however, must be modified, as indicated below.

(a) The 32 treatment combinations for the Resolution III design given in the first part of this section are in the proper order for the computer program; that is,

Trial		
Trial	1	aeg
	2	ab
	3	abcfg
	4	abcd
		⋮
	31	deg
	32	(1)

(b) For analysis of the data, type the statement

32 FAN 6

Note the *6*, rather than the true number of variables, which is seven.

(c) The computer program will type out the effects of A, B, C, D, E, and F. However, we know that E is confounded with FG, and F is confounded with EG. It is a good procedure, therefore, to add (+FG) to the heading E; i.e., label E as E + FG. Likewise, label F as F + EG.

(d) All of the two-factor interaction columns are correct estimates of the column headings except EF, which really measures EF + G.

(e) Four of the two-factor interactions, AG, BG, CG, and DG, are not part of the labeling under main factors or two-factor interactions. They are found in the estimate-of-error section.

Contrast 9 is an estimate of BG.
Contrast 12 is an estimate of DG.
Contrast 13 is an estimate of AG.
Contrast 28 is an estimate of CG.

(f) The eight trials of the second part of the experiment are analyzed using the computer statement

8 FAN 3

This requires no special treatment except that the computer printout will be labeled A, B, AB, etc., rather than E, F, EF, etc.

The criterion for the variables and interactions that are measured by the first part of the design is

$$t_\alpha S\sqrt{\tfrac{1}{16}+\tfrac{1}{16}} \qquad \text{or} \qquad U_\alpha \sigma\sqrt{\tfrac{1}{16}+\tfrac{1}{16}},$$

whereas the criterion for E, F, and G, and their interactions, is

$$t_\alpha S\sqrt{\tfrac{1}{4}+\tfrac{1}{4}} \qquad \text{or} \qquad U_\alpha \sigma\sqrt{\tfrac{1}{4}+\tfrac{1}{4}}.$$

Therefore, the criterion for the same α will be twice as large for E, F, and G and their interactions, which are measured in the second part of the design, as for A, B, C, D, etc., which are measured in the first part of the design. Therefore, with such designs, the variables that are expected to have the largest effect or largest two-factor interaction should be labeled E, F, and G.

Eight-Variable, 46-Trial Design

The Resolution III part of the design is the 32-trial, 8-variable design with the defining contrast I = ABCDEF = ABCDG = ABCEH = EFG = DFH = DEGH = ABCFGH and with the treatment combinations:

aeg	abefh	dfg	abceg
ab	bc	efh	bcdfg
abcfgh	acdfg	afgh	acdegh
abcdh	bdef	bfgh	abdegh
abcdef	ceg	ac	bcefh
bcdegh	adh	bdh	cdh
cdef	beg	acefh	degh
adef	cfgh	abdfg	(1)

The following main factors and two-factor interactions are confounded:

D = FH
E = FG
F = EG = DH
G = EF
H = DF
DE = GH
DG = EH

Since the confounding involves only five variables, these confoundings can be resolved with a 16-trial Resolution V design with A, B, and C low and the defining contrast I = DEFGH. The treatment combinations are:

dh	efgh	degh	eh
de	dfgh	ef	fh
defh	eg	fg	gh
defg	df	dg	(1)

The treatment combinations degh and (1) occur in both parts of the design; therefore, the total number of trials is 32 + 14 = 46.

The criterion for testing the significance of the main factors and two-factor interactions estimated by the first part of the design is

$$t_\alpha S\sqrt{\tfrac{1}{16}+\tfrac{1}{16}} \quad \text{or} \quad U_\alpha \sigma\sqrt{\tfrac{1}{16}+\tfrac{1}{16}}.$$

For the second part of the design, the criterion is

$$t_\alpha S\sqrt{\tfrac{1}{8}+\tfrac{1}{8}} \quad \text{or} \quad U_\alpha \sigma\sqrt{\tfrac{1}{8}+\tfrac{1}{8}}.$$

To use the computer program described in Chapter 13 for the analysis of the first part of this design, use the statement

32 FAN 6

Note the *6* rather than the true number of variables, which is eight.

The output analysis sheet must be partially relabeled:

D is relabeled D + FH.
E is relabeled E + FG.
F is relabeled F + EG + DH.
EF is relabeled EF + G.
DF is relabeled DF + H.

The following interactions will be found under estimates of error:

AG Contrast 13
AH Contrast 29
BG Contrast 9
BH Contrast 23
CG Contrast 28
CH Contrast 18

The computer program is used in the second part of this design in the conventional fashion, starting with the statement

16 FAN 5

All main factors and interactions will be properly printed out except, of course, in terms of A, B, C, D, and E instead of D, E, F, G, and H. The user must, therefore, relabel.

Nine-Variable, 71-Trial Design

The Resolution V design for nine variables in 71 trials consists of the 64-trial Resolution III design with the defining contrast I = ABCFG = ACDEH = BDEFJ = GHJ = BDEFGH = ACDEGJ = ABCFHJ. The only confoundings of main factors and/or two-factor interactions are:

G = HJ

H = GJ

J = GH

(Note, the ninth variable is labeled J rather than I to prevent confusion with the identity I.) These, of course, can be estimated with an 8-trial complete factorial in variables G, H, and J with A, B, C, D, E, and F set at their low levels.

The treatment combinations for the first part of the design are:

agh	abdfgj	bcefgj	defgj	bdgh	dhj
abhj	bde	acdfgh	aef	ceg	ehj
abcgj	cdf	bdegj	bf	dfgh	fgj
abcdgh	adegh	cef	ac	adgj	(1)
abcdegj	abefgj	adf	abd	adfgh	
abcdef	bcfgh	begh	abcegh	bchj	
bcdefgh	acdhj	cfhj	abcdfhj	cdg	
acdefgj	abdehj	adgj	bcdehj	de	
bdef	abcefhj	abe	cedfhj	efgh	
acefgh	bcdfgj	abcf	adefhj	afhj	
bdfhj	acde	bcd	befhj	bgj	
acehj	abdefgh	cdegh	acfgj	cgh	

The treatment combinations for the second part of the design are:

g	gj
gh	h
ghj	j
hj	(1)

There is only one treatment combination in common, namely (1). Therefore, the total number of trials is 64 + 7 = 71.

The computer program described in Chapter 13 can be used for analysis of this design. The Resolution III part of the design requires the command

64 FAN 6

Note the use of the *6* instead of the actual number of variables, which is nine.

The labeling will be correct for all the variables and interactions that are printed. The interactions of A, B, C, D, E, and F with G, H, and J, however, will appear under estimates of error:

Interaction	Contrast column
AG	18
BG	32
CG	12
DG	41
EG	44
FG	27
AH	29
BH	57
CH	46
DH	53
EH	49
FH	60
AJ	22
BJ	30
CJ	58
DJ	47
EJ	54
FJ	50

The second part of the design is analyzed with the conventional command

8 FAN 3

Twelve-Variable, 143-Trial Design

The Resolution III part of the design is obtained from the 128-trial design for 12 variables with the defining contrast I = ABCD(H) = ACEF(J) = ADEG(K) = BDEF(L) = BCEG(M) = HJL = HKM = JKLM = (a variety of other fifth-order and higher elements). This design is supplemented with an additional 16-trial Resolution V design in H, J, K, L, and M. There is one common trial, (1); therefore, the total number of trials is 128 + 15 = 143.

The reader should carefully note that the Resolution III part of this design cannot be obtained with the computer program of Chapter 13. Part of the design (for the first seven variables) can be obtained with the command

128 DES 7

This matrix must then be supplemented with the contrast columns ABCD, ACEF, ADEG, BDEF, and BCEG, which are labeled H, J, K, L, and M, respectively.

Analysis of the data from the Resolution III part of the design starts with the statement

128 FAN 7

Estimable interactions will appear in the following contrast columns under estimates of error:

AH	58	AJ	93	AK	94	AL	101	AM	77
BH	91	BJ	80	BK	42	BL	60	BM	115
CH	32	CJ	88	CK	86	CL	49	CM	45
DH	57	DJ	83	DK	54	DL	110	DM	119
EH	48	EJ	96	EK	28	EL	114	EM	21
FH	118	FJ	113	FK	124	FL	92	FM	102
GH	38	GJ	52	GK	109	GL	84	GM	33

Class AB Designs

A third class of Resolution V designs can be obtained by combining the Class A and Class B principle. Only one such design has been obtained, namely a 58-trial design for nine variables.

The first part of the design is a Resolution III design for nine variables in 32 trials, with the following added variables.

F = ABCDE
G = ABCD
H = ABCE
J = DE

These lead to the primary identity elements

$$I = ABCDEF = ABCDG = ABCEH = DEJ.$$

By multiplication, the following three and four letter words are generated:

$$I = EFG = DFH = DEGH = EFHJ = GHJ = DFGJ = DEFJ.$$

The following main effects and interactions can be estimated from these 32 trials using the program of Chapter 13.

A AB BC CD
B AC BD CE
C AD BE CF
AE BF CG
AF BG CH
AG BH CJ
AH BJ
AJ

The second part of the design is the Resolution V Class A design for five variables D, E, F, G, and H in 16 trials with the identity element I = DEFGH, combined with the 16-trial Resolution IV design for six variables D, E, F, G, H, and J. The total trials in the second part is 28.

The 16-trial Resolution V part can estimate

D DE EF FG GH
E DF EG FH
F DG EH
G DH
H

The 16-trial Resolution IV part of the design has the defining contrast I = DEGH + DEFJ + FGHJ. From this part, one can estimate by difference

D $DE + GH + \underline{FJ}$
E $EF + \underline{DJ}$
F $FG + \underline{HJ}$
G $DG + \underline{EJ}$
H $FH + \underline{GJ}$
J

Two of the trials required for the Class A part of the design are part of the original 32-trial Resolution III. Therefore, the total number of trials is

$$32 + 28 - 2 = 58$$

The criterion for testing the significance of the variables and interactions in the first part of the design is

$$|\overline{X}_{high} - \overline{X}_{low}|^* = U_\alpha \sigma \sqrt{\frac{1}{16} + \frac{1}{16}}$$

or

$$= t_\alpha s \sqrt{\frac{1}{16} + \frac{1}{16}}$$

Variables D, E, F, G, and H are the averages of the two parts of the Class A design and have the criterion

$$|\overline{X}_{high} - \overline{X}_{low}|^* = U_\alpha \sigma \ (0.40)$$

or

$$= t_\alpha S \ (0.40)$$

Interactions DE, DF, etc. have the criterion

$$|\overline{X}_{high} - \overline{X}_{low}|^* = U_\alpha \sigma \sqrt{\frac{1}{8} + \frac{1}{8}}$$

or

$$= t_\alpha S \sqrt{\frac{1}{8} + \frac{1}{8}}$$

The criterion for the interactions with J are too complicated for this text. Consult a qualified statistician.

EXERCISES

1. A class B 38-trial, 7-variable experiment with $\alpha = 0.05$, $\beta = 0.01$, and $\sigma = 4$ is conducted on the following variables in a development project, the object of which is to obtain an improved tape recording medium:

	Variable	Low level	High level
A	Percent iron	5	7
B	Shape	Regular	Elongated
C	Film thickness	2 mils	2.5 mils
D	Coating method	Doctor blade	Extrusion
E	Curing time	5 minutes	7 minutes
F	Curing temperature	250°F	300°F
G	Posttreatment	Regular	Burnishing

The following results are obtained:

Treatment combinations	Results, uncharged bits/100,000 bits
aeg	26
ab	32
abcfg	13
abcd	31
abcdef	24
bcdeg	17
cdef	25
adef	27
abef	26
bc	34
acdfg	23
bdef	28
ceg	24
ad	31
beg	18
cfg	24
dfg	26
ef	28
afg	27
bfg	17
ac	35
bd	33
acef	25
abdfg	14
abceg	15
bcdfg	16
acdeg	22
abdeg	16
bcef	27
cd	32
deg	26
(1)	36
abcde	29
abcdf	26
abcdg	18
abcdeg	15
abcdfg	14
abcdefg	12

- Calculate the criterion.
- Analyze the data completely.
- Make engineering decisions.
- Recommend management decisions.

2. Use the data in the example of page 168 and verify the mean effects stated in E.

10 Resolution V Designs with Efficiency = 1

In previous chapters, various Resolution V Designs were described. Prior to the discovery of the John's $\frac{3}{4}$ fractional factorials, only 2^{p-k} designs were available, and most of these precluded their use because of the large number of trials required when the number of variables was large. For example, with nine variables, 128 trials were required; the John's designs reduced the number of trials to 96; the unbalance designs of Chapter 9 further reduced the number of trials to 54.

The efficiency of a design is defined as:

$$\text{Efficiency} = \frac{\text{No. of Contrasts}}{\text{No. of Trials} - 1}$$

Thus, in the above examples,

$$\text{Prior to John's } \frac{3}{4} \text{ Fractional Factorials} = \frac{45}{128} = .35$$

$$\text{John's } \frac{3}{4} \text{ Fractional Factorials} = \frac{45}{96} = .47$$

$$\text{AB Design} = \frac{45}{58} = .78$$

$$\text{Ultimate Design} = \frac{45}{46 - 1} = 1.00$$

All of the following designs are Resolution V, with efficiency of 1.00.

By definition, a 2-factor noninteraction of A and B can be drawn as follows:

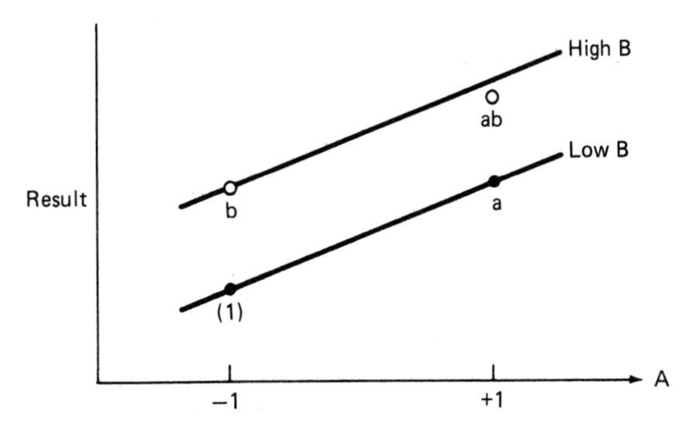

$$\text{Interaction AB} = \frac{\text{b} + \text{a} - \text{ab} - (1)}{2}$$

If the lines are substantially not parallel, then the above calculation of the AB interaction will be a number, either positive or negative, greater than the criterion.

To determine the interaction of any two variables, it is only required to have four treatment combinations; in the above case (1), a, b, and ab. For the BG interaction with seven or more variables, only (1), b, g, and bg are required to determine the BG interaction.

Three-Variable Designs

With three variables, the six contrasts that must be estimated for a Resolution V Design are A, B, C, AB, AC, and BC.

The seven treatment combinations required are (1), a, b, c, ab, ac, and bc.

$$\text{The effect of A} = \frac{\text{a} - (1) + \text{ab} - \text{b} + \text{ac} - \text{c}}{3}$$

The criteria for acceptance of the null or alternative Hypothesis is:

$$|\bar{X}_{\text{high}} - \bar{X}_{\text{low}}|^* = t_\alpha S \sqrt{\frac{1}{N_{\text{high}}} + \frac{1}{N_{\text{low}}}}$$

$$= t_\alpha S \sqrt{\frac{1}{3} + \frac{1}{3}}$$

$$= t_\alpha S\ (.816)$$

The effect of B and C are computed in similar fashion and they have the same criteria.

The effect of the AB interaction is calculated:

$$\frac{\text{a} + \text{b} - \text{ab} - 1}{2}$$

The criterion for acceptance of the null or alternative hypothesis for interactions is:

$$|X_{\text{high}} - X_{\text{low}}|^* = t_\alpha S \sqrt{\frac{1}{N_{\text{high}}} + \frac{1}{N_{\text{low}}}}$$

$$= t_\alpha S \sqrt{\frac{1}{2} + \frac{1}{2}}$$

$$= t_\alpha S \ (1)$$

The effect of AC and BC are computed in similar fashion and they will have the same criterion.

The efficiency of design is:

$$E = \frac{\text{No. of Contrast}}{\text{No. of trial} - 1} = \frac{6}{7-1} = 1.00$$

Four-Variable Designs

There are ten contrasts to be estimated. A, B, C, D, AB, AC, AD, BC, BD, and CD. The treatment combinations required are all of those in the three-variable designs plus the additional treatment combinations d, ad, bd, and cd. The total number of treatments for 4 variables is 7 + 4 = 11

[(1), a, b, c, ab, ac, bc] + [d, ad, bd, cd]

$$\text{The effect of A} = \frac{\text{a} - (1) + \text{ab} - \text{b} + \text{ac} - \text{c} + \text{ad} - \text{d}}{4}$$

$$\text{The effect of BD} = \frac{\text{b} + \text{d} - \text{bd} - 1}{2}$$

The criterion for the main effects is:

$$|\overline{X}_{\text{high}} - \overline{X}_{\text{low}}|^* = t_\alpha S \sqrt{\frac{1}{4} + \frac{1}{4}}$$

$$= t_\alpha S \ (.7)$$

The criterion for the interactions is exactly the same as for the three-variable case. In fact, the criterion for interaction will be the same regardless of the number of variables.

$$|\overline{X}_{\text{high}} - \overline{X}_{\text{low}}|^* = t_\alpha S.$$

The efficiency of the 4 factor design is:

$$E = \frac{10}{11 - 1} = 1.00$$

Five-Variable Designs

The regular 16 trial Hadamard matrix with I = ABCD is already at efficiency = 1. Using the principle described for three and four variables, a new Resolution V design is obtained with the 16 treatment combinations.

(1), a, b, c, d, c, e, ab, ac, ad, ae, bc, bd, be, cd, ce, and de.

$$\text{Effect of A} = \frac{a - (1) + ab - b + ac - c + ad - d + ae - e}{5}$$

The criterion is:

$$|\overline{X}_{\text{high}} - \overline{X}_{\text{low}}|^* = t_\alpha S \sqrt{\frac{1}{5} + \frac{1}{5}} = t_\alpha S\ (.63)$$

This compares to the criterion for the conventional Resolution V design for both main effects and interactions

$$|\overline{X}_{\text{high}} - \overline{X}_{\text{low}}|^* = t_\alpha S \sqrt{\frac{1}{8} + \frac{1}{8}} = t_\alpha S\ (.5)$$

Thus, the conventional design has a smaller variance and is preferable to these new designs.

Six and Higher Numbers of Variables

The treatment combinations required for any number of variables are:

- Treatment combination (1), that is all variables at the low level.
- All treatment combinations with a single letter, that is, all treatment combinations at the high level of one variable and the low level of all other variables.
- All treatment combinations with two letters, that is, all treatment combinations with two variables at the high level and the low levels of all other variables.

As the number of variables increase, the variance of the estimates of the main effects will decrease.

$$\text{Standard Error of Main Effects} = \sqrt{\frac{2}{\text{No. of Variables}}}\ \sigma$$

The variance of the estimate of the interactions is the same regardless of the number of variables.

Standard Error of Interactions $= \sigma$

Summary of Design; Efficiency = 1

No. of Variables	No. of Contrasts	No. of trials Required	σ for Main Effects
3	6	7	.82
4	10	11	.70
5	15	16	.63
6	21	22	.58
7	28	29	.53
8	36	37	.50
9	45	46	.47
10	55	56	.45
11	66	67	.42

It should be noted that the efficiency of these designs is one. These designs require fewer trials than the John's or unbalanced designs of Chapter 9. However, the variance of the estimates of the main effects will be greater than the John's or unbalanced designs. For example, the John's 96 trial designs for 9, 10, or 11 variables will have a standard deviation of $.25\sigma$, whereas these designs have, at best, a standard deviation of $.42\sigma$. The standard deviation of the interactions is $.25\sigma$ for the John's design, but the much larger σ for these designs.

11 Hadamard Matrix Designs for Binomial and Poisson Responses

In all prior examples, the responses of interest to the experiment were such qualities as the power of a laser or the tensile strength of a weld. In this chapter, the Hadamard matrix will be applied to problems where the response of interest is a binomial response such as the yield of a process or the failure rate of a mechanism or a poisson response such as the number of blisters on a painted part.

Binomial

All of the Hadamard matrix designs previously discussed can be used when the response of interest is binomial. The choice of the design however will require some different rules as will be explained later.

It is no longer necessary to calculate the number of Hadamard trials required. Rather, the experimenter must determine the total number of devices to be tested, or the number of sheets of paper that must be fed into a printer to determine failure rate, etc. The proper sample size for matrix experiments can be approximated by the following formula.

$$N_{\text{high}} = N_{\text{low}} = 1 \left\{ \frac{(U_\alpha)\,[(\rho_\alpha)\,(1 - \rho_\alpha)]^{1/2} + (U_\beta)\,[(\rho_\beta)\,(1 - \rho_\alpha)]^{1/2}}{\rho_\alpha - \rho_\beta} \right\}^2$$

Where U_α is the value from table 1 or 2 depending upon whether H_a is single- or double-sided, U_β is the value from table 1, ρ_α is the failure rate of which α applies and ρ_β is the failure rate which β applies.

Suppose that a new copier design has a specification of less than one paper jam per 1000 copies. The above formula is used to calculate the number of sheets of paper that must be fed at the high level of each variable and the number of sheets that must be fed at the low level of each variable. This number is independent of the number of variables and the desired resolution of the experiment and the chosen matrix size.

If the sample size (number of sheets) for example is:

$$N_{\text{high}} = N_{\text{low}} = 4000$$

and the experimenter chooses a complete matrix with three variables in eight trials, then the number of sheets fed into the machine at each treatment combination is

$$\frac{4000 = N_{\text{high}}}{4 = \text{No. of high trials}} = 1000 \text{ sheets/treatment combination.}$$

If a 16-trial Resolution IV design with 8 variables was chosen, the same 4000 sheets must be fed at the high level of each variable. The number of sheets at each treatment combination, however, will be

$$\frac{4000 = N_{\text{high}}}{8 = \text{No. of high trials}} = 500 \text{ sheets/treatment combination.}$$

EXAMPLE

A new latching mechanism has been designed and a sample of eight mechanisms can be built for product assurance testing. The product planning department has specified that the mechanism must have fewer than 1.5 latching failures per 1000 closings.

Four parameters of the mechanism are thought to influence the latching performance. These parameters and their proposed tolerances and specified values for the experiment are:

Variable	Parameter	Specification	High Level	Low Level
A	Spring tension	10 ± 0.5 #	10.5	9.5
B	Notch angle	20 ± 2 degrees	22	18
C	Striker length	2 ± 0.1 inch	2.1	1.9
D	Striker width	0.5 ± 0.05 inch	0.55	0.45

It is expected that there might be 2-factor interactions, but probably not 3-factor or higher interactions. A Resolution V design would be desirable, but since there are four variables and only eight devices can be made, a Resolution IV design is mandated.

A. State object of the experiment.

$$(H_0)_A: p_{10.5} = p_{9.5}$$
$$(H_a)_A: p_{10.5} \neq p_{9.5}$$
$$(H_0)_B: p_{220} = p_{180}$$
$$(H_a)_B: p_{220} \neq p_{180}$$
$$\vdots$$

$$(H_0)_{BD+AC}:(\rho_{BD+AC}) = 0$$
$$(H_a)_{BD+AC}:(\rho_{BD+AC}) > 0$$
etc.

B. Determine treatment combinations for the eight mechanisms with the computer program.

Trial	Treatment Combination	A	B	C	D
1	ad	10.5	18	1.9	0.55
2	ab	10.5	22	1.9	0.45
3	abcd	10.5	22	2.1	0.55
4	bc	9.5	22	2.1	0.45
5	ac	10.5	18	2.1	0.45
6	bc	9.5	22	1.9	0.55
7	cd	9.5	18	2.1	0.55
8	(1)	9.5	18	1.9	0.45

C. Choose α, β, ρ_α, and ρ_β. Note: ρ_β is chosen as some value higher than the specification (.0015), and ρ_α is chosen as some value lower than the specification.

$$\alpha = .10$$
$$\beta = .05$$
$$\rho_\alpha = .00075$$
$$\rho_\beta = .00225$$

D. Calculate sample size.

$$N_{\text{high}} = N_{\text{low}} = 2\left\{\frac{(U_\alpha)[(\rho_\alpha)(\rho_1 - \rho_\alpha)]^{1/2} + (U_\beta)[(\rho_\beta)(1 - \rho_\beta)]^{1/2}}{\rho_\beta - \rho_\alpha}\right\}^2$$

$$N_{\text{high}} - N_{\text{low}} = 2\left\{\frac{(1.645)[(0.00075)(1 - 0.00075)]^{1/2} + (1.445)[(0.00225)(1 - 0.00225)]^{1/2}}{0.00225 - .00075}\right\}^2 = 13{,}442$$

$$\frac{\text{Sample size}}{\text{Treatment combination}} = \frac{N_{\text{high}}}{\text{No. of high trials}} \qquad \frac{\text{No. of latchings}}{\text{treatment combination}} = \frac{13443}{4} = 3360$$

E. Make the eight specified variations of the latch mechanism and test each for 3360 cycles and count the number of failures to latch.

Trials	Failures	Failure Rate
1	32	0.0095
2	8	0.0024
3	4	0.0012
4	0	0
5	8	0.0024
6	12	0.0036
7	8	0.0024
8	30	0.0089

F. Calculate failure rate ($\hat{p}$) over all configurations of the device.

$$\hat{p} = \frac{32 + 8 + 4 + 0 + 8 + 12 + 18 + 30}{26,886}$$

$$\hat{p} = 0.0038$$

This estimated failure rate is much higher than the specification ($p = 0.0015$). Therefore, the data must be analyzed to determine the levels of the variable and interactions that are causing the poor performance.

G. Determine the effects of the variable and interaction pairs using the sums in the columns of the Hadamard matrix with the test results

A+BCD	B+ACD	C+ABD	D+ABC
+32	−32	−32	+32
+ 8	+ 8	− 8	− 8
+ 4	+ 4	+ 4	+ 4
− 0	+ 0	+ 0	− 0
+ 8	− 8	+ 8	− 8
−12	+12	−12	+12
− 8	− 8	+ 8	+ 8
−30	−30	−30	−30
+ 2	−54	−62	+10

−CD−AB	−AD−BC	−BD−AC
+32	−32	+32
− 8	+ 8	+ 8
− 4	− 4	− 4
+ 0	− 0	+ 0
+ 8	+ 8	− 8
+12	+12	−12
− 8	+ 8	+ 8
−30	−30	−30
+ 2	−30	− 6

Note carefully: These total results are *not* divided by 4. These numbers are the difference between 13,443 High level results and 13,443 Low level results.

H. Compute criterion

Note: $N_{\text{high}} = 13,443$

$$|p_{\text{high}} - p_{\text{low}}|^* = \sqrt{N_{\text{high}}}\, U_\alpha \sqrt{(p_\alpha)(1 - p_\alpha) + (p_\beta)(1 - p_\beta)}$$

$$|p_{\text{high}} - p_{\text{low}}|^* = \sqrt{13,443}\,(1.645)\,\sqrt{(0.00075)(1 - 0.00075) + (.00225)(1 - .00225)}$$

$$= 10.44 \text{ failures to latch}$$

I. Compare criterion with the sums of the Hadamard Contrast Column

$[(\rho_{\text{low B}} - \rho_{\text{high B}}) = 54]$
$> [(\rho_{\text{high B}} - \rho_{\text{low B}})^* = 10.44)]$
$[(\rho_{\text{low C}} - \rho_{\text{high C}}) = 60]$
$> [(\rho_{\text{high B}} - \rho_{\text{low B}})^* = 10.44)]$
$[(\rho_{\text{AD+BC}}) = 30]$
$> [(\rho_{\text{high}} - \rho_{\text{low}})^* = 10.44)]$

J. From theoretical considerations, it is reasonable to expect a BC interaction. Make up a two way BC table.

	1	c
1	32 30 $\bar{X} = 31$	8 8 $\bar{X} = 8$
b	8 12 $\bar{X} = 10$	4 0 $\bar{X} = 2$

K. Make an engineering decision.

The present specifications for the components of the latch are faulty. The product should be rejected until development has improved the product. It should be recommended that they increase the nominal notch angle and the nominal striker length.

Poisson

All of the Hadamard matrix designs can be used with Poisson responses. As with the binomial, it is no longer required to calculate the number of Hadamard trials required. Rather, the experiments must determine the total area to be examined, for example, for blisters, or the amount of time to observe some phenomenon.

The approximate sample size (number of units, or units of areas) for Hadamard matrix experiments is given by the following formula:

$$N_{\text{high}} = N_{\text{low}} = 2\left\{\frac{(U_\alpha)(\lambda_\alpha)^{1/2} + (U_\beta)(\lambda_\beta)^{1/2}}{\lambda_\beta - \lambda_\alpha}\right\}^2$$

Where λ_α is the number of defects/unit at which α applies, λ_β is the number of defect/unit at which β applies, and U_α is from table 1 or 2, depending upon whether the alternative hypothesis is single- or double-sided and U_β is obtained from table 1.

EXAMPLE

The specification on a photoconductor film is that it shall have less than 2 blister defects/100 square feet.

Suppose $\alpha = 0.05$, $\beta = 0.05$, $\lambda_\alpha = 1/100$, and $\lambda_\beta = 3/100$, and the variables and specifications are as shown on the following table.

	Variable	Specification	Levels for validation of process	
			Low	High
A	Temperature	150 ± 10 degrees	140	160
B	Tension	20 ± 1	19	21
C	Airflow	50 ± 5 cfm	45	55
D	Speed	10 + 1 fpm	9	11

Only eight runs can be made to validate the process. Therefore, an 8-trial Resolution IV Hadamard matrix is required.

A. State the object of the experiment.

$(H_0)_A: \lambda_{\text{high A}} = \lambda_{\text{low A}}$
$(H_a)_B: \lambda_{\text{high A}} = \lambda_{\text{low A}}$

⋮

$(H_0)_{AC+BD}: |\lambda_{AC+BD}| = 0$
$(H_a)_{AC+BD}: |\lambda_{AC+BD}| > 0$

B. Calculate $N_{\text{high}} = N_{\text{low}}$.

$$N_{\text{high}} = N_{\text{low}} = 2\left\{\frac{(U_\alpha)(\lambda_\alpha)^{\frac{1}{2}} + (U_\beta)(\lambda_\beta)^{\frac{1}{2}}}{\lambda_\beta - \lambda_\alpha}\right\}^2$$

$$N_{\text{high}} = N_{\text{low}} = 2\left\{\frac{(1.96)(.01)^{\frac{1}{2}} + (1.645)(.03)^{\frac{1}{2}}}{.03 - .01}\right\}^2$$

$= 1156$ sq ft at high level
$= 1156$ sq ft at low level

Calculate area at each treatment combination.

$$\frac{1156}{\text{4 trials at high}} = 289 \text{ sq ft/trial}$$

C. Compute Criterion

$$|\lambda_{\text{high}} - \lambda_{\text{low}}|^* = U_\alpha \sqrt{N}\sqrt{\lambda_\alpha + \lambda_\beta}$$

$$|\lambda_{\text{high}} - \lambda_{\text{low}}|^* = (1.96)\sqrt{1156}\sqrt{.01 + .03} = 13.33$$

D. Determine treatment combinations and make 289 sq ft at each trial. Count the number of blisters at each trial.

Trial	Treatment combination	Results
1	ad	8
2	ab	4
3	abcd	16
4	bc	1
5	ac	4
6	bc	4
7	cd	3
8	(1)	1
	Total	41

E. Compute the overall estimate of λ.

$\lambda = \dfrac{41}{2312} = 0.0177$ blisters/sq ft, which is less than the specifications of 0.02blister/sq ft

F. The process will mostly supply good photoconductor with the present process specification. Analyze the data to determine which variables and which levels will produce the best photoconductor.

A	B	C	D
+8	−8	−8	+8
+4	+4	−4	−4
+16	+16	+16	+16
−1	+1	+1	−1
+4	−4	+4	−4
−4	+4	−4	+4
−3	−3	+3	+3
−1	−1	−1	−1
23	9	7	21

$[(X_{high} - X_{low})_A = 23] > [|X_{high} - X_{low}|^* = 13.33]$

$[(X_{high} - X_{low})_D = 21] > [|X_{high} - X_{low}|^* = 13.33]$

G. Make an engineering decision.

High speed and high temperatures are causing blisters.

H. Make up two-way table.

	1	d
1	1 1 $\overline{X} = 1$	4 3 $\overline{X} = 3.5$
a	4 4 $\overline{X} = 4$	8 16 $\overline{X} = 12$

At low temperature (140°) and low speed (9 ft/m) the failure rate is 2/578 = .35/100 sq ft, which is better than the specification. At high speed (11 ft/m) and low temperature (140°), the failure rate is 8/578 = 1.4/100 sq ft, which is better than the specification.

I. Make an engineering decision.

Operate the process with 140° nominal temperature and keep all other variable at present nominal values and tolerances.

Note: lowering the speed will result in a better product, but yield will be reduced by 10 percent.

EXERCISES

1. A 16-trial Resolution IV design with eight variables is to be conducted. $\rho_\alpha = 0.05$; $\rho_\beta = 0.15$; $\alpha = \beta = 0.05$. How many parts must be tested at each treatment combination? What is the criterion?

2. An 8-trial Resolution III design with seven variables is to be conducted. $\lambda_\alpha = 2/100$ sq. ft., $\lambda_\beta = 4/100$ sq. ft., $\alpha = \beta = 0.10$. How many square feet of material must be examined at each treatment combination? What is the criterion?

12 Summary of Two-Level Matrix Designs

The purpose of this chapter is to condense in one place all the designs that are available to the experimenter for two-level multivariable experiments. It is also the purpose of this chapter to convince the experimenter that all the apparently complex discussion in the previous five chapters is really not difficult to apply.

- Before choosing a design, the experimenter must determine the number of variables that can possibly influence the responses of interest and therefore the number of variables that must be included in the experiment.
- The experimenter must choose α, β, and δ.
- The experimenter must calculate $N_{\text{high}} = N_{\text{low}}$. Only two possible formulas are involved:

$$N_{\text{high}} = N_{\text{low}} = 2(U_\alpha + U_\beta)^2 \frac{\sigma^2}{\delta^2} \quad \text{if } \sigma^2 \text{ is known.}$$

$$N_{\text{high}} = N_{\text{low}} = 2(t_\alpha + t_\beta)^2 \frac{S^2}{\delta^2} \quad \text{if } \sigma^2 \text{ is unknown.}$$

- The experimenter must calculate a criterion for decision making. Only two possible formulas are involved:

$$|\bar{X}_{\text{high}} - \bar{X}_{\text{low}}|^* = U_\alpha \sigma \sqrt{\frac{1}{N_{\text{high}}} + \frac{1}{N_{\text{low}}}} \quad \text{if } \sigma^2 \text{ is known.}$$

$$|\bar{X}_{\text{high}} - \bar{X}_{\text{low}}|^* = t_\alpha S \sqrt{\frac{1}{N_{\text{high}}} + \frac{1}{N_{\text{low}}}} \quad \text{if } \sigma^2 \text{ is unknown.}$$

- The experimenter must determine what resolution is desired. Only three possibilities exist in most practical problems:

(1) Resolution V: all main factors and two-factor interactions can be estimated.

(2) Resolution IV: all main factors and groups of two-factor interactions can be estimated.
(3) Resolution III: only main factors can be estimated.

Having computed or made decisions on the above, the experimenter can go to the following summary of designs, find the single Hadamard matrix that meets all the requirements of the experiment, and then go to the referenced pages in this text for the details of that design. Suppose, for example, that $N_{high} = N_{low}$ is calculated to be 15 for some α, β, and δ. The number of variables to be studied in the experiment is 12, and a Resolution IV design is desired.

- Go down the number of variables column to 12.
- Within the list of 12-variable experiments, go down the resolution column to the four possible designs of Resolution IV:

Number of variables	Resolution	$N_{high} = N_{low}$		Order of matrix
12	III	≤8		16
12	III	>8	≤10	20
12	IV	≤8		24
12	IV	{>8	≤16	32
12	IV	>16	≤32	64
12	IV	>32	≤64	128
12	V	≤64		192

- Of the four designs that are Resolution IV, choose the one that is correct for $N_{high} = N_{low} = 15$. There is only one such design matrix, and that is of order 32.
- The order-32 Hadamard matrix is the only matrix for this experiment. It meets all the requirements.
  (1) 12 variables
  (2) Resolution IV
  (3) $8 \le (N_{high} = N_{low} = 15) \le 16$
- Go to the computer program (Chapter 13) and type in 32 DES 12.

Summary of Two-Level Matrix Designs

Number of variables	Resolution	$N_{high} = N_{low}$ for desired α, β, and δ		Order of matrix	Matrix type	Page
2	V	≤2		4	Hadamard	—
2	V	>2	≤4	8	Hadamard	98
2	V	>4	≤8	16	Hadamard	126
2	V	>8	≤16	32	Hadamard	127
2	V	>16	≤32	64	Hadamard	128
2	V	>32	≤64	128	Hadamard	129
3	III	≤2		4	Hadamard	—
3	IV	≤2		6	John's	155
3	V+	≤4		8	Hadamard	104
3	V+	>4	≤8	16	Hadamard	126

Number of variables	Resolution	$N_{high} = N_{low}$ for desired α, β, and δ		Order of matrix	Matrix type	Page
3	V+	>8	≤16	32	Hadamard	127
3	V+	>16	≤32	64	Hadamard	128
3	V+	>32	≤64	128	Hadamard	129
4	III	≤2		6	John's	143
4	IV	≤4		8	Hadamard	110
4	V	≤4		12	John's	144
4	V+	>4	≤8	16	Hadamard	126
4	V+	>8	≤16	32	Hadamard	127
4	V+	>16	≤32	64	Hadamard	128
4	V+	>32	≤64	128	Hadamard	129
5	III	≤2		6	John's	—
5	III	>2	≤4	8	Hadamard	116
5	IV	≤4		12	John's	156
5	V	≤8		16	Hadamard	126
5	V+	>8	≤16	32	Hadamard	127
5	V+	>16	≤32	64	Hadamard	128
5	V+	>32	≤64	128	Hadamard	129
6	III	≤4		8	Hadamard	116
6	IV	≤4		12	John's	156
6	IV	>4	≤8	16	Hadamard	116
6	V+	≤16		32	Hadamard, unbalanced	127, 168
6	V+	>16	≤32	64	Hadamard	128
6	V+	>32	≤64	128	Hadamard	129
7	III	≤4		8	Hadamard	116
7	III	>4	≤6	12	Irregular Hadamard	132
7	IV	≤8		16	Hadamard	126
7	IV	>8	≤16	32	Hadamard	127
7	V	≤16		48	John's, unbalanced	151, 173, 177
7	V+	>16	≤32	64	Hadamard	128
7	V+	>32	≤64	128	Hadamard	129
8	III	≤4		12	John's	—
8	III	>4	≤6	12	Irregular Hadamard	132
8	IV	≤8		16	Hadamard	126
8	IV	>8	≤16	32	Hadamard	127
8	V	≤16		48	John's, unbalanced	151, 175, 179
8	V	>16	≤32	64	Hadamard	128
8	V+	>32	≤64	128	Hadamard	129
9	III	≤4		12	John's	—
9	III	>4	≤6	12	Irregular Hadamard	132
9	III	>6	≤8	16	Hadamard	127
9	III	>8	≤10	20	Irregular Hadamard	132
9	IV	≤8		24	John's	158
9	IV	>8	≤16	32	Hadamard	127
9	IV	>16	≤32	64	Hadamard	129
9	V	≤32		96	John's, unbalanced	154, 176, 180, 182
9	V	>32	≤64	128	Hadamard	129
10	III	≤4		12	John's	—
10	III	>4	≤6	12	Irregular Hadamard	132
10	III	>6	≤8	16	Hadamard	127
10	III	>8	≤10	20	Irregular Hadamard	132
10	IV	≤8		24	John's	158
10	IV	>8	≤16	32	Hadamard	127
10	IV	>16	≤32	64	Hadamard	129
10	V	≤32		96	John's	154
10	V	>32	≤64	128	Hadamard, unbalanced	129, 176
11	III	≤4		12	John's	—
11	III	>4	≤8	16	Hadamard	127

Number of variables	Resolution	$N_{high} = N_{low}$ for desired α, β, and δ		Order of matrix	Matrix type	Page
11	III	>8	≤10	20	Irregular Hadamard	132
11	IV	≤8		24	John's	158
11	IV	>8	≤16	32	Hadamard	127
11	IV	>16	≤32	64	Hadamard	129
11	V	≤32		96	John's, unbalanced	154, 176
11	V	>32	≤64	128	Hadamard	129
12	III	≤8		16	Hadamard	127
12	III	>8	≤10	20	Irregular Hadamard	132
12	IV	≤8		24	John's	158
12	IV	>8	≤16	32	Hadamard	127
12	IV	>16	≤32	64	Hadamard	129
12	IV	>32	≤64	128	Hadamard	129
12	V	≤64		192	John's, unbalanced	182
13	III	≤8		16	Hadamard	127
13	III	>8	≤10	20	Irregular Hadamard	132
13	III	>10	≤12	24	Irregular Hadamard	132
13	III	>12	≤14	28	Irregular Hadamard	132
13	IV	≤16		32	Hadamard	127
13	IV	>16	≤32	64	Hadamard	129
13	IV	>32	≤64	128	Hadamard	129
13	V	≤64		192	John's	128
14	III	≤8		16	Hadamard	127
14	III	>8	≤10	20	Irregular Hadamard	132
14	III	>10	≤12	24	Irregular Hadamard	132
14	III	>12	≤14	28	Irregular Hadamard	132
14	IV	≤16		32	Hadamard	127
14	IV	>16	≤32	64	Hadamard	129
14	IV	>32	≤64	128	Hadamard	129
14	V	≤64		192	John's	132
15	III	≤8		16	Hadamard	127
15	III	>8	≤10	20	Irregular Hadamard	132
15	III	>10	≤12	24	Irregular Hadamard	132
15	III	>12	≤14	28	Irregular Hadamard	132
15	IV	≤16		32	Hadamard	127
15	IV	>16	≤32	64	Hadamard	129
15	IV	>32	≤64	128	Hadamard	129
15	V	≤64		192	John's	132
16	III	≤10		20	Irregular Hadamard	132
16	III	>10	≤12	24	Irregular Hadamard	132
16	III	>12	≤14	28	Irregular Hadamard	132
16	IV	≤16		32	Hadamard	127
16	IV	>16	≤32	64	Hadamard	129
16	IV	>32	≤64	128	Hadamard	129
17	III	≤10		20	Irregular Hadamard	132
17	III	>10	≤12	24	Irregular Hadamard	132
17	III	>12	≤14	28	Irregular Hadamard	132
17	III	>14	≤16	32	Hadamard	128
17	III	>16	≤18	36	Irregular Hadamard	132
17	III	>18	≤20	40	Irregular Hadamard	132
17	III	>20	≤22	44	Irregular Hadamard	132
17	IV	≤16		48	John's	159
17	IV	>16	≤32	64	Hadamard	129
17	IV	>32	≤64	128	Hadamard	129
18	III	≤10		20	Irregular Hadamard	132
18	III	>10	≤12	24	Irregular Hadamard	132
18	III	>12	≤14	28	Irregular Hadamard	132
18	III	>14	≤16	32	Hadamard	128

Number of variables	Resolution	$N_{high} = N_{low}$ for desired α, β, and δ		Order of matrix	Matrix type	Page
18	III	>16	≤ 18	36	Irregular Hadamard	132
18	III	>18	≤ 20	40	Irregular Hadamard	132
18	III	>20	≤ 22	44	Irregular Hadamard	132
18	IV	≤ 16		48	John's	159
18	IV	>16	≤ 32	64	Hadamard	129
18	IV	>32	≤ 64	128	Hadamard	129
19	III	≤ 10		20	Irregular Hadamard	132
19	III	>10	≤ 12	24	Irregular Hadamard	132
19	III	>12	≤ 14	28	Irregular Hadamard	132
19	III	>14	≤ 16	32	Hadamard	128
19	III	>16	≤ 18	36	Irregular Hadamard	132
19	III	>18	≤ 20	40	Irregular Hadamard	132
19	III	>20	≤ 22	44	Irregular Hadamard	132
19	IV	≤ 16		48	John's	159
19	IV	>16	≤ 32	64	Hadamard	129
19	IV	>32	≤ 64	128	Hadamard	129
20	III	≤ 12		24	Irregular Hadamard	132
20	III	>12	≤ 14	28	Irregular Hadamard	132
20	III	>14	≤ 16	32	Hadamard	128
20	III	>16	≤ 18	36	Irregular Hadamard	132
20	III	>18	≤ 20	40	Irregular Hadamard	132
20	III	>20	≤ 22	44	Irregular Hadamard	132
20	IV	≤ 16		48	John's	159
20	IV	>16	≤ 32	64	Hadamard	129
20	IV	>32	≤ 64	128	Hadamard	129
21	III	≤ 12		24	Irregular Hadamard	132
21	III	>12	≤ 14	28	Irregular Hadamard	132
21	III	>14	≤ 16	32	Hadamard	128
21	III	>16	≤ 18	36	Irregular Hadamard	132
21	III	>18	≤ 20	40	Irregular Hadamard	132
21	III	>20	≤ 22	44	Irregular Hadamard	132
21	IV	≤ 16		48	John's	159
21	IV	>16	≤ 32	64	Hadamard	129
21	IV	>32	≤ 64	128	Hadamard	131
22	III	≤ 12		24	Irregular Hadamard	132
22	III	>12	≤ 14	28	Irregular Hadamard	132
22	III	>14	≤ 16	32	Hadamard	128
22	III	>16	≤ 18	36	Irregular Hadamard	132
22	III	>18	≤ 20	40	Irregular Hadamard	132
22	III	>20	≤ 22	44	Irregular Hadamard	132
22	IV	≤ 16		48	John's	159
22	IV	>16	≤ 32	64	Hadamard	129
22	IV	>32	≤ 64	128	Hadamard	129
23	III	≤ 12		24	Irregular Hadamard	132
23	III	>12	≤ 14	28	Irregular Hadamard	132
23	III	>14	≤ 16	32	Hadamard	128
23	III	>16	≤ 18	36	Irregular Hadamard	132
23	III	>18	≤ 20	40	Irregular Hadamard	132
23	III	>20	≤ 22	44	Irregular Hadamard	132
23	IV	≤ 16		48	John's	159
23	IV	>16	≤ 32	64	Hadamard	129
23	IV	>32	≤ 64	128	Hadamard	129
24	III	≤ 14		28	Irregular Hadamard	132
24	III	>14	≤ 16	32	Hadamard	128
24	III	>16	≤ 18	36	Irregular Hadamard	132
24	III	>18	≤ 20	40	Irregular Hadamard	132
24	III	>20	≤ 22	44	Irregular Hadamard	132
24	IV	≤ 16		48	John's	159

Number of variables	Resolution	$N_{high} = N_{low}$ for desired α, β, and δ		Order of matrix	Matrix type	Page
24	IV	>16	≤32	64	Hadamard	129
24	IV	>32	≤64	128	Hadamard	129
25	III	≤14		28	Irregular Hadamard	132
25	III	>14	≤16	32	Hadamard	128
25	III	>16	≤18	36	Irregular Hadamard	132
25	III	>18	≤20	40	Irregular Hadamard	132
25	III	>20	≤22	44	Irregular Hadamard	132
25	III	>22	≤24	48	Irregular Hadamard	132
25	III	>24	≤26	52	Irregular Hadamard	132
25	III	>26	≤28	56	Irregular Hadamard	132
25	III	>28	≤30	60	Irregular Hadamard	132
25	IV	≤32		64	Hadamard	129
25	IV	>32	≤64	128	Hadamard	129

Unbalanced Resolution V Designs

Class A Designs

No. of variables	No. of trials
6	28
7	36
8	44
9	69
10	81
11	93

Class B Designs

No. of variables	No. of trials
7	38
8	46
9	71
12	143

Class AB Design

No. of variables	No. of trials
9	58

EXERCISES

1. Continue the summary of designs for from 26 to 32 variables.
2. What is the correct matrix size for:
   a. 21 variables, Resolution III, $N_{high} = N_{low} = 21$
   b. 15 variables, Resolution IV, $N_{high} = N_{low} = 18$
   c. 13 variables, Resolution V, $N_{high} = N_{low} = 55$
   d. 5 variables, Resolution V, $N_{high} = N_{low} = 42$

13 A Computer Program for Generating Hadamard Matrix Designs and Analyzing the Data from Such Designs

General library programs for design of experiments have not been available because of the multitude of concepts regarding the various designs available (e.g., three-quarter fractional factorials, one-half fractional factorials, saturated designs, etc.). On the other hand, many statisticians have written individual programs, for their private use, for various classes of designs. The unification of all the two-level designs through the Hadamard matrix concepts made feasible the general APL computer program described in this chapter.

The APL computer program listing is attached as Chapter 13 Appendix 1 at the end of this chapter. The use of the program and the computing methods are described in the following pages.

The user of the program can obtain a short description of the program by entering the word DESCRIBE on the keyboard. The most important part of the output received from this instruction is the table of experiment designs that are programmed. The smallest design is the Resolution IV design in six trials, and the largest design is the Resolution III design for 127 variables in 128 trials. By inference, the designs that are not on the list are not programmed. For example, one could certainly obtain a Resolution V design for five variables in 24 trials, but it is not programmed because a better design can be obtained with only 16 trials. If one should request a design that is not programmed, the computer program will type a message to that effect.

The user can control the output format of the design matrix and/or the format of the data report by use of the word START. After this word is typed in, the computer will request the number of characters to be included in a page width. After the user replies to this request with some number up to 120, the program will request the number of columns across the page. In the design phase, this is usually the number of variables. It is not necessary to use this START program, as there is a built-in system that is quite adequate for most problems.

DESCRIBE

THIS WORKSPACE PROVIDES DESIGNS AND/OR ANALYSIS OF TWO LEVEL FACTORIAL EXPERIMENTS. THE POSSIBLE EXPERIMENTS THAT CAN BE DESIGNED ARE SHOWN IN THE TABLE BELOW

NUMBER OF VARIABLES FOR A GIVEN RESOLUTION

NUMBER OF TRIALS	WORKABLE RANGE RESOLUTION V+	V	IV	III	SUGGESTED RANGE RESOLUTION V+	V	IV	III
6			3				3	
8	3		4	5-7	3		4	5-7
12		4	5-6	7-11		4	5-6	8-11
16	4	5	6-8	9-15	4	5	7-8	12-15
20				11-19				16-19
24			9-12	13-23			9-12	20-23
32	5	6	7-16	17-31	5	6	13-16	24-31
36				19-35				32-35
44				23-43				36-43
48		7-8	17-24	25-47		7-8	17-24	44-47
60				31-59				48-59
64	6	7-8	9-32	33-63	6		25-32	60-63
68				35-67				64-67
72				37-71				68-71
80				41-79				72-79
84				43-83				80-83
96		9-11	33-48			9-11	33-48	
128	7	8-11	12-64	65-127	7		49-64	84-127

RESOLUTION
V+ : COMPLETE FACTORIAL

V : ALL MAIN EFFECTS AND TWO FACTOR INTERACTIONS ARE UNCONFOUNDED WITH OTHER MAIN EFFECTS OR TWO FACTOR INTERACTIONS

IV : ALL MAIN EFFECTS ARE UNCONFOUNDED BY TWO FACTOR INTERACTIONS TWO FACTOR INTERACTIONS ARE CONFOUNDED WITH EACH OTHER

III: MAIN EFFECTS CONFOUNDED WITH TWO FACTOR INTERACTIONS

TO USE THIS WORKSPACE TYPE START TO SET PAGE FORMAT CONTROLS
THEN TYPE NT DES NV FOR DESIGNS NT FAN NV FOR ANALYSIS
NT=NO. OF TRIALS NV=NO. OF VARIABLES

Designing the Experiment

To design an experiment, the user enters the statement NT DES NV; where NT is the number of trials and NV is the number of variables. For example, 16 DES 5 will produce the experiment design for a Resolution V experiment with five variables in 16 trials; the identity element will be I = ABCDE.

After the user types in the requested design, the computer will request the lower levels of the variables. Two alternatives are available for supplying this information. With, for example, an 8-trial experiment with seven variables (Resolution III), the user could either

- Type in the actual numerical value of the low level of each variable (e.g., 100, 25, 10, 60, 50, 1.2, and 10) or
- Type in the number of zeros that represent the low level, 7ρ0 or NVρ0 where ρ is the uppercase *R*.

With the first alternative, the design will be given in actual values of the variables; with the second alternative, the design will be given in zeros and ones.

After the user supplies the low levels of the variables, the computer will request the high levels of the variables. The same two alternatives as above are available; type in either the actual values or 7ρ1 or NVρ1.

The following subsections give examples of the experiment design outputs that will be received, depending on the experiment design requested and the options used by the experimenter.

Example of Saturated Resolution III Designs. 15 Variables in 16 Trials, Not Using the START Program, and Coding the Levels of the Variables as 0 or 1

```
       16 DES 15
    ENTER LOWER LEVEL (BV) OR NV ρ0
⎕:
       NVρ0
ENTER UPPER LEVEL (PS) OR NV ρ1
⎕:
       NVρ1

 NT ABCDEFGHIJKLMNO
  1 100010011010111
  2 110001001101011
  3 111000100110101
  4 111100010011010
  5 011110001001101
  6 101111000100110
  7 010111100010011
  8 101011110001001
  9 110101111000100
 10 011010111100010
 11 001101011110001
 12 100110101111000
 13 010011010111100
 14 001001101011110
 15 000100110101111
 16 000000000000000
```

Note: The vector of the lower levels of the variables is stored in the computer under the name (BV), and the vector of the upper level of the variables is stored under the name (PS). Thus, if, after obtaining the above design, the user desired the design for a 32-trial experiment with the same variables and levels, the following procedure is used:

- Type 32 DES 15.
- The computer requests lower levels. Type BV.
- The computer requests upper levels. Type PS.

Example of Saturated Resolution IV Designs. John's Three-Quarter Fractional Factorial with 6 Variables in 12 Trials, Not Using the START Program, and Coding the Levels of the Variables as 0 or 1

```
      12 DES 6
   ENTER LOWER LEVEL (BV) OR NV ρ0
⎕:
      NVρ0
ENTER UPPER LEVEL (PS) OR NV ρ1
⎕:
      NVρ1

 NT XT ABCDEF
  1  1 100010
  2  3 111000
  3  4 111111
  4  5 011101
  5  6 101110
  6  7 010110
  7  8 101001
  8 10 011010
  9 12 100101
 10 13 010001
 11 15 000111
 12 16 000000
```

In the above design, the first column on the left (NT) is the number of the trials to be conducted (i.e., trial 1 has the treatment combination 100010 or ae). The second column from the left (XT) is the number of the trial in the complete Hadamard matrix from which the John's design was derived. The numbers that are missing in this column are those trials that were in block *i* and were, therefore, eliminated. For this example, the John's design consists of the trials in the 16 × 16 Hadamard matrix except for trials 2, 9, 11, and 14. This column is added so that if, after completing the experiment, one should desire to complete the entire matrix, it is easy to determine which trials were not conducted.

Example of Resolution IV Designs. John's Three-Quarter Fractional Factorial with 9 Variables in 24 Trials, Using the START Program, and Using the Actual Levels of the Variables

```
        START
ENTER PAGE WIDTH
⎕:
        60
ENTER NUMBER OF COLUMNS ACROSS PAGE
⎕:
        10
ENTER NT DES NV FOR DESIGN OR NT FAN NV FOR ANALYSIS
NT = NO OF TRIALS    NV = NO OF VARIABLES

        24 DES 9
    ENTER LOWER LEVEL (BV) OR NV ρ0
⎕:
        0 0 0 0 0 0 0 0 0
ENTER UPPER LEVEL (PS) OR NV ρ1
⎕:
        1 2 3 4 5 6 7 8 9
```

NT	XT	A	B	C	D	E	F	G	H	I
1	1	1.0	.0	.0	.0	5.0	6.0	7.0	.0	.0
2	3	1.0	2.0	3.0	.0	.0	6.0	.0	.0	9.0
3	4	1.0	2.0	3.0	4.0	.0	.0	7.0	.0	.0
4	5	1.0	2.0	3.0	4.0	5.0	6.0	7.0	8.0	9.0
5	6	.0	2.0	3.0	4.0	5.0	6.0	.0	.0	.0
6	8	1.0	.0	.0	4.0	5.0	.0	.0	.0	9.0
7	10	.0	2.0	3.0	.0	.0	6.0	7.0	8.0	.0
8	11	1.0	.0	3.0	4.0	.0	.0	7.0	.0	.0
9	12	.0	2.0	.0	4.0	5.0	.0	7.0	8.0	.0
10	14	1.0	.0	.0	4.0	.0	6.0	.0	8.0	.0
11	15	.0	2.0	.0	.0	5.0	6.0	.0	8.0	9.0
12	17	.0	.0	.0	4.0	.0	6.0	7.0	.0	9.0
13	18	.0	.0	.0	.0	5.0	6.0	.0	8.0	9.0
14	19	1.0	.0	.0	.0	.0	.0	7.0	8.0	9.0
15	20	.0	2.0	.0	.0	.0	.0	.0	.0	.0
16	21	1.0	.0	3.0	.0	.0	6.0	.0	.0	9.0
17	22	.0	2.0	.0	4.0	.0	6.0	7.0	.0	9.0
18	23	1.0	.0	3.0	.0	5.0	.0	.0	8.0	.0
19	25	1.0	2.0	3.0	.0	5.0	.0	.0	8.0	.0
20	26	.0	2.0	3.0	4.0	.0	.0	.0	8.0	9.0
21	27	1.0	.0	3.0	4.0	5.0	6.0	7.0	8.0	9.0
22	29	.0	2.0	3.0	.0	5.0	.0	7.0	.0	9.0
23	31	.0	.0	.0	4.0	5.0	.0	7.0	8.0	.0
24	32	0	0	0	0	0	0	0	0	0

If, in the above example, the page width was specified as some smaller number of spaces (e.g., 30), then the data would be outputted as, for example,

```
A B C D E F G H I
1 0 0 0 5 6 7 0 0
```

On the other hand, if one wanted the levels specified to two decimal places, the page width would have to be increased to, say, 80 or 90 characters wide.

Example of Saturated Resolution III Designs. 15 Variables in 16 Trials, Not Using the START Program, and Using the Actual Levels of the Variables

```
      16 DES 15
   ENTER LOWER LEVEL (BV) OR NV ρ0
⎕:
      2 2 2 2 2 2 2 2 2 2 2 2 2 2 2
ENTER UPPER LEVEL (PS) OR NV ρ1
⎕:
      4 4 4 4 4 4 4 4 4 4 4 4 4 4 4
```

NT	A	B	C	D	E	F	G	H
1	4.00	2.00	2.00	2.00	4.00	2.00	2.00	4.00
2	4.00	4.00	2.00	2.00	2.00	4.00	2.00	2.00
3	4.00	4.00	4.00	2.00	2.00	2.00	4.00	2.00
4	4.00	4.00	4.00	4.00	2.00	2.00	2.00	4.00
5	2.00	4.00	4.00	4.00	4.00	2.00	2.00	2.00
6	4.00	2.00	4.00	4.00	4.00	4.00	2.00	2.00
7	2.00	4.00	2.00	4.00	4.00	4.00	4.00	2.00
8	4.00	2.00	4.00	2.00	4.00	4.00	4.00	4.00
9	4.00	4.00	2.00	4.00	2.00	4.00	4.00	4.00
10	2.00	4.00	4.00	2.00	4.00	2.00	4.00	4.00
11	2.00	2.00	4.00	4.00	2.00	4.00	2.00	4.00
12	4.00	2.00	2.00	4.00	4.00	2.00	4.00	2.00
13	2.00	4.00	2.00	2.00	4.00	4.00	2.00	4.00
14	2.00	2.00	4.00	2.00	2.00	4.00	4.00	2.00
15	2.00	2.00	2.00	4.00	2.00	2.00	4.00	4.00
16	2.00	2.00	2.00	2.00	2.00	2.00	2.00	2.00

NT	I	J	K	L	M	N	O
1	4.00	2.00	4.00	2.00	4.00	4.00	4.00
2	4.00	4.00	2.00	4.00	2.00	4.00	4.00
3	2.00	4.00	4.00	2.00	4.00	2.00	4.00
4	2.00	2.00	4.00	4.00	2.00	4.00	2.00
5	4.00	2.00	2.00	4.00	4.00	2.00	4.00
6	2.00	4.00	2.00	2.00	4.00	4.00	2.00
7	2.00	2.00	4.00	2.00	2.00	4.00	4.00
8	2.00	2.00	2.00	4.00	2.00	2.00	4.00
9	4.00	2.00	2.00	2.00	4.00	2.00	2.00
10	4.00	4.00	2.00	2.00	2.00	4.00	2.00
11	4.00	4.00	4.00	2.00	2.00	2.00	4.00
12	4.00	4.00	4.00	4.00	2.00	2.00	2.00
13	2.00	4.00	4.00	4.00	4.00	2.00	2.00
14	4.00	2.00	4.00	4.00	4.00	4.00	2.00
15	2.00	4.00	2.00	4.00	4.00	4.00	4.00
16	2.00	2.00	2.00	2.00	2.00	2.00	2.00

Without START, the program automatically sets the level numbers to two decimal places, which requires that variables I through O be placed below variables A through H.

It is, of course, easy to put all the levels of the trials on one line by cutting off the bottom section and pasting it beside the last column of the top section.

Analyzing the Data

After the experiment has been conducted, the data are analyzed by use of the statement NT FAN NV. For example, if the design used in the experiment was obtained with the statement 16 DES 5, then the analysis of the data must be made with the statement 16 FAN 5. After the user types in the proper statement, the computer program will request the results in standard order. The standard order is the order in which the trials were specified by the design program; i.e., the first result entered is the result obtained by the treatment combination that was given first in the design phase. If all the results cannot be entered on one line, the user depresses the carriage return button, and the computer will again request the results. The user types in the letter W followed by a comma, and then continues with the data.

In all cases, the output of the analysis program consists of the average of all the data followed by two numbers for each main effect, interaction, or estimate of error. The first number that is calculated is the average (AVE), which is nothing more than $(\bar{X}_{\text{high}} - \bar{X}_{\text{low}})$. The second number that is calculated, but not used in most cases, is called SS, which is $(\sum X_{\text{high}} - \sum X_{\text{low}})^2/T$, and which is, under certain circumstances, an estimate of σ^2.

The following subsections give examples of the outputs that can be obtained.

Example of Resolution V Designs. 4 Variables in 16 Trials

```
        16 FAN 4
ENTER RESULTS (W) IN STANDARD ORDER
⎕:
        2.45 4 8.55 13.9 10.55 9.35 10.95 7.1 9.45 5.1
ENTER RESULTS (W) IN STANDARD ORDER
⎕:
        W, 6 4.95 5.65 3.55 6.4 4.05

AVE   7.00000

MAIN EFFECTS

           A          B          C          D
AVE    .93750    3.03750    2.02500    3.88750
S S    3.5156    36.9056    16.4025    60.4506

ESTIMATES OF TWO FACTOR INTERACTIONS

           AB         AC         BC         AD         BD         CD
AVE    .02500   ¯2.48750     .01250     .00000   ¯1.50000     .01250
S S     .0025    24.7506      .0006      .0000     9.0000      .0006

ESTIMATES OF ERROR
           13          8         11         12         14
AVE¯.0125000  .0125000  .0000000  .0250000¯.0750000
S S .0006250  .0006250  .0000000  .0025000  .0225000
```

The average of all the data, $\sum X_i/T$, is 7.00. The mean effect of A, i.e., $(\bar{X}_{\text{high A}} - \bar{X}_{\text{low A}})$ is 0.9375. The SS number under A is ignored. (Note: for the reader who understands ANOVA, this is the sum of squares for A, which is in the numerator of the F-test.) If the sign for AVE A is plus, this signifies that the high level of A gives a higher response than the low level of A. If the sign is minus, it signifies that the low level of A gives a higher response than the high level of A. Of course, this value of A must be compared to the criterion to determine if H_0 or H_a should be accepted.

The estimates of the two-factor interactions also use only the AVE values. In this example, $(\bar{X}_{\text{high AB}} - \bar{X}_{\text{low AB}})$ is 0.0250. This number is compared to the criterion, and the decision is made that there is or is not a true AB interaction. Actually, these columns should be labeled as −AB, −AC, etc. However, the sign of the interaction average is not used for any purpose. Some experimenters try to interpret, for example, −2.4875 for the interaction AC as a bad interaction with a negative effect. This is not true; the nature of the AC interaction can only be found by making a two-way chart.

Again, the *SS* value under an interaction is of no value to the reader of this text. If, however, the reader knows ANOVA, this sum-of-square term is put in the numerator of the F-test.

The data under ESTIMATES OF ERROR is treated in one of two ways, depending upon the assumptions that were made prior to conducting the experiment:

- If the experimenter had a prior estimate of the variance and conducted the experiment with the intention of estimating all three- and higher-order interactions, then the numbers (13, 8, 11, 12, and 14 in this example) refer to the columns of the Hadamard matrix that measure the ABCD, ABD, ABC, BCD, and ACD interactions; the AVE value is the mean effect of the interaction; and the *SS* value is not used. The interaction that is measured by each contrast column number can be found in figure 7-1.
- If the experimenter conducted the experiment with the presumption that interactions of higher order than two were insignificant and that all such contrast columns would be used as an estimate of variance, then the *SS* numbers are averaged to give an average estimate of the variance, with five degrees of freedom in this case.

Example of Saturated Resolution IV Designs. 8 Variables in 16 Trials

```
      16 FAN 8
ENTER RESULTS (W) IN STANDARD ORDER
⎕:
      26 25 20 36 22 21 27 23 21 32 25 32 25 33 32 0

AVE  25.0

MAIN EFFECTS

        A       B       C       D       E       F       G       H
AVE  1.00    2.00    3.00    4.00    5.00    6.00    7.00    8.00
S S     4      16      36      64     100     144     196     256

ESTIMATES OF TWO FACTOR INTERACTIONS

        5       6       7       9      10      13      15
AVE  2.00     .00    5.00    4.00    3.00     .00     .00
S S    16       0     100      64      36       0       0
```

The first thing that is printed is the average of all the data. In this case, the average of the 16 trials is 25.0. The ($\bar{X}_{high} - \bar{X}_{low}$) for each variable is given in the row of AVE; thus, ($\bar{X}_{high\ A} - \bar{X}_{low\ A}$) is 1.00. Since the two-factor interactions are confounded with each other, the program does not specify any interactions under the estimates of two-factor interaction. Instead, the AVE for a given contrast column is printed.

To determine what two-factor interactions are included in any contrast column that is found to be significant, one must refer to the proper subsection in Chapter 13 Appendix 2. Suppose in this case that contrast column 7 is significantly greater than the criterion. Go to the subsection concerning 16-trial designs with six to eight variables. Go down the left column to contrast 7; all the interactions in row 7 out to column 8 are measured by contrast column 7. In this case, (CD + EF + BG + AH) would be significant with $(1 - \alpha)(100)$ percent confidence. If there had been only seven variables in the experiment, then contrast column 7 would measure CD + EF + BG. Note also that, if there are only seven variables, that contrast column 14 does not measure any two-factor interactions or main factors. Therefore, the *SS* value for this contrast column can be used as an estimate of variance with one degree of freedom and would be so printed in the output.

Example of Resolution III Designs. 13 Variables in 16 Trials

```
        16 FAN 13
ENTER RESULTS (W) IN STANDARD ORDER
⎕:
        1 2 3 4 5 6 7 8 9 1 2 3 4 5 6 7


AVE   4.5625

MAIN EFFECTS

            A         B         C         D         E         F         G
AVE     ¯.1250    ¯.3750    ¯.6250    1.3750    ¯.3750    1.6250    1.3750
S S      .063      .563     1.563     7.563      .563    10.563     7.563

            H         I         J         K         L         M
AVE     ¯.3750   ¯2.1250   ¯2.3750   ¯1.8750     .1250     .6250
S S      .563    18.063    22.563    14.063      .063     1.563

ESTIMATES OF ERROR

           14        15
AVE   ¯1.1250    ¯.6250
S S    5.0625    1.5625
```

With Resolution III designs, the output is given either as an average estimate of main effects or as an estimate of variance for all unused columns. No attempt is made to determine the confounding, since the assumption in such designs is that all interactions are insignificant. The only numbers of interest under main effects are those printed in the AVE rows; the number 1.6250 under *F* is ($\bar{X}_{high\ F} - \bar{X}_{low\ F}$). The *SS* numbers under ESTIMATES OF ERROR can be used as estimates of variance, but usually they are biased-high estimates of σ^2 in these designs.

Example of John's Three-Quarter Fractional Factorial. 9 Variables in 24 Trials

```
      START
ENTER PAGE WIDTH
⎕:
      60
ENTER NUMBER OF COLUMNS ACROSS PAGE
⎕:
      6
ENTER NT DES NV FOR DESIGN OR NT FAN NV FOR ANALYSIS
NT = NO OF TRIALS    NV = NO OF VARIABLES

      24 FAN 9
ENTER RESULTS (W) IN STANDARD ORDER
⎕:
      1 2 3 4 5 6 7 8 9 1 2 3 4 5 6 7 8 9 1 2 3 4 5 6


AVE  4.62500
AVE2  3.87500 AVE3  6.50000 AVE4  3.50000

MAIN EFFECTS

           A         B         C         D         E         F
AVE ¯1.62500 ¯1.25000    .87500    .00000   ¯.25000  ¯1.25000
S S  10.5625    6.2500    3.0625     .0000     .2500    6.2500

           G         H         I
AVE  .750000  ¯.500000  ¯.500000
S S  2.25000   1.00000   1.00000

ESTIMATES OF TWO FACTOR INTERACTIONS

           1         2         3         4         5         6
AVE ¯2.62500    .37500   3.00000   2.37500  ¯1.37500  ¯2.50000
S S  27.5625     .5625   36.0000   22.5625    7.5625   25.0000

           7         8         9        10        11        12
AVE  1.12500   ¯.12500   1.00000    .12500   ¯.12500    .00000
S S  5.06250    .06250   4.00000    .06250    .06250    .00000

          13        14        15        16        17        18
AVE  ¯.12500    .12500    .00000  ¯1.62500   1.12500  ¯1.00000
S S    .0625     .0625     .0000   10.5625    5.0625    4.0000

          19        20        21
AVE  2.12500  ¯2.62500   ¯.50000
S S  18.0625   27.5625    1.0000

ESTIMATES OF ERROR

          10        11        12
AVE  ¯.25000    .75000   1.25000
S S   .25000   2.25000   6.25000
```

The first difference in output of these designs is that, in addition to the grand average of all the data, the printout also includes the average of each of the three blocks of data. These are of no value to the experimenter.

The main effects use only the AVE values, as in the previous cases. The *SS* value is not used. The estimates of two-factor interactions are labeled with numbers. If a two-factor interaction column is significant, then the actual confounding is obtained from the proper subsection in Chapter 13 Appendix 2. (Note: in the John's three-quarter fractional factorial analysis, the numbers associated with various interactions are *not* contrast columns; they are simply identifying numbers.)

In the example above, there are three contrast columns that are not used for main effects or interactions. These three columns are, therefore, listed under ESTIMATES OF ERROR, and the average of the three *SS* numbers can be used as an estimate of σ^2.

EXERCISES

1. Obtain the design matrix for a Resolution V design for seven variables using 0 and 1 for the low and high levels, respectively. Write treatment combinations.
2. Obtain the design matrix for a Resolution IV design for eight variables.

	A	B	C	D	E	F	G	H
High	2	4	6	8	1	2	3	4
low	6	2	4	10	3	5	1	2

3. Obtain the design matrix for a Resolution III design for seven variables using 0 and 1, respectively, for the low and high levels of the variables. Write treatment combinations.
4. Analyze the following data using the statement 16 FAN 7.

   1, 2, 3, 4, 5, 6, 7, 8, 1, 2, 3, 4, 5, 6, 7, 8

Chapter 13 Appendix 1
APL Computer Program Coding

Note: This computer program is not an IBM product and is not guaranteed or maintained by IBM.

```
    ∇ AAA
[1]   LB←LB,LB,LB,LB,LB←'ABCDEFGHIJKLMNOPQRSTUVWXYZ'
[2]   LC←'1234567890'
[3]   →(XT= 8 12 16 20 24 32 36 44 48 60 64 68 72 80 84 128)/L0,L1,L2,L3,L4,L5,L6,L7,L8,L9,M0,M1,M2,M3,M4,M5
[4]   →0,INO←ρA← 1 1 0
[5]  L0:A← 1 1 1 0 1 0 0
[6]   →0,INO← 4 7 5
[7]  L1:→0,A← 1 1 0 1 1 1 0 0 0 1 0
[8]  L2:A← 1 1 1 0 1 0 1 1 0 0 1 0 0 0
[9]   →0,INO← 5 9 6 15 10 7 12 14 8 11
[10] L3:→0,A← 1 1 0 0 1 1 1 1 0 1 0 1 0 0 0 0 1 1 0
[11] L4:→0,A← 1 1 1 1 1 0 1 0 1 1 0 0 1 1 0 0 1 0 1 0 0 0 0
[12] L5:A← 1 1 1 1 1 0 0 1 1 0 1 0 0 1 0 0 0 0 1 0 1 0 1 1 1 0 1 1 0 0 0
[13]  →0,INO← 19 6 20 30 7 21 11 31 8 22 25 15 17 27 24
[14] L6:→0,A← 0 1 0 1 1 1 0 0 0 1 1 1 1 1 0 1 1 1 0 0 1 0 0 0 0 1 0 1 0 1 1 0 0 1 0
[15] L7:→0,A← 1 1 0 0 1 0 1 0 0 1 1 1 0 1 1 1 1 1 0 0 0 1 0 1 1 1 0 0 0 0 0 1 0 0 0 1 1 0 1 0 1 1 0
[16] L8:→0,A← 1 1 1 1 1 0 1 1 1 1 0 0 1 0 1 0 1 1 1 0 0 1 0 0 1 1 0 1 1 0 0 0 1 0 1 0 1 1 0 0 0 0 1 0 0 0 0
[17] L9:A← 1 1 0 1 1 1 0 1 0 1 0 0 1 0 0 1 1 1 0 1 1 1 1 0 0 1 1 1 1 1 0 0 0 0 0 1 1 0 0 0 0 1 0 0 0 1 1 0
[18]  →0,A←A, 1 1 0 1 0 1 0 0 0 1 0
[19] M0:A← 1 1 1 1 1 1 0 1 0 1 0 1 1 0 0 1 1 0 1 1 1 0 1 1 0 1 0 0 1 0 0 1 1 1 0 0 0 1 0 1 1 1 1 0 0 1 0 1 0 0
[20]  A←A, 0 1 1 0 0 0 0 1 0 0 0 0 0
[21]  →0,INO← 7 13 8 33 14 9 25 34 15 10 63 26 35 16 11 28 49 36 27 57 41 51 44 62 60 32 53 43
[22] M1:A← 1 1 0 0 1 0 1 0 0 1 1 0 0 0 1 1 1 1 0 1 0 1 1 1 1 1 1 0 0 1 0 0 0 1 0 1 1 1 0 1 1 0 0 0 0 0 0 1 0 1
[23]  →0,A←A, 0 0 0 0 1 1 1 0 0 1 1 0 1 0 1 1 0
[24] M2:A← 1 1 1 1 1 1 1 0 1 1 1 0 1 0 0 1 1 0 1 1 1 0 0 0 1 1 0 1 0 1 1 0 1 0 0 0 1 1 1 0 1 0 0 1 0 1 0 0 1 1
[25]  →0,A←A, 1 0 0 0 1 0 0 1 1 0 1 0 0 0 1 0 0 0 0 0 0
[26] M3:A← 1 1 1 0 1 1 0 0 1 1 1 1 0 1 0 0 1 0 1 1 1 1 1 1 0 1 1 0 0 0 0 1 1 0 0 0 1 0 1 0 1 0 1 0 1 1 1 0 0 1
[27]  →0,A←A, 1 1 1 0 0 1 0 0 0 0 0 0 1 0 1 1 0 1 0 0 0 0 1 1 0 0 1 0 0
[28] M4:A← 1 1 0 1 1 0 0 1 0 1 1 1 1 0 0 0 1 1 0 0 0 1 0 1 0 1 1 1 1 1 1 1 0 1 0 0 1 1 1 0 1 1 0 0 1 0 0 0 1 1
[29]  →0,A←A, 0 1 0 0 0 0 0 0 0 1 0 1 0 1 1 1 0 0 1 1 1 0 0 0 0 1 0 1 1 0 0 1 0
[30] M5:A← 1 1 1 1 1 1 1 0 1 0 1 0 1 0 0 1 1 0 0 1 1 1 0 1 1 1 0 1 0 0 1 0 1 1 0 0 0 1 1 0 1 1 1 1 0 1 1 0 1 0
[31]  A←A, 1 1 0 1 1 0 0 1 0 0 1 0 0 0 1 1 1 0 0 0 0 1 0 1 1 1 1 1 0 0 1 0 1 0 1 1 1 0 0 1 1 0 1 0 0
[32]  A←A, 0 1 0 0 1 1 1 1 0 0 0 1 0 1 0 0 0 0 1 1 0 0 0 0 0 1 0 0 0 0 0 0
[33]  INO← 8 15 9 64 16 10 29 65 17 11 55 30 66 18 12 127 56 31 67 19 13 48 59 92 33 58 49 119 117 43 95 82
[34]  INO←INO, 90 85 93 28 40 105 120 36 116 71 51 96 73 61 26 52 124 126 54 88 103 69 97
    ∇
    ∇ CHECK
[1]   →(NV≥NT)/L1
[2]   E←NV≥ 2 3 3 4 4 11 9 5 19 23 7 31 6 35 37 41 43 9 7
[3]   →(0=+/(NT= 4 6 8 12 16 20 24 32 36 44 48 60 64 68 72 80 84 96 128)×E)/L1
[4]   →(+/(NT= 48 96)×(NV≤ 16 32)∧NV≥ 9 12)/L1
[5]   →(+/(NT= 6 96)×NV> 3 48)/L1
[6]   →0,CK←3
[7]  L1:'NOT PROGRAMMED FOR ';NT;' TRIALS ';NV;' VARIABLES'
[8]   →CK←0
    ∇
    ∇ DATA
[1]  L1:'   ENTER LOWER LEVEL (BV) OR NV ρ0'
[2]   →(NV≠ρBV←⎕)/L1
[3]  L2:'ENTER UPPER LEVEL (PS) OR NV ρ1'
[4]   →(NV≠ρPS←⎕)/L2
[5]   LL←' NT '
[6]   E←P
[7]   F←S
[8]   →((2×NV)≠(+/0=BV)++/1=PS)/M2
```

```
[9]    LA←(E←NV)↑LB
[10]   LAA←NV↑(26ρ' '),(26ρ'1'),(26ρ'2'),(26ρ'3'),26ρ'4'
[11]   →(XT=NT)/M1
[12]   LAA←'    ',LAA
[13] M1:'      ',LAA
[14]   →L3,F←1
[15] M2:LAB F
[16] L3:→(XT=NT)/L4
[17]   LL←LL,'XT '
[18] L4:NO←0
[19] M0:L←E⌊NV-NO
[20]   LL,LA[(F×NO)+⍳F×L]
[21]   J←I←0
[22] L5:VV←BV
[23]   →(XT=I←I+1)/L6
[24]   →(0=+/B=I)/L5
[25]   T←(-I)⌽⍉A
[26]   VV←(BV×~T[VNO])+T[VNO]×PS
[27] L6:K←I
[28]   →(XT=NT)/L7
[29]   K←(J←J+1),I
[30] L7:(3 0 ⍕K),' ',F PD VV[NO+⍳L]
[31]   →(XT>I)/L5
[32]   ' '
[33]   ' '
[34]   →((NO←NO+L)≠NV)/M0
     ∇
     ∇ NT DES NV
[1]    CHECK
[2]    →CK
[3]    VNO←NV↑VS←⍳NT-1+TE←0
[4]    B←⍳XT←NT
[5]    →(NT<2×NV)/L1
[6]    →(0=+/NT= 6 12 24 48 96)/L1
[7]    XT←NT×4÷3
[8]  L1:AAA
[9]    →(NT<2×NV)/M4
[10]   B←((~(H←(¯1⌽A)≠¯2⌽A)×G←A≠¯2⌽A)/⍳XT-1),XT
[11]   →(NT= 6 12 24 48 96)/L3,L5,L7,L8,M1
[12]   B←⍳XT
[13]   VNO←NV↑VS←((-(⍟XT)÷⍟2)⌽⍉A)/⍳XT-1
[14]   S LBD TE←(~(⍳NT-1)∊VS)/⍳NT-1
[15]   →((NT= 8 16 32 64 128)×NV> 3 5 6 8 11)/M4
[16]   TE←(+/⍳NV-1)↑INO
[17]   LBI S
[18]   →(NT= 4 8 32 64 128)/M4,L4,L2,M0,M3
[19]   →M4,VNO←NV↑VS← 1 2 3 4 13
[20] L2:→M4,VNO←NV↑VS← 1 2 3 4 5 16
[21] L3:CN← 6 4
[22] L4:→M4,VNO←NV↑VS← 1 2 3
[23] L5:→(NV=4)/L6
[24]   VNO←NV↑VS← 1 2 3 4 14 12
[25]   →M4,CN← 11 8 5 13
[26] L6:B←((~(G←¯4⌽A)×H←¯11⌽A)/⍳XT-1),XT
[27]   →M4,VNO←VS← 1 2 3 4
```

```
[28] L7:VNO←NV↑VS← 1 2 3 4 5 14 9 23 26 13 29 10
[29]  →M4,CN← 12 28 18 16 19 24 27 17
[30] L8:→(NV≤8)/L9
[31]  VNO←NV↑VS← 1 2 3 4 5 6 29 38 51 30 49 53 32 46 24 62 60 28 37 18 50 54 47 58
[32]  →M4,CN← 27 36 17 12 57 41 44 22 7 19 55 48 39 52 31 59
[33] L9:B←((~(G←¯16⌽A)×H←¯37⌽A)/⍳XT-1),XT
[34] M0:→M4,VNO←NV↑VS← 1 2 3 4 5 6 19 23
[35] M1:→(NV≤11)/M2
[36]  VS← 1 2 3 4 5 6 7 59 34 70 93 115 111 60 35 94 61 50 91 113 96 63 109 44 28 88 54 126 83 104
[37]  VNO←NV↑VS←VS, 86 52 124 58 33 69 21 92 114 97 110 45 89 49 119 81 102 84
[38]  CN← 57 32 68 20 14 48 118 38 80 77 107 101 74 42 26 122 8 22 98 46 90 71 116
[39]  →M4,CN←CN, 112 36 95 62 120 82 103 85 51
[40] M2:B←((~(G←¯7⌽A)×H←¯49⌽A)/⍳XT-1),XT
[41] M3:VNO←NV↑VS← 1 2 3 4 5 6 7 23 77 101 125
[42] M4:→(Q=2)/0
[43]  DATA
    ∇
    ∇ DESCRIBE
[1]   'THIS WORKSPACE PROVIDES DESIGNS AND/OR ANALYSIS OF TWO LEVEL'
[2]   'FACTORIAL EXPERIMENTS.  THE POSSIBLE EXPERIMENTS THAT CAN BE '
[3]   'DESIGNED ARE SHOWN IN THE TABLE BELOW'
[4]   ' '
[5]   '       NUMBER OF VARIABLES FOR A GIVEN RESOLUTION'
[6]   'NUMBER    WORKABLE RANGE        SUGGESTED RANGE'
[7]   '  OF        RESOLUTION             RESOLUTION'
[8]   'TRIALS V+  V    IV     III    V+  V    IV     III'
[9]   '    4   2                3     2                3'
[10]  '    6             3                      3'
[11]  '    8   3         4    5-7     3         4    5-7'
[12]  '   12      4    5-6   7-11        4    5-6   8-11'
[13]  '   16   4  5    6-8   9-15     4  5    7-8  12-15'
[14]  '   20                11-19                  16-19'
[15]  '   24          9-12  13-23             9-12 20-23'
[16]  '   32   5  6    7-16 17-31     5  6   13-16 24-31'
[17]  '   36                19-35                  32-35'
[18]  '   44                23-43                  36-43'
[19]  '   48     7-8  17-24 25-47       7-8  17-24 44-47'
[20]  '   60                31-59                  48-59'
[21]  '   64   6 7-8   9-32 33-63     6      25-32 60-63'
[22]  '   68                35-67                  64-67'
[23]  '   72                37-71                  68-71'
[24]  '   80                41-79                  72-79'
[25]  '   84                43-83                  80-83'
[26]  '   96     9-11 33-48             9-11 33-48'
[27]  '  128   7 8-11 12-64 65-127    7      49-64 84-127'
[28]  ' '
[29]  'RESOLUTION'
[30]  'V+ :  COMPLETE FACTORIAL'
[31]  ' '
[32]  'V  :  ALL MAIN EFFECTS AND TWO FACTOR INTERACTIONS ARE UNCONFOUNDED'
[33]  '      WITH OTHER MAIN EFFECTS OR TWO FACTOR INTERACTIONS'
[34]  ' '
[35]  'IV :  ALL MAIN EFFECTS ARE UNCONFOUNDED BY TWO FACTOR INTERACTIONS'
[36]  '      TWO FACTOR INTERACTIONS ARE CONFOUNDED WITH EACH OTHER'
[37]  ' '
```

```
[38]  'III:  MAIN EFFECTS CONFOUNDED WITH TWO FACTOR INTERACTIONS'
[39]  ' '
[40]  'TO USE THIS WORKSPACE TYPE START TO SET PAGE FORMAT CONTROLS'
[41]  'THEN TYPE  NT DES NV  FOR DESIGNS  NT FAN NV  FOR ANALYSIS'
[42]  'NT=NO. OF TRIALS  NV=NO. OF VARIABLES'
[43]  ' '
[44]  'FOR FURTHER DESCRIPTION AND TABLES OF CONTRAST ASSIGNMENTS '
[45]  'SEE TR 77.0007'
[46]  'CONTACT FRANCIS CLARKE 263-4930 DEPT 509 BLD 023 BOULDER'
[47]  '   OR   BILL DIAMOND   263-5730 DEPT 509 BLD 023 BOULDER'
[48]  'FOR HELP, PROBLEMS, OR SUGGESTIONS'
[49]  'CHANGED MAR. 17 1975, CHANGED APRIL 8,1975'
    ∇
    ∇ NT FAN NV
[1]    CHECK
[2]    →CK,SM←SI←NVρ1
[3]   L1:'ENTER RESULTS (W) IN STANDARD ORDER'
[4]    →(NT≠ρW←⎕)/L1
[5]    Q←2
[6]    NT DES NV
[7]    Q←1
[8]    'AVE',S PO+/W÷NT
[9]    LX←ρVS
[10]   →(NT=XT)/L2
[11]   XB2←(B∊B2←(G×~H)/⍳XT-1)/⍳NT
[12]   XB3←(B∊B3←(H×~G)/⍳XT-1)/⍳NT
[13]   XB4←(B∊B4←(((~G)×~H)/⍳XT-1),XT)/⍳NT
[14]   'AVE2',(S PO+/W[XB2]÷XT÷4),' AVE3',(S PO+/W[XB3]÷XT÷4),' AVE4',S PO+/W[XB4]÷XT÷4
[15]   JAN
[16]   SM←NV↑SM
[17]   SI←LY↑SI
[18]   →L6
[19]  L2:TF←R←⍳I←0
[20]  L3:T←(I⌽A),0
[21]   R←R,(+/W×T-~T)÷NT÷2
[22]   →((1-NT)<I←I-1)/L3
[23]   LY←ρ,TE
[24]   RR←R[VS]
[25]   →(TE=0)/L5
[26]   RR←RR,R[TE]
[27]  L4:→(LX=NT÷2)/L5
[28]   RR←RR,R[TF←(~(⍳NT-1)∊VS,TE)/⍳NT-1]
[29]  L5:R←RR
[30]  L6:PJ
    ∇
    ∇ JAN
[1]    S LBD⍳LY←NT-3+J←0
[2]    →(+/(NT= 6 12 24 48 96)×NV> 0 4 0 8 11)/L5
[3]    LY←+/⍳NV-1
[4]    SM←11ρ1
[5]    SI←55ρ1
[6]    LBI S
[7]    →(NT= 48 96)/L1,L3
[8]    RET 13 11 8 14
[9]    R←R[2 1 6 9 11 8 7 5 4 3 12]
```

```
[10]  SI[1]←4÷3
[11]  →0,R[5]←R[5]+0.5×R[11]
[12] L1:RET 58 56 22 39 21 31 12 18 37 61 40 52 48 20 42 59
[13]  RR←R
[14]  RR[E]←R[E← 2 29 35 9 33 42]-0.5×R[F← 3 30 36 8 32 41]
[15]  RR[E]←R[E← 1 28 34 7 31 40 20 21 23 24 26 27]+0.5×R[F, 19 19 22 22 25 25]
[16]  SM[1 3 7 8]←4÷3
[17]  SI[2 3 4 5 8 9 11 14 15 17 19 26 27 28]←4÷3
[18]  →(NV=8)/L2
[19]  RR[E]←RR[E← 4 10 37 43 5 12 38 45]+0.5×RR[F],-RR[F← 6 11 39 44]
[20]  RR[E]←RR[E← 23 27 28 31]+((-RR[24 26]),RR[29 33])÷3
[21]  RR[E]←R[E← 24 26 29 33]
[22] L2:R←RR[1 5 7 12 14 18 20 24 13 27 28 31 23 16 4 2 34 37 40 43 10 9 21 45 42 38 35 17 15]
[23]  SM[2 4]←4÷3
[24]  SI[6+ι13]←4÷3
[25]  →0,R←R,RR[39 11 44 6 33 29 26 3 8 19 22 25 30 32 36 41 47 48]
[26] L3:CN← 121 57 32 68 20 14 39 115 60 106 84 102 22 81 42 107 104 46 38 83 112 118 62
[27]  RET CN, 109 113 44 72 63 79 45 110 122
[28]  RR←R
[29]  RR[(F+1),F+2]←R[(F+1),F+2]+0.5×R[F],R[F← 4 7 13 16 19 28]
[30]  RR[(F-2),F-1]←R[(F-2),F-1]+0.5×R[F],-R[F← 3 42 45 48 57 66 78 81 84 87 96]
[31]  RR[F+1]←R[F+1]+(R[F-1]+R[F← 53 62 71 74 89 92])÷3
[32]  SM[1 2 3 4 6 7 8 11]←4÷3
[33]  SI[1 7 8 9 10 15 20 29 30 31 32 34 35 36 37 38 39 40 42 43 44 50 54 55]←4÷3
[34]  SI[13 19 24 25 51 52]←1.5
[35]  →(NV=11)/L4
[36]  RR[F]←R[F← 10 31 34 12 32 35]+0.5×R[E],-R[E← 11 33 36]
[37]  RR[F]←R[F← 79 85 21 39 29]+(R[80 86 19 37 28]+R[81 87],-R[20 38 30])÷3
[38]  RR[F]←R[F← 20 30 80 86 90 93]
[39]  SM[5 8]←(4÷3),1.5
[40]  SI[2 3 4 5 45]←4÷3
[41]  SI[6 26 36 44]←1.5
[42]  →(NV=10)/L4
[43]  RR[F]←R[F← 23 24 25 26]+0.5×R[22 22 27],-R[27]
[44]  RR[F]←R[F← 46 55 6 12 18]+(R[47 56 4 10 16]+R[48 57],-R[5 11 17])÷3
[45]  RR[F]←R[F← 5 10 11 17 47 56 82 83 85 86]
[46]  SM[9 3 5 7]←(4÷3), 1.5 1.5 1.5
[47]  SI[14 18 33 10 15]←(4÷3),(4÷3),(4÷3),1.5,1.5
[48] L4:R←RR[1 2 6 9 12 15 18 21 24 27 30 95 31 32 34 35 39 40 41 43 46 50 49 54 23 55 59 58 26 63 64 69]
[49]  R←R,RR[68 67 72 75 29 60 51 76 77 65 14 25 8 44 79 82 83 17 56 22 47 5 85 10 37 38 36 33 20 90 93 11]
[50]  →0,R←R,RR[86 80 3 4 7 13 16 19 28 42 45 48 52 53 57 61 62 66 70 71 73 74 78 81 84 87 88 89 91 92 96]
[51] L5:RET CN
[52] L6:→(((NT÷2)-NV)<J←J+1)/L7
[53]  R[E-1]←R[E-1]-0.5×R[E←(NT÷2)-3×J-1]
[54]  →L6,R[E-2]←R[E-2]+0.5×R[E]
[55] L7:RR←(ρVS)ρJ←0
[56]  RR[3,3+ι(XT÷8)-1]←R[¯2+E←3×ιXT÷8]
[57]  RR[1,2+(XT÷8)+ι(XT÷8)-1]←R[E-1]
[58]  RR[2,3+(2×¯1+XT÷8)+ι(XT÷8)-1]←R[E]
[59]  RR←RR,R[LX+2]-R[LX+3]
[60]  RR←RR,R[LX+1+ι2]
[61] M1:→(((LY-LX)÷3)<J←J+1)/M2
[62]  RR←RR,R[(E←LX+3×J)+ι3]
[63]  →M1,RR←RR,(R[E+2]-R[E+3]),R[E+1]+R[E+3],R[E+2]
[64] M2:R←RR
```

```
[65]   J←0
[66]  M3:→(((NT÷2)-NV)<J←J+1)/0
[67]   →M3,SM[((NT÷6)+3-J),(NT÷3)+2-J]←4÷3
     ∇
     ∇ LAB A
[1]    →(NV>26)/L1
[2]    LC←'
[3]   L1:LA←⍳J←0
[4]   L2:J←J+1
[5]    I←0
[6]   L3:→(26<I←I+1)/L2
[7]    LA←LA,((A-2)⍴' '),LB[I],LC[J]
[8]    →(NV>I+26×J-1)/L3
     ∇
     ∇ A LBD LZ
[1]    LD←⍳¯1+I←1
[2]   L1:LD←LD,((A-2+XT=128)⍴' '),((2+XT=128),0)⍕LZ[I]
[3]    →((⍴LZ)≥I←I+1)/L1
     ∇
     ∇ LBI A
[1]    LD←⍳¯2+I←2
[2]   L1:J←1
[3]   L2:LD←LD,((A-2)⍴' '),LB[J],LB[I]
[4]    →(I>J←J+1)/L2
[5]    →(NV≥I←I+1)/L1
     ∇
     ∇ Z←Y PD X;TT;RR;I
[1]    →(Y≤2)/L4
[2]    →(0=⌈/|X)/L4
[3]    RR←⍳I←0
[4]   L1:→((⍴,X)<I←I+1)/L3
[5]    TT←(Y,0)⍕X[I]
[6]    →(0=X[I])/L2
[7]    TT←(Y,(Y-2)⌊0⌈⌈Y-4+10⍟|X[I])⍕X[I]
[8]   L2:RR←RR,TT
[9]    →L1
[10]  L3:Z←RR
[11]   →0
[12]  L4:Z←(Y,0)⍕X
     ∇
     ∇ PJ
[1]    ' '
[2]    'MAIN EFFECTS'
[3]    LAB S
[4]    LA PRINT R[⍳NV]
[5]    ' '
[6]    →(NV=NT-1)/0
[7]    →(NV>NT÷2)/L1
[8]    'ESTIMATES OF TWO FACTOR INTERACTIONS'
[9]    LD PRINT R[LX+⍳LY]
[10]   ' '
[11]   →((NV=5)∧NT=16)/0
[12]   →(NV=NT÷2)/0
[13]  L1:'ESTIMATES OF ERROR'
[14]   TD←LZ←RR←⍳0
```

```
[15]   →(LX=NV)/L2
[16]   TD←(~VS∊VNO)/VS
[17]   RR←R[LZ←NV+⍳LX-NV]
[18]  L2:→(LX=NT÷2)/L3
[19]   →(LX=NT-1)/L5
[20]   RR←R[LZ←LZ,LX+LY+⍳NT-1+LX+LY]
[21]  L3:S LBD LZ
[22]   →(XT≠NT)/L4
[23]  L5:S LBD TD,TF
[24]  L4:LD PRINT RR
     ∇
     ∇ Z←Y PO X
[1]    →(Y≤2)/L1
[2]    →(0=⌈/|X)/L1
[3]    Z←(Y,(Y-2)⌊0⌈⌈Y-4+10⍟⌈/|X)⍕X←,X
[4]    →0
[5]   L1:Z←(Y,0)⍕X←,X
     ∇
     ∇ G PRINT F;J
[1]    SF←(⍴F)⍴1
[2]    →(NT=XT)/L2
[3]    →(NV≠⍴F)/L3
[4]    →L2,SF←SM
[5]   L3:→((⍴F)≠+/⍳NV-1)/L2
[6]    SF←SI
[7]   L2:NO←0
[8]   L1:J←P⌊(⍴F)-NO
[9]    ' '
[10]   '    ',G[(S×NO)+⍳S×J]
[11]   'AVE',S PO F[E←NO+⍳J]
[12]   'S S',S PO(F[E]*2)×SF[E]×XT÷4+4×XT≠NT
[13]   →((NO←NO+J)≠⍴F)/L1
     ∇
     ∇ RET E
[1]    R←⍳¯1+I←1
[2]   L1:T←((1-E[I])⌽A),0
[3]    R←R,(+/W[XB2,XB3]×(~T[B2,B3])-T[B2,B3])÷XT÷4
[4]    R←R,(+/W[XB2,XB4]×T[B2,B4]-~T[B2,B4])÷XT÷4
[5]    R←R,(+/W[XB3,XB4]×T[B3,B4]-~T[B3,B4])÷XT÷4
[6]    →((NT÷3)≥I←I+1)/L1
     ∇
     ∇ SAVE
     ∇
     ∇ START
[1]    ⎕PW←129+Q←1
[2]    'ENTER PAGE WIDTH'
[3]    F←⎕
[4]    'ENTER NUMBER OF COLUMNS ACROSS PAGE'
[5]    S←⌊(F-6)÷P←⎕
[6]    'ENTER NT DES NV FOR DESIGN OR NT FAN NV FOR ANALYSIS'
[7]    'NT = NO OF TRIALS    NV = NO OF VARIABLES'
     ∇
```

Chapter 13 Appendix **2**

Confounding of Two-Factor Interactions in Resolution IV Design

5 and 6 Variables in 12 Trials

Interaction Column	5 Variables	6 Variables
1	AC+DE−BC	−DF
2	AC+DE+AB	+EF
3	BC+AB	DF+EF
4	CD+AE−BE	BF−AF
5	AD+CE+BE	AF
6	BD+BE	CF+AF
7	AD+CE−BD	−CF
8	CD+AE+BD	BF+CF
9	CD+AE+AD	BF+CE

6 to 12 Variables in 24 Trials

Interaction Column	6 Variables	7 Variables	8 Variables	9 Variables	10 Variables	11 Variables	12 Variables
1	AC−BC	DG	EH	FI	−DJ	−EK	−FL
2	AC+AB	DG	EH	FI	GJ	HK	IL
3	BC+AB				DJ+GJ	EK+HK	FL+IL
4	CD+EF	AG−BG		HI	BJ−AJ	−IK	KL−HL
5	AD	CG+BG	FH	EI	AJ	IK	HL
6	BD	BG			CJ+AJ	FK+IK	EL+HL
7	AD−BD	CG	FH	EI	−CJ	−FK	−EL
8	CD+BD+EF	AG		HI	BJ+CJ	FK	KL+EL
9	CD+AD+EF	AG+CG	FH	HI+EI	BJ		KL
10	CE+DF		AH−BH	GI	−IJ	BK−AK	JL−GL
11	AE	FG	CH+BH	DI	IJ	AK	GL
12	BE		BH		FJ+IJ	CK+AK	DL+GL
13	AE−BE	FG	CH	DI	−FJ	−CK	−DL
14	CE+BE+DF		AH	GI	FJ	BK+CK	JL+DL
15	CE+AE+DF	FG	AH+CH	GI+DI		BK	JL
16	DE+CF		GH	AI−BI	−HJ	JK−GK	BL−AL
17	AF	EG	DH	CI+BI	HJ	GK	AL
18	BF			BI	EJ+HJ	DK+GK	CL+AL
19	AF−BF	EG	DH	CI	−EJ	−DK	−CL
20	DE+BF+CF		GH	AI	EJ	JK+DK	BL+CL
21	DE+AF+CF	EG	GH+DH	AI+CI		JK	BL

9 to 24 Variables in 48 Trials

Column	9 Var.	10 Var.	11 Var.	12 Var.	13 Var.	14.Var.	15 Var.	16 Var.	17 Var.	18 Var.	19 Var.	20 Var.	21 Var.	22 Var.	23 Var.	24 Var.
1	AC–BC		DK	EL	FM	GN	HO	IP	JQ	–DR	–ES	–FT	–GU	–HV	–IW	–JX
2	AC+AB		DK	EL	FM	GN	HO	IP	JQ	KR	LS	MT	NU	OV	PW	QX
3	BC+AB									DR+KR	+ES+LS	FT+MT	GU+NU	HV+OV	IW+PW	JX+QX
4	CD+EG+FH	IJ	AK–BK	LN			MO		PQ	BR–AR	–NS	–OT	SU–LU	TV–MV	–QW	WX–PX
5	AD		CK+BK	GL	HM	EN	FO	JP	IQ	AR	NS	OT	LU	MV	QW	PX
6	BD		BK							CR+AR	GS+NS	HT+OT	EU+LU	FV+MV	JW+QW	IX+PX
7	AD–BD		CK	GL	HM	EN	FO	JP	IQ	–CR	–GS	–HT	–EU	–FV	–JW	–IX
8	CD+BD+EG+FH	IJ	AK			LN	MO		PQ	BR+CR	GS	HT	SU+EU	TV+FV	JW	WX+IX
9	CD+AD+EG+FH	IJ	CK+AK	GL	HM	LN+EN	MO+FO	JP	PQ+IQ	BR			SU	TV		WX
10	CE+DG+FI	HJ		AL–BL		KN		MP	OQ	–NR	BS–AS	–PT	RU–KU	–QV	TW–MW	VX–OX
11	AE		GK	CL+BL	IM	DN	JO	FP	HQ	+NR	AS	PT	KU	QV	MW	OX
12	BE			BL						GR+NR	CS+AS	IT+PT	DU+KU	JV+QV	FW+MW	HX+OX
13	AE–BE		GK	CL	IM	DN	JO	FP	HQ	–GR	–CS	–IT	–DU	–JV	–FW	–HX
14	CE+BE+DG+FI	HJ		AL		KN		MP	OQ	GR	BS+CS	IT	RU+DU	JV	TW+FW	VX+HX
15	CE+AE+DG+FI	HJ	GK	AL+CL	IM	KN+DN	JO	MP+FP	OQ+HQ		BS		RU		TW	VX
16	CF+DH+EI	GJ			AM–BM		KO	LP	NQ	–OR	–PS	BT–AT	–QU	RV–KV	SW–LW	UX–NX
17	AF		HK	IL	CM+BM	JN	DO	EP	GQ	+OR	+PS	AT	QU	KV	LW	NX
18	BF				BM					HR+OR	IS+PS	CT+AT	JU+QU	DV+KV	EW+LW	GX+NX
19	AF–BF		HK	IL	CM	JN	DO	EP	GQ	–HR	–IS	–CT	–JU	–DV	–EW	–GX
20	CF+BF+DH+EI	GJ			AM		KO	LP	NQ	HR	IS	BT+CT	JU	RV+DV	SW+EW	UX+GX
21	CF+AF+DH+EI	GJ	HK	IL	AM+CM	JN	KO+DO	LP+EP	NQ+GQ			BT		RV	SW	UX
22	DE+CG+HI	FJ		KL		AN–BN		OP	MQ	–LR	RS–KS	–QT	BU–AU	–PV	VW–OW	TX–MX
23	AG		EK	DL	JM	CN+BN	IO	HP	FQ	LR	KS	QT	AU	PV	OW	MX
24	BG					BN				ER+LR	DS+KS	JT+QT	CU+AU	IV+PV	HW+OW	FX+MX
25	AG–BG		EK	DL	JM	CN	IO	HP	FQ	–ER	–DS	–JT	–CU	–IV	–HW	–FX
26	DE+CG+HI+BG	FJ		KL		AN		OP	MQ	ER	RS+DS	JT	BU+CU	IV	VW+HW	TX+FX
27	DE+CG+HI+AG	FJ	EK	KL+DL	JM	AN+CN	IO	OP+HP	MQ+FQ		RS		BU		VW	TX
28	DF+CH+GI	EJ			KM		AO–BO	NP	LQ	–MR	–QS	RT–KT	–PU	BV–AV	UW–NW	SX–LX
29	AH		FK	JL	DM	IN	CO+BO	GP	EQ	MR	QS	KT	PU	AV	NW	LX
30	BH						BO			FR+MR	JS+QS	DT+KT	IU+PU	CV+AV	GW+NW	EX+LX
31	AH–BH		FK	JL	DM	IN	CO	GP	EQ	–FR	–JS	–DT	–IU	–CV	–GW	–EX
32	DF+CH+GI+BH	EJ			KM		AO	NP	LQ	FR	JS	RT+DT	IU	BV+CV	UW+GW	SX+EX
33	DF+CH+GI+AH	EJ	FK	JL	KM+DM	IN	AO+CO	NP+GP	LQ+EQ			RT		BV	UW	SX
34	EF+GH+CI	DJ			LM		NO	AP–BP	KQ	–QR	–MS	ST–LT	–OU	UV–NV	BW–AW	RX–KX
35	AI		JK	FL	EM	HN	GO	CP+BP	DQ	+QR	MS	LT	OU	NV	AW	KX
36	BI							BP		JR+QR	FS+MS	ET+LT	HU+OU	GV+NV	CW+AW	DX+KX
37	AI–BI		JK	FL	EM	HN	GO	CP	DQ	–JR	–FS	–ET	–HU	–GV	–CW	–DX
38	EF+GH+CI+BI	DJ			LM		NO	AP	KQ	JR	FS	ST+ET	HU	UV+GV	BW+CW	RX+DX
39	EF+GH+CI+AI	DJ	JK	FL	LM+EM	HN	NO+GO	AP+CP	KQ+DQ			ST		UV	BW	RX
40	FG+EH+DI	CJ				MN	LO	KP	AQ–BQ	–PR	–OS	–NT	TU–MU	SV–LV	RW–KW	BX–AX
41	*	AJ	IK	HL	GM	FN	EO	DP	CQ+BQ	PR	OS	NT	MU	LV	KW	AX
42	*	BJ							BQ	IR+PR	HS+OS	GT+NT	FU+MU	EV+LV	DW+KW	CX+AX
43	*	AJ–BJ	IK	HL	GM	FN	EO	DP	CQ	–IR	–HS	–GT	–FU	–EV	–DW	–CX
44	FG+EH+DI	CJ+BJ				MN	LO	KP	AQ	IR	HS	GT	TU+FU	SV+EV	RW+DW	BX+CX
45	FG+EH+DI	CJ+AJ	IK	HL	GM	MN+FN	LO+EO	KP+DP	AQ+CQ				TU	SV	RW	BX

12 to 48 Variables in 96 Trials

		M	N	O	P	Q	R	S	T	U	V
1	AC–BC							DS	ET	FU	GV
2	AC+AB							DS	ET	FU	GV
3	BC+AB										
4	CD+EH+FI+GJ		KN	LO	MP		QR	AS–BS			
5	AD							CS+BS	HT	IU	JV
6	BD							BS			
7	AD–BD							CS	HT	IU	JV
8	CD+EH+FI+GJ+BD		KN	LO	MP		QR	AS			
9	CD+EH+FI+GJ+AD		KN	LO	MP		QR	AS+CS	HT	IU	JV
10	CE+DH+FK+GL		IN	JO		MQ	PR		AT–BT		
11	AE							HS	CT+BT	KU	LV
12	BE								BT		
13	AE–BE							HS	CT	KU	LV
14	CE+DH+FK+GL+BE		IN	JO		MQ	PR		AT		
15	CE+DH+FK+GL+AE		IN	JO		MQ	PR	HS	AT+CT	KU	LV
16	CF+DI+EK	GM	HN		JP	LQ	OR			AU–BU	
17	AF							IS	KT	CU+BU	MV
18	BF									BU	
19	AF–BF							IS	KT	CU	MV
20	CF+DI+EK+BF	GM	HN		JP	LQ	OR			AU	
21	CF+DI+EK+AF	GM	HN		JP	LQ	OR	IS	KT	AU+CU	MV
22	CG+DJ+EL	FM		HO	IP	KQ	NR				AV–BV
23	AG							JS	LT	MU	CV+BV
24	BG										BV
25	AG–BG							JS	LT	MU	CV
26	CG+DJ+EL+BG	FM		HO	IP	KQ	NR				AV
27	CG+DJ+EL+AG	FM		HO	IP	KQ	NR	JS	LT	MU	AV+CV
28	DE+CH+IK+JL		FN	GO		PQ	MR		ST		
29	AH							ES	DT	NU	OV
30	BH										
31	AH–BH							ES	DT	NU	OV
32	DE+CH+IK+JL+BH		FN	GO		PQ	MR		ST		
33	DE+CH+IK+JL+AH		FN	GO		PQ	MR	ES	ST+DT	NU	OV
34	DF+CI+HK	JM	EN		GP	OQ	LR			SU	
35	AI							FS	NT	DU	PV
36	BI										
37	AI–BI							FS	NT	DU	PV
38	DF+CI+HK+BI	JM	EN		GP	OQ	LR			SU	
39	DF+CI+HK+AI	JM	EN		GP	OQ	LR	FS	NT	SU+DU	PV
40	DG+GJ+HL	IM		EO	FP	NQ	KR				SV
41	AJ							GS	OT	PU	DV
42	BJ										
43	AJ–BJ							GS	OT	PU	DV
44	DG+CJ+HL+BJ	IM		EO	FP	NQ	KR				SV
45	DG+CJ+HL+AJ	IM		EO	FP	NQ	KR	GS	OT	PU	SV+DV
46	EF+HI+CK	LM	DN		OP	GQ	JR			TU	
47	AK							NS	FT	EU	QV
48	BK										
49	AK–BK							NS	FT	EU	QV
50	EF+HI+CK+BK	LM	DN		OP	GQ	TR			TU	
51	EF+HI+CK+AK	LM	DN		OP	GQ	JR	NS	FT	TU+EU	QV
52	EG+HJ+CL	KM		DO	NP	FQ	IR				TV
53	AL							OS	GT	QU	EV
54	BL										
55	AL–BL							OS	GT	QU	EV
56	EG+HJ+CL+BL	KM		DO	NP	FQ	IR				TV
57	EG+HJ+CL+AL	KM		DO	NP	FQ	IR	OS	GT	QU	TV+EV
58	FG+IJ+KL	CM		NO	DP	EQ	HR				UV
59	•	AM						PS	QT	GU	FV
60	•	BM									
61	•	AM–BM						PS	QT	GU	FV
62	FG+IJ+KL	CM+BM		NO	DP	EQ	HR				UV
63	FG+IJ+KL	CM+AM		NO	DP	EQ	HR	PS	QT	GU	UV+FV
64	FH+EI+DK		CN	MO	LP	JQ	GR				
65	•		AN					KS	IT	HU	RV
66	•		BN								
67	•		AN–BN					KS	IT	HU	RV
68	FH+EI+DK		CN+BN	MO	LP	JQ	GR				
69	FH+EI+DK		CN+AN	MO	LP	JQ	GR	KS	IT	HU	RV
70	GH+EJ+DL		MN	CO	KP	IQ	FR				
71	•			AO				LS	JT	RU	HV
72	•			BO							
73	•			AO–BO				LS	JT	RU	HV
74	GH+EJ+DL		MN	CO+BO	KP	IQ	FR				
75	GH+EJ+DL		MN	CO+AO	KP	IQ	FR	LS	JT	RU	HV
76	GI+FJ	DM	LN	KO	CP	HQ	ER				
77	•				AP			MS	RT	JU	IV
78	•				BP						
79	•				AP–BP			MS	RT	JU	IV
80	GI+FJ	DM	LN	KO	CP+BP	HQ	ER				
81	GI+FJ	DM	LN	KO	CP+AP	HQ	ER	MS	RT	JU	IV
82	GK+FL	EM	JN	IO	HP	CQ	DR				
83	•					AQ		RS	MT	LU	KV
84	•					BQ					
85	•					AQ–BQ		RS	MT	LU	KV
86	GK+FL	EM	JN	IO	HP	CQ+BQ	DR				
87	GK+FL	EM	JN	IO	HP	CQ+AQ	DR	RS	MT	LU	KV
88	JK+IL	HM	GN	FO	EP	DQ	CR				
89	•						AR	QS	PT	OU	NV
90	•						BR				
91	•						AR–BR	QS	PT	OU	NV
92	JK+IL	HM	GN	FO	EP	DQ	CR+BR				
93	JK+IL	HM	GN	FO	EP	DQ	CR+AR	OS	PT	OU	NV

12 to 48 Variables in 96 Trials (*continued*)

	W	X	Y	Z	A*	B*	C*	D*	E*	F*	G*	H*	I*
1	HW	IX	JY	KZ	LA*	MB*	NC*	OD*	PE*	QF*	RG*	−DH*	−EI*
2	HW	IX	JY	KZ	LA*	MB*	NC*	OD*	PE*	QF*	RG*	SH*	TI*
3												DH*+SH*	EI*+TI*
4	TW	UX	VY				ZC*	A*D*	B*E*		F*G*	BH*−AH*	−WI*
5	EW	FX	GY	NZ	OA*	PB*	KC*	LD*	ME*	RF*	QG*	AH*	WI*
6												CH*+AH*	HI*+WI*
7	EW	FX	GY	NZ	OA*	PB*	KC*	LD*	ME*	RF*	QG*	−CH*	−HI*
8	TW	UX	VY				ZC*	A*D*	B*E*		F*G*	BH*+CH*	HI*
9	TW+EW	UX+FX	VY+GY	NZ	OA*	PB*	ZC*+KC*	A*D*+LD*	B*E*+ME*	RF*	F*G*+QG*	BH*	
10	SW			UZ	VA*		XC*	YD*		B*F*	E*G*	−WH*	BI*−AI*
11	DW	NX	OY	FZ	GA*	QB*	IC*	JD*	RE*	MF*	PG*	WH*	AI*
12												HH*+WH*	CI*+AI*
13	DW	NX	OY	FZ	GA*	QB*	IC*	JD*	RE*	MF*	PG*	−HH*	−CI*
14	SW			UZ	VA*		XC*	YD*		B*F*	E*G*	HH*	BI*+CI*
15	SW+DW	NX	OY	UZ+FZ	VA*+GA*	QB*	XC*+IC*	YD*+JD*	RE*	B*F*+MF*	E*G*+PG*		BI*
16		SX		TZ		VB*	WC*		YE*	A*F*	D*G*	−XH*	−ZI*
17	NW	DX	PY	EZ	QA*	GB*	HC*	RD*	JE*	LF*	OG*	XH*	ZI*
18												IH*+XH*	KI*+ZI*
19	NW	DX	PY	EZ	QA*	GB*	HC*	RD*	JE*	LF*	OG*	−IH*	−KI*
20		SX		TZ		VB*	WC*		YE*	A*F*	D*G*	IH*	KI*
21	NW	SX+DX	PY	TZ+EZ	QA*	VB*+GB*	WC*+HC*	RD*	YE*+JE*	A*F*+LF*	D*G*+OG*		
22			SY		TA*	UB*		WD*	XE*	ZF*	C*G*	−YH*	−A*I*
23	OW	PX	DY	QZ	EA*	FB*	RC*	HD*	IE*	KF*	NG*	+YH*	A*I*
24												JH*+YH*	LI*+A*I*
25	OW	PX	DY	QZ	EA*	FB*	RC*	HD*	IE*	KF*	NG*	−JH*	−LI*
26			SY		TA*	UB*		WD*	XE*	ZF*	C*G*	JH*	LI*
27	OW	PX	SY+DY	QZ	TA*+EA*	UB*+FB*	RC*	WD*+HD*	XE*+IE*	ZF*+KF*	C*G*+NG*		
28	AW−BW			XZ	YA*		UC*	VD*		E*F*	B*G*	−TH*	H*I*−SI*
29	CW+BW	KX	LY	IZ	JA*	RB*	FC*	GD*	QE*	PF*	MG*	TH*	SI*
30	BW											EH*+TH*	DI*+SI*
31	CW	KX	LY	IZ	JA*	RB*	FC*	GD*	QE*	PF*	MG*	−EH*	−DI*
32	AW			XZ	YA*		UC*	VD*		E*F*	B*G*	EH*	H*I*+DI*
33	AW+CW	KX	LY	XZ+IZ	YA*+JA*	RB*	UC*+FC*	VD*+GD*	QE*	E*F*+PF*	B*G*+MG*		H*I*
34		AX−BX		WZ		YB*	TC*		VE*	D*F*	A*G*	−UH*	−C*I*
35	KW	CX+BX	MY	HZ	RA*	JB*	EC*	QD*	GE*	DF*	LG*	UH*	C*I*
36		BX										FH*+UH*	NI*+CI*
37	KW	CX	MY	HZ	RA*	JB*	EC*	QD*	GE*	OF*	LG*	−FH*	−NI*
38		AX		WZ		YB*	TC*		VE*	DF*	A*G*	FH*	NI*
39	KW	AX+CX	MY	WZ+HZ	RA*	YB*+JB*	TC+EC*	QD*	VE*+GE*	DF*+OF*	A*G*+LG*		
40			AY−BY		WA*	XB*		TD*	UE*	C*F*	ZG*	−VH*	−D*I*
41	LW	MX	CY+BY	RZ	HA*	IB*	QC*	ED*	FE*	NF*	KG*	VH*	D*I*
42			BY									GH*+VH*	OI*+D*I*
43	LW	MX	CY	RZ	HA*	IB*	QC*	ED*	FE*	NF*	KG*	−GH*	−OI*
44			AY		WA*	XB*		TD*	UE*	C*F*	ZG*	GH*	OI*
45	LW	MX	AY+CY	RZ	WA*+HA*	XB*+IB*	QC*	TD*+ED*	UE*+FE*	C*F*+NF*	ZG*+KG*		
46		WX		AZ−BZ		A*B*	SC*		D*E*	VF*	YG*	−C*H*	−UI*
47	IW	HX	RY	CZ+BZ	MA*	LB*	DC*	PD*	OE*	GF*	JG*	C*H*	UI*
48				BZ								NH*+C*H*	FI*+UI*
49	IW	HX	RY	CZ	MA*	LB*	DC*	PD*	OE*	GF*	JG*	−NH*	−FI*
50		WX		AZ		A*B*	SC*		D*E*	VF*	YG*	NH*	FI*
51	IW	WX+HX	RY	AZ+CZ	MA*	A*B*+LB*	SC*+DC*	PD*	D*E*+OE*	YF*+GF*	YG*+JG*		
52			WY		AA*−BA*	ZB*		SD*	C*E*	UF*	XG*	−D*H*	−VI*
53	JW	RX	HY	MZ	CA*+BA*	KB*	PC*	DD*	NE*	FF*	IG*	D*H*	VI*
54					BA*							OH*+D*H*	GI*+VI*
55	JW	RX	HY	MZ	CA*	KB*	PC*	DD*	NE*	FF*	IG*	−OH*	−GI*
56			WY		AA*	ZB*		SD*	C*E*	UF*	XG*	OH*	GI*
57	JW	RX	WY+HY	MZ	AA*+CA*	ZB*+KB*	PC*	SD*+DD*	C*E*+NE*	UF*+FF*	XG*+IG*		
58			XY		ZA*	AB*−BB*		C*D*	SE*	TF*	WG*	−E*H*	−F*I*
59	RW	JX	IY	LZ	KA*	CB*+BB*	OC*	ND*	DE*	EF*	HG*	E*H*	F*I*
60						BB*						PH*+E*H*	QI*+F*I*
61	RW	JX	IY	LZ	KA*	CB*	OC*	ND*	DE*	EF*	HG*	−PH*	−QI*
62			XY		ZA*	AB*		C*D*	SE*	TF*	WG*	PH*	QI*
63	RW	JX	XY+IY	LZ	ZA*+KA*	AB*+CB*	OC*	C*D*+ND	SE*+DE*	TF*+EF*	WG*+HG*		
64	UW	TX		SZ			AC*−BC*	B*D*	A*E*	YF*	VG*	−ZH*	−XI*
65	FW	EX	QY	DZ	PA*	OB*	CC*+BC*	MD*	LE*	JF*	GG*	ZH*	XI*
66							BC*					KH*+ZH*	II*+XI*
67	FW	EX	QY	DZ	PA*	OB*	CC*	MD*	LE*	JF*	GG*	−KH*	−II*
68	UW	TX		SZ			AC*	B*D*	A*E*	YF*	VG*	+KH*	II*
69	UW+FW	TX+EX	QY	SZ+DZ	PA*	OB*	AC*+CC*	B*D*+MD*	A*E*+LE*	YF*+JF*	VG*+GG*		
70	VW		TY		SA*		B*C*	A*D*−BD*	ZE*	XF*	UG*	−A*H*	−YI*
71	GW	QX	EY	PZ	DA*	NB*	MC*	CD*+BD*	KE*	IF*	FG*	A*H*	YI*
72								BD*				LH*+A*H*	JI*+YI*
73	GW	QX	EY	PZ	DA*	NB*	MC*	CD*	KE*	IF*	FG*	−LH*	−JI*
74	VW		TY		SA*		B*C*	A*D*	ZE*	XF*	UG*	LH*	JI*
75	VW+GW	QX	TY+EY	PZ	SA*+DA*	NB*	B*C*+MC*	A*D*+CD*	ZE*+KE*	XF*+IF*	UG*+FG*		
76		VX	UY			SB*	A*C*	ZD*	AE*−BE*	WF*	TG*	−B*H*	−G*I*
77	QW	GX	FY	OZ	NA*	DB*	LC*	KD*	CE*+BE*	HF*	EG*	B*H*	G*I*
78									BE*			MH*+B*H*	RI*+G*I*
79	QW	GX	FY	OZ	NA*	DB*	LC*	KD*	CE*	HF*	EG*	−MH*	−RI*
80		VX	UY			SB*	A*C*	ZD*	AE*	WF*	TG*	+MH*	RI*
81	QW	VX+GX	UY+FY	OZ	NA*	SB*+DB*	A*C*+LC*	ZD*+KD*	AE*+CE*	WF*+HF*	TG*+EG*		
82				VZ	UA*	TB*	YC*	XD*	WE*	AF*−BF*	SG*	−GH*	−B*I*
83	PW	OX	NY	GZ	FA*	EB*	JC*	ID*	HE*	CF*+BF*	DG*	GH*	B*I*
84										BF*		RH*+GH*	MI*+B*I*
85	PW	OX	NY	GZ	FA*	EB*	JC*	ID*	HE*	CF*	DG*	−RH*	−MI*
86				VZ	UA*	TB*	YC*	XD*	WE*	AF*	SG*	RH*	MI*
87	PW	OX	NY	VZ+GZ	UA*+FA*	TB*+EB*	YC*+JC*	XD*+ID*	WE*+HE*	AF*+CF*	SG*+DG*		
88				YZ	XA*	WB*	VC*	UD*	TE*	SF*	AG*−BG*	−F*H*	−E*I*
89	MW	LX	KY	JZ	IA*	HB*	GC*	FD*	EE*	DF*	CG*+BG*	F*H*	E*I*
90											BG*	QH*+F*H*	PI*+E*I*
91	MW	LX	KY	JZ	IA*	HB*	GC*	FD*	EE*	DF*	CG*	−QH*	−PI*
92				YZ	XA*	WB*	VC*	UD*	TE*	SF*	AG*	QH*	PI*
93	MW	LX	KY	YZ+JZ	XA*+IA*	WB*+HB*	VC*+GC*	UD*+FD*	TE*+EE*	SF*+DF*	AG*+CG*		

12 to 48 Variables in 96 Trials (*continued*)

	J*	K*	L*	M*	N*	O*	P*	Q*	R*	S*	T*	U*	V*
1	−FJ*	−GK*	−HL*	−IM*	−JN*	−KO*	−LP*	−MQ*	−NR*	−OS*	−PT*	−QU*	−RV*
2	UJ*	VK*	WL*	XM*	YN*	ZO*	A*P*	B*Q*	C*R*	D*S*	E*T*	F*U*	G*V*
3	FJ*+UJ*	GK*+VK*	HL*+WL*	IM*+XM*	JN*+YN*	KO*+ZO*	LP*+A*P*	MQ*+B*Q*	NR*+C*R*	OS*+D*S*	PT*+E*T*	QU*+F*U*	RV*+G*V*
4	−XJ*	−YK*	I*L*−TL*	J*M*−UM*	K*N*−VN*	−C*O*	−D*P*	−E*Q*	O*R*−ZR*	P*S*−A*S*	Q*T*−B*T*	−G*U*	U*V*−F*V*
5	XJ*	YK*	TL*	UM*	VN*	C*O*	D*P*	E*Q*	ZR*	A*S*	B*T*	G*U*	F*V*
6	IJ*+XJ*	JK*+YK*	EL*+TL*	FM*+UM*	GN*+VN*	NO*+C*O*	OP*+D*P*	PQ*+E*Q*	KR*+ZR*	LS*+A*S*	MT*+B*T*	RU*+G*U*	QV*+F*V*
7	−IJ*	−JK*	−EL*	−FM*	−GN*	−NO*	−OP*	−PQ*	−KR*	−LS*	−MT*	−RU*	−QV*
8	IJ*	JK*	I*L*+EL*	J*M*+FM*	K*N*+GN	NO*	OP*	PQ*	O*R*+KR*	P*S*+LS*	Q*T*+MT*	RU*	U*V*+QV*
9			I*L*	J*M*	K*N*				O*R*	P*S*	Q*T*		U*V*
10	−ZJ*	−A*K*	H*L*−SL*	−C*M*	−D*N*	J*O*−UO*	K*P*−VP*	−F*Q*	M*R*−XR*	N*S*−YS*	−G*T*	Q*U*−B*U*	T*V*−E*V*
11	ZJ*	A*K*	SL*	C*M*	D*N*	UO*	VP*	F*Q*	XR*	YS*	G*T*	B*U*	E*V*
12	KJ*+ZJ*	LK*+A*K*	DL*+SL*	NM*+C*M*	ON*+D*N*	FO*+UO*	GP*+VP*	QQ*+F*Q*	IR*+XR*	JS*+YS*	RT*+G*T*	MU*+B*U*	PV*+E*V*
13	−KJ*	−LK*	−DL*	−NM*	−ON*	−FO*	−GP*	−QQ*	−IR*	−JS*	−RT*	−MU*	−PV*
14	KJ*	LK*	H*L*+DL*	NM*	ON*	J*O*+FO*	K*P*+GP*	QQ*	M*R*+IR*	N*S*+JS*	RT*	Q*U*+MU*	T*V*+PV*
15			H*L*			J*O*	K*P*		M*R*	N*S*		Q*U*	T*V*
16	B*J*−AJ*	−B*K*	−C*L*	H*M*−SM*	−E*N*	I*O*−TO*	−F*P*	K*Q*−VQ*	L*R*−WR*	−G*S*	N*T*−YT*	P*U*−A*U*	S*V*−D*V*
17	AJ*	B*K*	C*L*	SM*	E*N*	TO*	F*P*	VQ*	WR*	G*S*	YT*	A*U*	D*V*
18	CJ*+AJ*	MK*+B*K*	NL*+C*L*	DM*+SM*	PN*+E*N*	EO*+TO*	QP*+F*P*	GQ*+VQ*	HR*+WR*	RS*+G*S*	JT*+YT*	LU*+A*U*	OV*+D*V*
19	−CJ*	−MK*	−NL*	−DM*	−PN*	−EO*	−QP*	−GQ*	−HR*	−RS*	−JT*	−LU*	−OV*
20	B*J*+CJ*	MK*	NL*	H*M*+DM*	PN*	I*O*+EO*	QP*	K*Q*+GQ*	L*R*+HR*	RS*	N*T*+JT*	P*U*+LU*	S*V*+OV*
21	B*J*			H*M*		I*O*		K*Q*	L*R*		N*T*	P*U*	S*V*
22	−B*J*	BK*−AK*	−D*L*	−E*M*	H*N*−SN*	−F*O*	T*P*−TP*	J*Q*−UQ*	−G*R*	L*S*−WS*	M*T*−XT*	O*U*−ZU*	R*V*−C*V*
23	B*J*	AK*	D*L*	E*M*	SN*	F*O*	TP*	UQ*	G*R*	WS*	XT*	ZU*	C*V*
24	MJ*+B*J*	CK*+AK*	OL*+D*L*	PM*+E*M*	DN*+SN*	QO*+F*O*	EP*+TP*	FQ*+UQ*	RR*+G*R*	HS*+WS*	IT*+XT*	KU*+ZU*	NV*+C*V*
25	−MJ*	−CK*	−OL*	−PM*	−DN*	−QO*	−EP*	−FQ*	−RR*	−HS*	−IT*	−KU*	−NV*
26	MJ*	BK*+CK*	OL*	PM*	H*N*+DN*	QO*	T*P*+EP*	J*Q*+FQ*	RR*	L*S*+HS*	M*T*+IT*	O*U*+KU*	R*V*+NV*
27		BK*			H*N*		T*P*	J*Q*		L*S*	M*T*	O*U*	R*V*
28	−C*J*	−D*K*	BL*−AL*	−ZM*	−A*N*	M*O*−XO*	N*P*−YP*	−G*Q*	J*R*−UR*	K*S*−VS*	−F*T*	T*U*−E*U*	Q*V*−B*V*
29	C*J*	D*K*	AL*	ZM*	A*N*	XO*	YP*	G*Q*	UR*	VS*	F*T*	E*U*	B*V*
30	NJ*+C*J*	OK*+D*K*	CL*+AL*	KM*+ZM*	LN*+A*N*	IO*+XO*	JP*+YP*	RQ*+G*Q*	FR*+UR*	GS*+VS*	QT*+F*T*	PU*+E*U*	MV*+B*V*
31	−NJ*	−OK*	−CL*	−KM*	−LN*	−IO*	−JP*	−RQ*	−FR*	−GS*	−QT*	−PU*	−MV*
32	NJ*	OK*	BL*+CL*	KM*	LN*	M*O*+IO*	N*P*+JP*	RQ*	J*R*+FR*	K*S*+GS*	QT*	T*U*+PU*	Q*V*+MV*
33			BL*			M*O*	N*P*		J*R*	K*S*		T*U*	Q*V*
34	H*J*−SJ*	−E*K*	−ZL*	BM*−AM*	−B*N*	L*O*−WO*	−G*P*	N*Q*−YQ*	I*R*−TR*	−F*S*	K*T*−VT*	S*U*−D*U*	P*V*−A*V*
35	SJ*	E*K*	ZL*	AM*	B*N*	WO*	G*P*	YQ*	TR*	F*S*	VT*	D*U*	A*V*
36	DJ*+SJ*	PK*+E*K*	KL*+ZL*	CM*+AM*	MN*+B*N*	HO*+WO*	RP*+G*P*	JQ*+YQ*	ER*+TR*	QS*+F*S*	GT*+VT*	OU*+D*U*	LV*+A*V*
37	−DJ*	−PK*	−KL*	−CM*	−MN*	−HO*	−RP*	−JQ*	−ER*	−QS*	−GT*	−OU*	−LV*
38	H*J*+DJ*	PK*	KL*	BM*+CM*	MN*	L*O*+HO*	RP*	N*Q*+JQ*	I*R*+ER*	QS*	K*T*+GT*	S*U*+OU*	P*V*+LV*
39	H*J*			BM*		L*O*		N*Q*	I*R*		K*T*	S*U*	P*V*
40	−E*J*	H*K*−SK*	−A*L*	−B*M*	BN*−AN*	−G*O*	L*P*−WP*	M*Q*−XQ*	−F*R*	I*S*−TS*	J*T*−UT*	R*U*−C*U*	O*V*−ZV*
41	E*J*	SK*	A*L*	B*M*	AN*	G*O*	WP*	XQ*	F*R*	TS*	UT*	C*U*	ZV*
42	PJ*+E*J*	DK*+SK*	LL*+A*L*	MM*+B*M*	CN*+AN*	RO*+G*O*	HP*+WP*	IQ*+XQ*	QR*+F*R*	ES*+TS*	FT*+UT*	NU*+C*U*	KV*+ZV*
43	−PJ*	−DK*	−LL*	−MM*	−CN*	−RO*	−HP*	−IQ*	−QR*	−ES*	−FT*	−NU*	−KV*
44	PJ*	H*K*+DK*	LL*	MM*	BN*+CN*	+RO*	L*P*+HP*	M*Q*+IQ*	QR*	I*S*+ES*	J*T*+FT*	R*U*+NU*	O*V*+KV*
45		H*K*			BN*		L*P*	M*Q*		I*S*	J*T*	R*U*	O*V*
46	I*J*−TJ*	−F*K*	−XL*	L*M*−WM*	−G*N*	BO*−AO*	−B*P*	P*Q*−A*Q*	H*R*−SR*	−E*S*	S*T*−D*T*	K*U*−VU*	N*V*−YV*
47	TJ*	F*K*	XL*	WM*	G*N*	AO*	B*P*	A*Q*	SR*	E*S*	D*T*	VU*	YV*
48	EJ*+TJ*	QK*+F*K*	IL*+XL*	HM*+WM*	RN*+G*N*	CO*+AO*	MP*+B*P*	LQ*+A*Q*	DR*+SR*	PS*+E*S*	OT*+D*T*	GU*+VU*	JV*+YV*
49	−EJ*	−QK*	−IL*	−HM*	−RN*	−CO*	−MP*	−LQ*	−DR*	−PS*	−OT*	−GU*	−JV*
50	I*J*+EJ*	QK*	IL*	L*M*+HM*	RN*	BO*+CO*	MP*	P*Q*+LQ*	H*R*+DR*	PS*	S*T*+OT*	K*U*+GU*	N*V*+JV*
51	I*J*			L*M*		BO*		P*Q*	H*R*		S*T*	K*U*	N*V*
52	−F*J*	I*K*−TK*	−YL*	−G*M*	L*N*−WN*	−B*O*	BP*−AP*	O*Q*−ZQ	−E*R*	H*S*−SS*	R*T*−C*T*	J*U*−UU*	M*V*−XV*
53	F*J*	TK*	YL*	G*M*	WN*	B*O*	AP*	ZQ*	E*R*	SS*	C*T*	UU*	XV*
54	QJ*+F*J*	EK*+TK*	JL*+YL*	RM*+G*M*	HN*+WN*	MO*+B*O*	CP*+AP*	KQ*+ZQ*	PR*+E*R*	DS*+SS*	NT*+C*T*	FU*+UU*	IV*+XV*
55	−QJ*	−EK*	−JL*	−RM*	−HN*	−MO*	−CP*	−KQ*	−PR*	−DS*	−NT*	−FU*	−IV*
56	QJ*	I*K*+EK*	JL*	RM*	L*N*+HN*	MO*	BP*+CP*	O*Q*+KQ*	PR*	H*S*+DS*	R*T*+NT*	J*U*+FU*	M*V*+IV*
57		I*K*			L*N*		BP*	O*Q*		H*S*	R*T*	J*U*	M*V*
58	−VJ*	J*K*−UK*	−G*L*	−YM*	M*N*−XN*	−A*O*	O*P*−ZP*	BQ*−AQ*	−D*R*	R*S*−C*S*	H*T*−ST*	I*U*−TU*	L*V*−WV*
59	VJ*	UK*	G*L*	YM*	XN*	A*O*	ZP*	AQ*	D*R*	C*S*	ST*	TU*	WV*
60	GJ*+VJ*	FK*+UK*	RL*+G*L*	JM*+YM*	IN*+XN*	LO*+A*O*	KP*+ZP*	CQ*+AQ*	OR*+D*R*	NS*+C*S*	DT*+ST*	EU*+TU*	HV*+WV*
61	−GJ*	−FK*	−RL*	−JM*	−IN*	−LO*	−KP*	−CQ*	−OR*	−NS*	−DT*	−EU*	−HV*
62	GJ*	J*K*+FK*	RL*	JM*	M*N*+IN*	LO*	O*P*+KP	BQ*+CQ*	OR*	R*S*+NS*	H*T*+DT*	I*U*+EU*	L*V*+HV*
63		J*K*			M*N*		O*P*	BQ*		R*S*	H*T*	I*U*	L*V*
64	−WJ*	−G*K*	J*L*−UL*	I*M*−TM*	−F*N*	H*O*−SO*	−E*P*	−D*Q*	BR*−AR*	Q*S*−B*R*	P*T*−A*T*	N*U*−YU*	K*V*−VV*
65	WJ*	G*K*	UL*	TM*	F*N*	SO*	E*P*	D*Q*	AR*	B*R*	A*T*	YU*	VV*
66	HJ*+WJ*	RK*+G*K*	FL*+UL*	EM*+TM*	QN*+F*N*	DO*+SO*	PP*+E*P*	OQ*+D*Q*	CR*+AR*	MS*+B*R*	LT*+A*T*	JU*+YU*	GV*+VV*
67	−HJ*	−RK*	−FL*	−EM*	−QN*	−DO*	−PP*	−OQ*	−CR*	−MS*	−LT*	−JU*	−GV*
68	HJ*	RK*	J*L*+FL*	I*M*+EM*	QN*	H*O*+DO*	PP*	OQ*	BR*+CR*	Q*S*+MS*	P*T*+LT*	N*U*+JU*	K*V*+GV*
69			J*L*	I*M*		H*O*			BR*	Q*S*	P*T*	N*U*	K*V*
70	−G*J*	−WK*	K*L*−VL*	−F*M*	IN*−TN*	−E*O*	H*P*−SP*	−C*Q*	Q*R*−B*R*	BS*−AS*	O*T*−ZT*	M*U*−XU*	J*V*−UV*
71	G*J*	WK*	VL*	F*M*	TN*	E*O*	SP*	C*Q*	B*R*	AS*	ZT*	XU*	UV*
72	RJ*+G*J*	HK*+WK*	GL*+VL*	QM*+F*M*	EN*+TN*	PO*+E*O*	DP*+SP*	NQ*+C*Q*	MR*+B*R*	CS*+AS*	KT*+ZT*	IU*+XU*	FV*+UV*
73	−RJ*	−HK*	−GL*	−QM*	−EN*	−PO*	−DP*	−NQ*	−MR*	−CS*	−KT*	−IU*	−FV*
74	RJ*	HK*	K*L*+GL*	QM*	I*N*+EN*	PO*	H*P*+DP*	NQ*	Q*R*+MR*	BS*+CS*	O*T*+KT*	M*U*+IU*	J*V*+FV*
75			K*L*		I*N*		H*P*		Q*R*	BS*	O*T*	M*U*	J*V*
76	−YJ*	−XK*	−F*L*	K*M*−VM*	J*N*−UN*	−D*O*	−C*P*	H*Q*−SQ*	P*R*−A*R*	O*S*−ZS*	BT*−AT*	L*U*−WU*	I*V*−TV*
77	YJ*	YK*	F*L*	VM*	UN*	D*O*	C*P*	SQ*	A*R*	ZS*	AT*	WU*	TV*
78	JJ*+YJ*	IK*+XK*	QL*+F*L*	GM*+VM*	FN*+UN*	OO*+D*O*	NP*+C*P*	DQ*+SQ*	LR*+A*R*	KS*+ZS*	CT*+AT*	HU*+WU*	EV*+TV*
79	−JJ*	−IK*	−QL*	−GM*	−FN*	−OO*	−NP*	−DQ*	−LR*	−KS*	−CT*	−HU*	−EV*
80	JJ*	IK*	QL*	K*M*+GM*	J*N*+FN*	OO*	NP*	H*Q*+DQ*	P*R*+LR*	O*S*+KS*	BT*+CT*	L*U*+HU*	I*V*+EV*
81				K*M*	J*N*			H*Q*	P*R*	O*S*	BT*	L*U*	I*V*
82	−A*J*	−ZK*	−E*L*	−D*M*	−C*N*	K*O*−VO*	J*P*−UP*	I*Q*−TQ*	N*R*−YR*	M*S*−XS*	L*T*−WT*	BU*−AU*	H*V*−SV*
83	A*J*	ZK*	E*L*	D*M*	C*N*	VO*	UP*	TQ*	YR*	XS*	WT*	AU*	SV*
84	LJ*+A*J*	KK*+ZK*	PL*+E*L*	OM*+D*M*	NN*+C*N*	GO*+VO*	FP*+UP*	EQ*+TQ*	JR*+YR*	IS*+XS*	HT*+WT*	CU*+AU*	DV*+SV*
85	−LJ*	−KK*	−PL*	−OM*	−NN*	−GO*	−FP*	−EQ*	−JR*	−IS*	−HT*	−CU*	−DV*
86	LJ*	KK*	PL*	OM*	NN*	K*O*+GO*	J*P*+FP	I*Q*+EQ*	N*R*+JR*	M*S*+IS*	L*T*+HT*	BU*+CU*	H*V*+DV*
87						K*O*	J*P*	I*Q*	N*R*	M*S*	L*T*	BU*	H*V*
88	−O*J*	−C*K*	−B*L*	−A*M*	−ZN*	N*O*−YO*	M*P*−XP*	L*Q*−WQ*	K*R*−VR*	J*S*−US*	I*T*−TT*	H*U*−SU*	BV*−AV*
89	O*J*	C*K*	B*L*	A*M*	ZN*	YO*	XP*	WQ*	VR*	US*	TT*	SU*	AV*
90	OJ*+O*J*	NK*+C*K*	ML*+B*L*	LM*+A*M*	KN*+ZN*	JO*+YO*	IP*+XP*	HQ*+WQ*	GR*+VR*	FS*+US*	ET*+TT*	DU*+SU*	CV*+AV*
91	−OJ*	−NK*	−ML*	−LM*	−KN*	−JO*	−IP*	−HQ*	−GR*	−FS*	−ET*	−DU*	−CV*
92	OJ*	NK*	ML*	LM*	KN*	NO*+JO*	MP*+IP*	L*Q*+HQ*	K*R*+GR*	J*S*+FS*	I*T*+ET*	H*U*+DU*	BV*+CV*
93						NO*	MP*	L*Q*	K*R*	J*S*	I*T*	H*U*	BV*

Estimates of Main Effects and Two-Factor Interactions for 16-Trial Hadamard Matrix Designs with 6 to 8 Variables

Contrast	Number of variables				
			6	7	8
1	A				
2	B				
3	C				
4	D				
5	AB	+ DE	+ CF		GH
6	BC		+ AF	DG	EH
7	CD		+ EF	BG	AH
8		E			
9	AC		+ BF	EG	DH
10	BD	+ AE		CG	FH
11	*	*	F		
12	*	*	*	G	
13	*	CE	+ DF	AG	BH
14	*	*	*	*	H
15	AD	+ BE		FG	CH

* Random error.

Note: All of the above interactions are shown with a plus sign only to keep the tables neat. Actually, all of these interactions should have minus signs in front of them. The same applies to all the following subsections. For example, contrast 7 really measures (−CD − EF − BG − AH).

Estimates of Main Effects and Two-Factor Interactions for 32-Trial Hadamard Matrix Designs with 7 to 16 Variables

			Number of Variables									
Contrast			7	8	9	10	11	12	13	14	15	16
1	A											
2	B											
3	C											
4	D											
5	E											
6	AC	+ DF		BH			GK		EM	JN	IO	LP
7	BD		+ EG	FH	CI				KM	LN	AO	JP
8	CE				GI	DJ		HL	AM	FN	KO	BP
9	*	F										
10	*	*	G									
11	AE					FJ	IK	BL	CM	DN	GO	HP
12	*	*	*	H								
13	*	*	*	*	I							
14	*	*	*	*	*	J						
15	*	EF		GH		AJ	BK	IL	DM	CN		OP
16	*	*	*	*	*	*	K					
17	*	*	AG			HJ	CK	DL	IM	BN	EO	FP
18	*	*	*	*	*	*	*	L				
19	AB			CH	FI		JK	EL		GN	DO	MP
20	BC			AH	DI	GJ			LM	KN	FO	EP
21	CD	+ AF			BI	EJ		KL		MN	HO	GP
22	DE		+ BG			CJ	HK		FM	AN	LO	IP
23	*	*	*	*	*	*	*	*	M			
24	*	BF		DH	AI		EK	JL	GM		CO	NP
25	*	*	CG		EI	BJ	AK	FL		HN	MO	DP
26	*	*	*	*	*	*	*	*	*	N		
27	*	*	FG	EH			DK	CL	BM	IN	JO	AP
28	*	*	*	*	*	*	*	*	*	*	O	
29	*	*	*	*	*	*	*	*	*	*	*	P
30	AD	+ CF			HI			GL	JM	EN	BO	KP
31	BE		+ DG			IJ	FK	AL	HM		NO	CP

* Random Error

Estimates of Main Effects and Two-Factor Interactions for 64-Trial Hadamard Matrix Designs with 9 to 32 Variables

Contrast	No. of Variables			9	10	11	12	13	14	15	16	17	18	19	20	21	22	23	24	25	26	27	28	29	30	31	32
1	A																										
2	B																										
3	C																										
4	D																										
5	E																										
6	F																										
7	AB	FG	EH				CL			JO	IP	DQ			ST				MX	VY	UZ	RA·	KB·	NC·		D·E·	WF·
8	BC			FI			AL	DM			GP		ER		KT				QX	NY	WZ	HA·	SB·	VC·	OD·	JE·	UF·
9	CD							BM	EN		KP	LQ		FS	GT	JU			AX	RY	OZ	VA·	IB·	HC·	WD·		E·F·
10	DE				GJ				CN	FO		HQ	MR			TU	AV			BY	SZ	XA·	WB·	LC·	ID·	PE·	KF·
11	EF		GH							DO		JQ	IR	NS		LU	KV	BW			CZ	PA·	YB·	TC·	MD·	XE·	AF·
12 ABF	*	G																									
13	AC			GI			BL				FP	MQ	HR	KS			NV	UW	DX			EA·	TB·	YC·	JD·	OE·	ZF·
14	BD					GK		CM				AQ	NR	IS	PT		HV	OW	LX	EY			FB·	A·C·	ZD·	UE·	JF·
15	CE						HL		DN				BR	OS	JT	GU		IW	VX	MY	FZ	AA·		QC·	B·D·	KE·	PF·
16	DF				HJ	AK		IM		EO		GQ		CS	LT				PX	WY	NZ		BB·	UC·	RD·	A·E·	VF·
17 ABE	*		H																								
18 BCF	*			I																							
19 ABCD						IK	DL	AM	HN			CQ		GS	FT	OU	RV		BX		JZ	YA·	PB·	EC·		WE·	D·F·
20 BCDE								EM	BN	IO	JP		DR			KU	LV	SW	HX	CY		QA·	ZB·	AC·	FD·	GE·	TF·
21 CDEF							JL		FN	CO				ES	HT	QU	PV	MW		IY	DZ	KA·	RB·	GC·	BD·	AE·	XF·
22 ABDEF	*				J																						
23 ACEF				HI				JM	KN		EP		GR			BU	SV	LW	OX	TY	AZ	FA·		B·C·	QD·	DE·	CF·
24 ADF	*					K																					
25	AE		BH							KO			LR			IU	DV	GW	NX	QY	PZ	CA·	JB·	MC·	TD·	SE·	FF·
26	BF	AG	CI								LP	KQ		MS			JV	EW	TX	OY	RZ	UA·	DB·		ND·	C·E·	HF·
27 ABC	*						L																				
28 BCD	*							M																			
29 CDE	*								N																		
30 DEF	*									O																	
31 ABEF		EG	FH		DJ							OQ	PR		NT	CU		AW		KY	LZ	IA·	VB·	SC·	XD·	ME·	BF·
32 ACF	*										P																
33	AD					FK		LM				BQ		PS	IT		EV	JW	CX	HY		NA·	GB·	RC·	UD·	ZE·	OF·
34	BE		AH			JK			MN				CR			PU	QV	FW		DY	IZ	LA·	OB·	XC·	SD·	TE·	GF·
35	CF			BI			GL			NO	AP			DS	QT	HU		RW	KX		EZ		MB·	JC·	YD·	VE·	A·F·
36 ABD	*											Q															
37 BCE	*												R														
38 CDF	*													S													
39 ABDE			DH		FJ				LN	GO		EQ				SU	BV	KW	RX	AY	TZ	MA·		CC·	PD·	IE·	B·F·
40 BCEF				EI						MO	HP		FR			AU	TV	CW	JX	SY	BZ	GA·	NB·	KC·	DD·	QE·	LF·
41 ABCDF	*														T												
42 ACDE					IJ			HM	AN		OP		QR				CV	TW	EX	LY	KZ	DA·	UB·	BC·	GD·	FE·	SF·
43 BDEF					AJ	HK			IN	BO				RS				GV	DW	UX	FY	MZ	TA·	EB·	PC·	CD·	LE·	QF·
44 ABCEF	*															U											
45 ACDF						CK		GM			DP	IQ	JR	AS	BT				FX	UY	VZ	OA·	LB·	WC·	HD·	EE·	NF·
46 ADE	*																V										
47 BEF	*																	W									
48 ABCF		CG		AI			FL	KM	JN		BP			QS	DT	EU					HZ	WA·	XB·	OC·	VD·	YE·	RF·
49 ACD	*																		X								
50 BDE	*																			Y							
51 CEF	*																				Z						
52 ABDF		DG			EJ	BK				HO	MP	FQ		LS	CT	NU		VW	IX				AB·	ZC·	A·D·	RE·	YF·
53 ACF	*																					A·					
54 BDF	*																						B·				
55 ABCE			CH				EL					NQ	AR	JS	OT	FU	MV	PW		XY	GZ	BA·		DC·	KD·	B·E·	IF·
56 BCDF				DI			KL	FM				PQ	OR	BS	AT		UV	NW	GX		YZ	JA·	CB·		ED·	HE·	C·F·
57 ABCDE	*																							C·			
58 BCDEF	*																								D·		
59 ABCDEF					CJ				GN	LO			KR	HS	ET	DU	IV		WX	PY	QZ		A·B·	FC·	AD·	BE·	MF·
60 ACDEF	*																									E·	
61 ADEF					BJ	EK				AO	NP				RT	MU	FV	QW		GY	XZ	SA	HB·	IC·	LD·	CE·	DF·
62 AEF	*																										F·
63	AF	BG				DK	IL				CP				MT	RU	OV	HW	SX	JY		ZA·	QB·		C·D·	NE·	EF·

Note to reader: the table for 128 trials with 12 to 64 variables is the foldout located elsewhere in the book.

14 ANOG—Analysis of Goodness

A variety of techniques have been derived for analyzing the data from orthogonal matrix experiments. In this text, the method of orthogonal contrast coefficients have been used. Other authors would use ANOVA—Analysis of Variance, or the Yates Algorithm or normal plots for all of the examples in this text. Their conclusions, for the most part, would be exactly the same as those given in this text. The method of orthogonal contrast coefficients, if applied to their data, would reach the same conclusions as they do.

It will be shown in this chapter that ANOG—Analysis of Goodness, as a method by itself or in conjunction with the more conventional methods of analysis of factorial experiment, will give additional information and in some cases, more correct, practical information to the experimenter. ANOG is particularly applicable to experiments where the response of interest is the yield of a process or the percent defectives from a process.

ANOG is based on two premises:

- The symmetry of the Hadamard Matrix
- That the laws of nature must be obeyed in all experiments

Suppose that the law of nature for a process is that high A and low B will yield a 95 percent product. Any other combination of A and B will yield a substantially lower quality product. The symmetry of the Hadamard Matrix is such that one-fourth of the trials are at high A and low B. Since the laws of nature must be obeyed, one-fourth of the trials will have a yield of about 95 percent and three-fourths of the trials will be substantially lower.

The converse is ANOG. Obtain the results of a Hadamard Matrix. List the results in ascending or decending order. Write the treatment combinations beside each trial and then examine the best two, or four, or eight trials for patterns of pluses or minuses. The pattern identifies the significant variables.

EXAMPLE

A process is producing a chemical with a yield of 70 percent. A second expensive step is required to remove the contaminating materials. A Resolution V experiment with 5 variables and 16 trials is proposed with the expectation that a higher-yielding process can be developed.

The variables and their levels were:

Variables	(−) Present	(+) Proposed
A Temperature (°C)	150	180
B Pressure (psi)	80	60
C Catalyst (percent)	1	2
D pH	7	9
E Charge (pounds)	10	8

The treatment combinations and results are as follows:

Trial	A	B	C	D	E	Result	Ranking
1	+	−	−	−	+	.65	
2	+	+	−	−	−	.85	5
3	+	+	+	−	+	.91	1
4	+	+	+	+	−	.89	3
5	−	+	+	+	+	.88	4
6	+	−	+	+	+	.67	
7	−	+	−	+	−	.83	7
8	+	−	+	−	−	.68	
9	+	+	−	+	+	.82	8
10	−	+	+	−	−	.90	2
11	−	−	+	+	−	.66	
12	+	−	−	+	−	.63	
13	−	+	−	−	+	.84	6
14	−	−	+	−	+	.68	
15	−	−	−	+	+	.63	
16	−	−	−	−	−	.64	

A. List the results in descending order with the treatment combination beside each result.

Trial	Result	Treatment Combination A	B	C	D	E
3	.91	+	+	+	−	+
10	.90	−	+	+	−	−
4	.89	+	+	+	+	−
5	.88	−	+	+	+	+
2	85	+	+	−	−	−
13	84	−	+	−	−	+
7	83	−	+	−	+	−
9	82	+	+	−	+	+

B. Make an engineering decision.

1) The most important variable is B. For a high yield (0.82), the high level of B (60 psi) or lower must be used.

2) At 60 psi pressure, a yield of at least .88 will be obtained at the high level of C (2 percent catalyst).
3) The third most important variable is D. The low level of D (pH = 7), the present level, is better than the proposed high level (pH = 9).
4) High A (180°) might be better than low A (150°).

C. Compute the effect of variables A, C, and D at high B.

$$A = \frac{.91 - .90 + .89 - .88 + .85 - .84 - .83 + .82}{4}$$
$$= .005$$
$$C = \frac{.91 + .90 + .89 + .88 - .85 - .84 - .83 - .82}{4}$$
$$= .06$$
$$D = \frac{-.91 - .90 + .89 + .88 - .85 - .84 + .83 + .82}{4}$$
$$= -.02$$

D. Compute a regression equation at high B. Note: The coefficients are $\frac{1}{2}$ the effects calculated in C.

Yield = .865 + .0025 (A) + 0.03(C) −0.01 (D)

Note: If all 16 results are used, the coefficient will be very different.

E. Propose a follow-up plan.

Operate the process with the variables set as follows:

A = 180°C
B = 60 psi
C = 2 percent catalyst
D = 7 pH
E = 10 pound charge.

The yield over several weeks should be about 0.90. Determine the S^2 of the process. After operating several weeks at the new standard conditions, design an EVOP experiment with four variables.

	Low Level	High Level
A	180°C	190°C
B	60 psi	55 psi
C	2%	2.3%
D	7	6
E	fix at 10 pounds	

Most texts on experiment design rarely mention a procedure for finding bad data in an experiment. ANOG will immediately highlight a bad datum.

EXAMPLE

A replicated 16 trial Resolution V Design with four variables at 2 levels was conducted. The response was the sharpness of printwork score for typewriter ribbons; the variables were manufacturing process settings.

The results and treatment combinations were:

Trial	A	B	C	D	Results	Sum of results
1	+	−	−	−	13 18	31
2	+	+	−	−	19 17	36
3	+	+	+	−	6 6	12
4	+	+	+	+	16 15	31
5	−	+	+	+	16 12	28
6	+	−	+	+	24 24	48
7	−	+	−	+	8 12	20
8	+	−	+	−	17 18	35
9	+	+	−	+	(18 25)	43
10	−	+	+	−	12 9	21
11	−	−	+	+	(12 23)	35
12	+	−	−	+	(16 23)	39
13	−	+	−	−	12 9	21
14	−	−	+	−	12 14	26
15	−	−	−	+	12 11	23
16	−	−	−	−	14 15	39

Note: circled replicates have excessive deviation and cast doubt on the whole experiment.

A. List the results in ascending order

Trial	Sum	A	B	C	D	
3	12	+	+	+	−	
7	20	−	+	−	+	
10	21	−	+	+	−	
13	21	−	+	−	−	
15	23	−	−	−	+	
14	26	−	−	+	−	
5	28	−	+	+	+	
11	29	−	−	+	+	* Bad replicates
1	31	+	−	−	−	
4	31	+	+	+	+	
16	35	−	−	−	−	
8	35	+	−	+	−	
2	36	+	+	−	−	
12	39	+	−	−	+	* Bad replicates
9	43	+	+	−	+	* Bad replicates
6	48	+	−	+	+	

It is immediately apparent that trial 3 doesn't fit the pattern for Variable A. Seven of the eight best results are at the low level, but trial 3, the best result of the experiment, is at high A.

B. List only the Low A results in ascending order

Trial	Sum	A	B	C	D
7	20	−	+	−	+
10	21	−	+	+	−
13	21	−	+	−	−
15	23	−	−	−	+
14	26	−	−	+	−
5	28	−	+	+	+
16	29	−	−	−	−
11	35	−	−	+	+

Variable B or C seem to be the second most important variable after A. Trial #5, if moved to the fourth row, would make the results consistant in both B and C.

C. Move trial 5 to the fourth position.

Trial	Sum	A	B	C	D	
7	20	−	+	−	+	
10	21	−	+	+	−	
13	21	−	+	−	−	Avg. sum
5	28	−	+	+	+	22.5
15	23	−	−	−	+	
14	26	−	−	+	−	Avg. sum
16	29	−	−	−	−	28.5
11	35	−	−	+	+	

Variables A, B, and C are all in a consistent pattern.

Conclusion: For good sharpness (low score) operate at Low A, High B, and Low C. A follow-up experiment at Lower A, Higher B, and Lower C should be considered to obtain an even lower result.

A three-way table of the results can be very informative.

	c		1	
	1	b	1	b
a	24–24 17–18 $\bar{X}$ = 20.75	~~6–6~~ 16–15 $\bar{X}$ = 15.50	13–18 16–23 $\bar{X}$ = 17.5	19–17 18–25 $\bar{X}$ = 19.75
1	12–23 12–14 $\bar{X}$ = 15.25	16–12 12–9 $\bar{X}$ = 12.25	12–11 14–15 $\bar{X}$ = 13.00	8–12 12–9 $\bar{X}$ = 10.25

The worst block is High A, Low B, High C. The best block is Low A, High B, Low C. Except for the crossed out data points, all of the data is consistent.

ANOG is particularly applicable to Resolution III designs. It has been stated earlier that Resolution III designs can only estimate main effects. The experimenter must assume all interactions are zero. It will now be shown that with ANOG and the sparsity of variables principle, Resolution III designs converge to Resolution IV designs in most real experiments, and in some cases to Resolution V.

EXAMPLE

The example on page 117, Chapter 6, can be analyzed by ANOG. List the results in ascending order.

Trial	Result (percent)	A	B	C	D	E
2	2.5	+	+	−	−	+
6	7.0	−	+	−	+	+
4	8.0	−	+	+	+	−
3	12.0	+	+	+	−	−
7	12.0	−	−	+	−	+
5	13.5	+	−	+	+	+
8	13.5	−	−	−	−	−
1	15.5	+	−	−	+	−

The combination of High B and High E are necessary for low results. Low C might be important, but it is probably the interaction BE, which is one of the interactions confounded with Variable C. These are exactly the same conclusions that were reached in Chapter 6 with several pages of calculations, with the exception that ANOG identified the BE interaction, even though the design was Resolution III.

15 Alternative Methods of Analysis

As mentioned many times throughout this book, there are a variety of methods of analysis for the results of Hadamard matrix experiments. Most of these methods will give exactly the same answer to the questions "What is the effect of the variables on the response of interest?" and/or "What interactions are significant?" This chapter will show how the data from one method of analysis converts to other methods of analysis.

EXAMPLE

A device had been in production for a relatively short period of time. However, it was obvious that the performance was not entirely satisfactory. Many of the mechanisms produced an annoying vibration. An engineer was placed in charge of the project to reduce the vibration. A vibration scale was devised; the level expected by the designer of the device was called zero: the worst device produced thus far was assigned a value of 20. The mechanism variation was estimated to be $S = 1.4$ with 9 degrees of freedom. The objective of the project was to reduce the vibration to a value of 3 or better.

Three groups of factors were involved in the problem.

- Product design factors (part dimensions, materials, etc.)
- Manufacturing process factors (recipes, material lots, vendors, etc.)
- User environment factors (temperature, humidity, supplies, etc.)

To evaluate user environment, eight testing systems were devised using the Addleman principle.

A	B	C	Test System
−1	−1	−1	0
−1	−1	+1	1
−1	+1	−1	2
−1	+1	+1	3
+1	−1	−1	4
+1	−1	+1	5
+1	+1	−1	6
+1	+1	+1	7

Five variables of the manufacturing process were included in the design.

D—Frame vendor: Smith versus Ortego
E—Welding method: Ultrasonic versus chemical
F—Shaft insertion: Before versus after welding
G—Amount of gear lubrication: Normal versus double
I—Amount of bearing lubrication: Normal versus double

A	B	C	D	E	F	G	H	I	J	K	L	M	N	O	P	
U	U	U	M1	M2	M3	M4	U	M5	P1	P2	P3	P4	P5	P6	P7	
1	0	0	0	1	1	1	1	0	1	0	0	0	0	1	1	15.4
1	1	0	0	0	1	1	0	1	0	0	0	1	1	0	1	20.4
1	1	1	0	0	0	1	1	0	1	1	0	0	1	0	0	7.1
1	1	1	1	0	1	0	1	1	0	0	0	0	0	1	0	16.3
1	1	1	1	1	1	1	1	1	1	1	1	1	1	1	1	−15.3
0	1	1	1	1	0	1	0	1	1	0	0	0	0	0	1	15.7
0	0	1	1	1	0	0	1	0	1	1	1	0	0	1	0	−15.3
1	0	0	1	1	0	0	1	1	0	1	0	0	1	0	1	6.1
1	1	0	0	1	1	0	0	1	1	1	1	0	0	0	0	−10.0
0	1	1	0	0	1	1	0	0	1	0	1	1	0	1	0	5.6
1	0	1	1	0	1	1	0	0	0	1	1	0	0	0	1	−9.0
0	1	0	1	1	1	1	1	0	0	1	0	1	0	0	0	5.8
0	0	1	0	1	1	1	1	1	0	0	1	0	1	0	0	8.6
1	0	0	1	0	0	1	1	1	1	0	1	1	0	0	0	4.9
0	1	0	0	1	0	0	1	1	1	0	0	1	1	1	0	15.3
0	0	1	0	0	1	0	1	1	1	1	0	1	0	0	1	5.3
0	0	0	1	0	1	1	0	1	1	1	0	0	1	1	0	3.5
0	0	0	0	1	0	1	0	0	1	1	1	1	1	0	1	−9.3
1	0	0	0	0	1	0	1	0	0	1	1	1	1	1	0	−12.9
0	1	0	0	0	0	1	1	1	0	1	1	0	0	1	1	−13.4
1	0	1	0	0	0	0	0	1	1	0	1	0	1	1	1	3.7
0	1	0	1	0	1	0	1	0	1	0	1	0	1	0	1	4.0
1	0	1	0	1	0	1	0	1	0	1	0	1	0	1	0	4.7
1	1	0	1	0	0	0	0	0	1	1	0	1	0	1	1	4.2
1	1	1	0	1	0	0	1	0	0	0	1	1	0	0	1	6.6
0	1	1	1	0	0	0	0	1	0	1	1	1	1	0	0	−12.9
1	0	1	1	1	1	0	0	0	1	0	0	1	1	0	0	17.7
1	1	0	1	1	0	1	0	0	0	0	1	0	1	1	0	7.2
0	1	1	0	1	1	0	0	0	0	1	0	0	1	1	1	5.7
0	0	1	1	0	0	1	1	0	0	0	0	1	1	1	1	14.7
0	0	0	1	1	1	0	0	1	0	0	1	1	0	1	1	2.1
0	0	0	0	0	0	0	0	0	0	0	0	0	0	0	0	17.8

Fig. 15-1. Experiment design and measured vibration

Note that the contrast column labeled H in the Hadamard matrix cannot be used because it is really the ABC interaction and really measures the test systems. Seven design factors were included in the design. J, K, L, and M were dimensional variables where the present nominals were contrasted with changes, either smaller or larger, suggested by engineering judgement.

N—Frame cover: None versus added cover
O—Shaft O-ring: None versus added shaft O-ring
P—Shaft material: Current versus new

The total number of variables was: 12 variables at 2 levels, and environment variables in 8 blocks, which required 3 of the contrast columns. The experimenters had two choices of design:

- A 16-trial Resolution III design
- A 32-trial Resolution IV design

They chose the latter. The design matrix and the results are shown in figure 15-1.

The analysis of the data using the Gem 2 program from Chapter 13 is shown in figure 15-2. Note that a lower number is better; therefore, a minus average effect means that the new level (+1) of the variables is better than the old level of the variables.

The variables with the biggest effect are K and L. D and O are also significant

```
AVE   3.75938

MAIN EFFECTS

           A          B          C        ** D          E          F
AVE    .86875     .26875    -.11875   -1.30625     .10625     .38125
S S    6.0378      .5778      .1128    13.6503      .0903     1.1628

           G          H          I          J        ** K       ** L
AVE     .8062     -.8688     -.6437     -.9563   -14.4813   -14.4438
S S      5.20       6.04       3.32       7.32    1677.65    1668.98

           M          N       ** O          P
AVE   -.40625     .43125   -2.33125    -.40625
S S    1.3203     1.4878    43.4778     1.3203

ESTIMATES OF TWO FACTOR INTERACTIONS

           6          7          8         11         15         17
AVE   .318750    .231250    .406250    .393750    .506250    .343750
S S    .81281     .42781    1.32031    1.24031    2.05031     .94531

          19         20       ** 21        22         24         25
AVE   -.46875     .46875    3.11875     .31875     .04375     .18125
S S    1.7578     1.7578    77.8128      .8128      .0153      .2628

          27         30         31
AVE  -.043750   -.943750    .143750
S S    .01531    7.12531     .16531
```

Fig. 15-2. FAN program output with significant contrasts indicated by **.

(due to the small variance), but the effect is small. Likewise, interaction 21 is small but significant. Interaction 21 measures CD + AF + BI + EJ + KL + MN + HO + GP

Significance at the five percent confidence level was determined by a comparison of the average value of the computer analysis sheet with the criterion.

$$|\bar{X}_{\text{high}} - \bar{X}_{\text{low}}|^* = t_\alpha S \sqrt{\frac{1}{N_{\text{high}}} + \frac{1}{N_{\text{low}}}};\ \alpha = .05 \text{ Double-sided}$$

$$= (1.4)(2.28)\sqrt{\frac{1}{16} + \frac{1}{16}} = 1.13$$

The same data was put into a regression program called Source and the result is shown in figure 15-3.

Line 31 is equivalent to the ABCDE interaction which was used in the design

Note: Under SOURCE, numbers are used in the place of factor letters. That is, 1 indicates factor A, 15 indicates the AE interaction, etc.

LINE	SOURCE	SUM SQUARES	D.F.	MEAN SQUARE
1	1	6.03781250	1	6.037812500
2	2	0.57781250	1	0.577812500
3	3	0.11281250	1	0.112812500
4	4	13.65031250	1	13.650312500
5	5	0.09031250	1	0.090312500
6	12	1.75781250	1	1.757812500
7	13	0.81281250	1	0.812812500
8	14	7.12531250	1	7.125312500
9	15	1.24031250	1	1.240312500
10	23	1.75781250	1	1.757812500
11	24	0.42781250	1	0.427812500
12	25	0.16531250	1	0.165312500
13	34	77.81281250	1	77.812812500
14	35	1.32031250	1	1.320312500
15	45	0.81281250	1	0.812812500
16	123	6.03781250	1	6.037812500
17	124	43.47781250	1	43.477812500
18	125	1668.97531250	1	1668.975312500
19	134	1.16281250	1	1.162812500
20	135	1.32031250	1	1.320312500
21	145	1.48781250	1	1.487812500
22	234	3.31531250	1	3.315312500
23	235	1.32031250	1	1.320312500
24	245	5.20031250	1	5.200312500
25	345	7.31531250	1	7.315312500
26	1234	0.01531250	1	0.015312500
27	1235	0.01531250	1	0.015312500
28	1345	2.05031250	1	2.050312500
29	1245	0.94531250	1	0.945312500
30	2345	0.26281250	1	0.262812500
31	12345	1677.65281250	1	1677.652812500
	TOTAL	3534.25718750	31	

Fig. 15-3. Analysis of variance table.

matrix for variable K. This is the biggest Sum of Squares, therefore the most important variable. Line 18, which is the ACD interaction, was used for variable L and also has a very large Sum of Squares value. The next largest Sum of Squares is associated with source 34, or the CD interaction. Note, Gem 2 shows that CD, AF and/or etc. was significant. The last two variables with large Sums of Squares are 124 and 4, which were used for variable O and variable D.

An F-test of the mean squares divided by S^2 would test significance.

$$F = \frac{\text{MSD}}{S^2} \text{ with } \phi = 1 \text{ in numerator and}$$

$$S^2 = 1.96 \text{ with } \phi = 9 \text{ in the denominator}$$

$$F = \frac{13.65}{1.96} = 6.96$$

$$F^*(.05,\ 1,\ 9) = 5.12$$

Therefore, high D is better than low D, with at least $(1 - {}^{\alpha}/_{2})$ 100 percent of confidence.

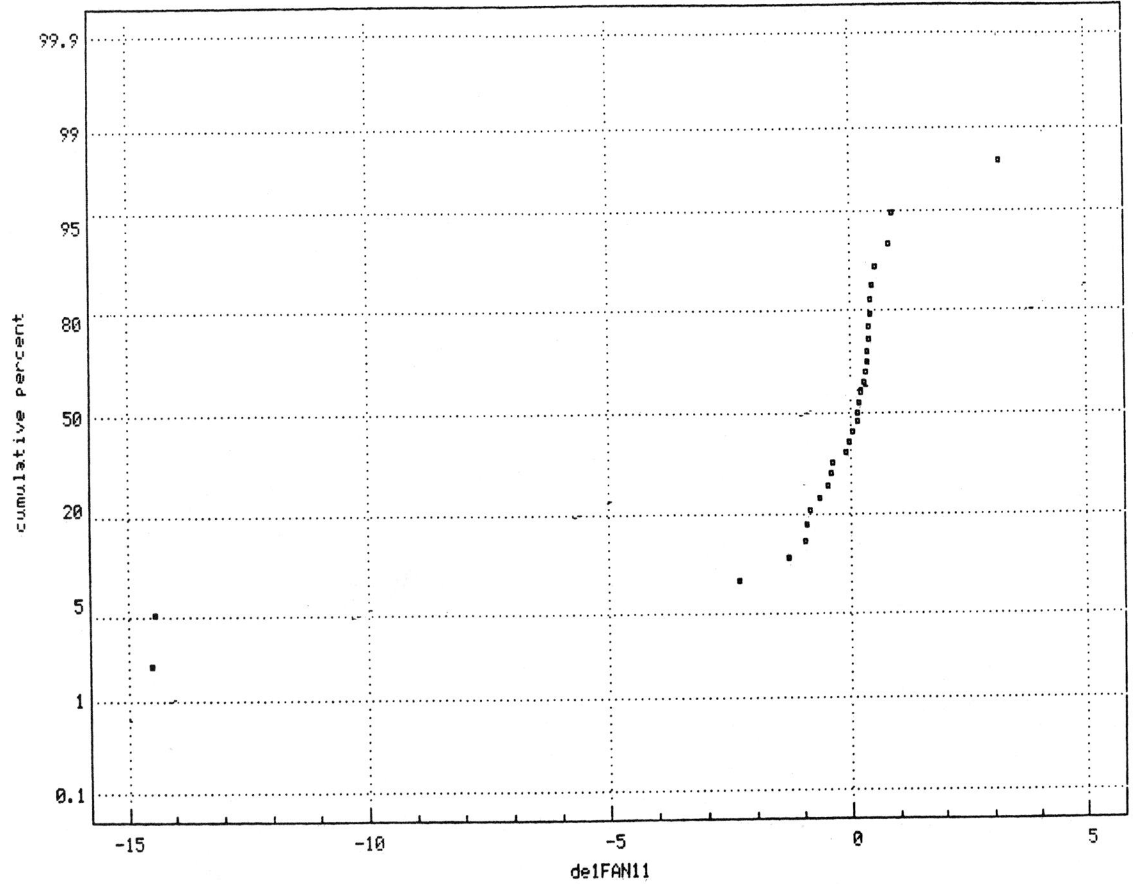

Fig. 15–4. Normal probability plot.

Some statisticians prefer analyzing the data by what is called Normal Probability Plots. The average data from figure 15-2 is plotted as shown in figure 15-4.

There are 31 contrast columns in a 32-trial Hadamard matrix

$$\text{probability/contrast column} = \frac{100}{32} = 3.125$$

Therefore the largest minus effect (−14.4813 for K) is plotted at the cumulative percent of 3.125. The next largest negative effect (−14.4438 for L) is plotted at 3.125 + 3.125 = 6.25. The final plot (the largest positive effect is plotted at the cumulative probability of 96.875.

From figure 15.4, an experimenter would conclude that the extreme points on the bottom left and the extreme point on the top right are tails of the normal distinction and therefore significant. These points correspond to K, L and the interactions of contrast column 21 (KL).

The simplest of all methods for analysis of a matrix experiment is ANOG. In figure 15-5, the data has been ordered from the best result to the worst result. The answer to the problem of how to reduce the vibration is unequivocal. The

1	1	1	1	1	1	1	1	1	1	1	1	1	1	1	1	−15.3
0	0	1	1	1	0	0	1	0	1	1	1	0	0	1	0	−15.3
0	1	0	0	0	0	1	1	1	0	1	1	0	0	1	1	−13.4
0	1	1	1	0	0	0	0	1	0	1	1	1	1	0	0	−12.9
1	0	0	0	0	1	0	1	0	0	1	1	1	1	1	0	−12.9
1	1	0	0	1	1	0	0	1	1	1	1	0	0	0	0	−10.0
0	0	0	0	1	0	1	0	0	1	1	1	1	1	0	1	−9.3
1	0	1	1	0	1	1	0	0	0	1	1	0	0	0	1	−9.0
0	0	0	1	1	1	0	0	1	0	0	1	1	0	1	1	2.1
0	·0	0	1	0	1	1	0	1	1	1	0	0	1	1	0	3.5
1	0	1	0	0	0	0	0	1	1	0	1	0	1	1	1	3.7
0	1	0	1	0	1	0	1	0	1	0	1	0	1	0	1	4.0
1	1	0	1	0	0	0	0	0	1	1	0	1	0	1	1	4.2
1	0	1	0	1	0	1	0	1	0	1	0	1	0	1	0	4.7
1	0	0	1	0	0	1	1	1	1	0	1	1	0	0	0	4.9
0	0	1	0	0	1	0	1	1	1	1	0	1	0	0	1	5.3
0	1	1	0	0	1	1	0	0	1	0	1	1	0	1	0	5.6
0	1	1	0	1	1	0	0	0	0	1	0	0	1	1	1	5.7
0	1	0	1	1	1	1	1	0	0	1	0	1	0	0	0	5.8
1	0	0	1	1	0	0	1	1	0	1	0	0	1	0	1	6.1
1	1	1	0	1	0	0	1	0	0	0	1	1	0	0	1	6.6
1	1	1	0	0	0	1	1	0	1	1	0	0	1	0	0	7.1
1	1	0	1	1	0	1	0	0	0	0	1	0	1	1	0	7.2
0	0	1	0	1	1	1	1	1	0	0	1	0	1	0	0	8.6
0	0	1	1	0	0	1	1	0	0	0	0	1	1	1	1	14.7
0	1	0	0	1	0	0	1	1	1	0	0	1	1	1	0	15.3
1	0	0	0	1	1	1	1	0	1	0	0	0	0	1	1	15.4
0	1	1	1	1	0	1	0	1	1	0	0	0	0	0	1	15.7
1	1	1	1	0	1	0	1	1	0	0	0	0	0	1	0	16.3
1	0	1	1	1	1	0	0	0	1	0	0	1	1	0	0	17.7
0	0	0	0	0	0	0	0	0	0	0	0	0	0	0	0	17.8
1	1	0	0	0	1	1	0	1	0	0	0	1	1	0	1	20.4

Fig. 15-5. Experiment design with observed data sorted on the data, best to worst.

eight best results from 9 to 15 units better than the design expectation are all at the new level of K and L, which are two of the dimensional variables. Just changing these dimensional specifications on the engineering drawing to the new dimensions will eliminate the problem. In fact, it will produce a device better than specified. Note, however, that two of the better results had exactly the same vibration value (12.9). If these two treatment combinations are reversed on the list, then it is noted that variable O has a pattern, with four plus results being better than four minuses. To have a product that ranged from −12.9 to −15.3 the device must have the new dimension for K and L and also a O-ring on the shaft. Leave off the O-ring and the vibration level will range from − 9 to −12.9.

The best way, by far, to present this data is in block form.

	(1)		k	
	(1)	1	(1)	1
(1)	20.4	4.8	5.3	−12.9
	17.8	4.9	5.8	−10.0 $\overline{X}$ = −10.3
	17.7	6.6	6.1	− 9.3
	15.7	8.6	7.1	− 9.0
0	16.3	2.4	3.5	−15.9
	15.4	3.7	4.2	−15.3 $\overline{X}$ = −14.23
	15.3	5.6	4.7	−13.4
	14.7	7.2	5.7	−12.9

In Chapter 24, it is shown that the four results in each block can be used to obtain a good estimate of the variance. It is not an unbiased estimate, but it is a good (probably biased high) estimate of σ^2. Each block gives an estimate with three degrees of freedom. The experimenter can use these S^2 values to determine whether the levels of K, L, and O have any effect on the variance. If it happens, as it does in this case, that there is no significant difference in the various blocks, then all of the estimates can be combined to give an average S^2 with 24 degrees of freedom which can be used in any future experiments on this device. If this exercise is carried out on these numbers, one will find that this average estimate agrees well with the prior estimate.

The experimenter should note that the list of confounded interactions that was significant included KL. The preceding chart confirms that a KL interaction, along with variable O, significantly explains all the data.

It should be noted that ANOG finds no significant pattern with variable D.

Having used any of the previous methods to determine that D, K, L, O, and KL are significant and important variables and interactions, one can now fit the data to a model of these variables and interactions.

Only the data matrix of the important variables is used. Thus, the first seven columns are used to test the seven environmental systems versus system O (figure 15-6). Then the columns are labeled for the levels of D, K, L, O, and KL.

Any typical regression analysis will produce an output similar to figure 15-7. One can now write a regression equation or a predictive equation for the experimental space. In this case, the first term in the equation is B_0, the mean of

T1	T2	T3	T4	T5	T6	T7	D	K	L	O	KL	
0	0	0	1	0	0	0	−1	−1	−1	1	1	15.4
0	0	0	0	0	1	0	−1	−1	−1	−1	1	20.4
0	0	0	0	0	0	1	−1	1	−1	−1	−1	7.1
0	0	0	0	0	0	1	1	−1	−1	1	1	16.3
0	0	0	0	0	0	1	1	1	1	1	1	−15.3
0	0	1	0	0	0	0	1	−1	−1	−1	1	15.7
1	0	0	0	0	0	0	1	1	1	1	1	−15.3
0	0	0	1	0	0	0	1	1	−1	−1	−1	6.1
0	0	0	0	0	1	0	−1	1	1	−1	1	−10.0
0	0	1	0	0	0	0	−1	−1	1	1	−1	5.6
0	0	0	0	1	0	0	1	1	1	−1	1	−9.0
0	1	0	0	0	0	0	1	1	−1	−1	−1	5.8
1	0	0	0	0	0	0	−1	−1	1	−1	−1	8.6
0	0	0	1	0	0	0	1	−1	1	−1	−1	4.9
0	1	0	0	0	0	0	−1	−1	−1	1	1	15.3
1	0	0	0	0	0	0	−1	1	−1	−1	−1	5.3
0	0	0	0	0	0	0	1	1	−1	1	−1	3.5
0	0	0	0	0	0	0	−1	1	1	−1	1	−9.3
0	0	0	1	0	0	0	−1	1	1	1	1	−12.9
0	1	0	0	0	0	0	−1	1	1	1	1	−13.4
0	0	0	0	1	0	0	−1	−1	1	1	−1	3.7
0	1	0	0	0	0	0	1	−1	1	−1	−1	4.0
0	0	0	0	1	0	0	−1	1	−1	1	−1	4.7
0	0	0	0	0	1	0	1	1	−1	1	−1	4.2
0	0	0	0	0	0	1	−1	−1	1	−1	−1	6.6
0	0	1	0	0	0	0	1	1	1	−1	1	−12.9
0	0	0	0	1	0	0	1	−1	−1	−1	1	17.7
0	0	0	0	0	1	0	1	−1	1	1	−1	7.2
0	0	1	0	0	0	0	−1	1	−1	1	−1	5.7
1	0	0	0	0	0	0	1	−1	−1	1	1	14.7
0	0	0	0	0	0	0	1	−1	1	1	−1	2.1
0	0	0	0	0	0	0	−1	−1	−1	−1	1	17.8

Fig. 15-6. Matrix of independent variables (columns 1–12) and observed data; input for regression analysis model 1.

all the data. B_1 through B_7 refer to the environmental system; (B_6) is almost significant but very small. B_8 through B_{12} are the coefficients of variables D, K, L, O, and KL, and they are significant with the F test. Therefore:

$$\text{Vibration} = 3.52 - 7.24\text{K} - 7.22\text{L} + 1.56\ \text{KL} - 1.17\ \text{“O”} - 0.65\ \text{D}.$$

The reader should also note several other important numbers on the analysis data of figure 15-7. The R^2 factor = .99 means that 99 percent of the variation in the data is accounted for by the equation with all of the B values calculated. The residual MS of 1.87334 is an estimate of σ^2, with 19 degrees of freedom.

An improved model can be calculated by using only the terms that were significant in the first regression analysis (figure 15-7).

Figure 15-8 shows a regression equation

$$\text{Vibration} = 3.52 - 7.24\text{K} - 7.22\text{L} + 1.56\text{KL} - .65\text{D} + 1.93(\text{System 6}) - 1.17\text{“O”}$$

Estimated Model Coefficients, Coefficient Standard Errors,
t Values and Confidence Intervals (alpha = 0.05)

	COEFFICIENT		B/SIGMA(B)	CONFIDENCE INTERVA	
TERM	B	SIGMA(B)	T	LOWER	UPPER
B0	3.5250	.6843	5.1509	2.0923	4.9577
B1	-.2000	.9678	-.2067	-2.2262	1.8262
B2	-.6000	.9678	-.6200	-2.6262	1.4262
B3	.0000	.9678	.0000	-2.0262	2.0262
B4	-.1500	.9678	-.1550	-2.1762	1.8762
B5	.7500	.9678	.7749	-1.2762	2.7762
B6	1.9250	.9678	1.9890	-.1012	3.9512
B7	.1500	.9678	.1550	-1.8762	2.1762
B8	-.6531	.2420	-2.6994	-1.1597	-.1466
B9	-7.2406	.2420	-29.9256	-7.7472	-6.7341
B10	-7.2219	.2420	-29.8481	-7.7284	-6.7153
B11	-1.1656	.2420	-4.8175	-1.6722	-.6591
B12	-1.5594	.2420	-6.4449	-2.0659	-1.0528

THE THEORETICAL VALUE FOR T AT THE 0.025 LEVEL AND 19 DF = 2.094

REGRESSION ANOVA TABLE

SOURCE	SS	DF	MS	F
TOTAL (CORRECTED)	3534.25719	31	114.00830	
REGRESSION (CORRECTED)	3498.66375	12	291.55531	155.6
RESIDUAL	35.59344	19	1.87334	
CORRECTION FACTOR	452.25281	1		

MULTIPLE CORRELATION COEFFICIENT = .995
R SQUARED FACTOR = .990

Fig. 15-7. Regression analysis results for model 1.

Estimated Model Coefficients, Coefficient Standard Errors,
t Values and Confidence Intervals (alpha = 0.05)

	COEFFICIENT		B/SIGMA(B)	CONFIDENCE INTERVA	
TERM	B	SIGMA(B)	T	LOWER	UPPER
B0	3.5179	.2379	14.7863	3.0277	4.0080
B1	1.9321	.6729	2.8713	.5459	3.3184
B2	-.6531	.2225	-2.9348	-1.1116	-.1947
B3	-7.2406	.2225	-32.5351	-7.6991	-6.7822
B4	-7.2219	.2225	-32.4509	-7.6803	-6.7634
B5	-1.1656	.2225	-5.2376	-1.6241	-.7072
B6	-1.5594	.2225	-7.0069	-2.0178	-1.1009

THE THEORETICAL VALUE FOR T AT THE 0.025 LEVEL AND 25 DF = 2.060

REGRESSION ANALYSIS TABLE

SOURCE	SS	DF	MS	F
TOTAL (CORRECTED)	3534.25719	31	114.00830	
REGRESSION (CORRECTED)	3494.63518	6	582.43920	367.4
RESIDUAL	39.62201	25	1.58488	
CORRECTION FACTOR	452.25281	1		

MULTIPLE CORRELATION COEFFICIENT = .994
R SQUARED FACTOR = .989

Fig. 15-8. Regression analysis results for model 2.

It is suggested that the reader compare figure 15-2 with figure 15-3. In figure 15-2, the SS numbers are Sums of Squares. The number under A is the Sum of Squares due to A. It is 6.0378. In figure 15-3, the first line is for variable 1 which is A. and the Sum of Squares is 6.03781250 with one degree of freedom.

The final engineering decision is to change the nominal dimensions of variables K and L. With these two changes, a satisfactory device is obtained. If an O-ring is added to the shaft, a further improvement will be achieved, but the cost of the device will be increased. The choice of vendor of the frames likewise has a small effect, but having two vendors that are almost equally good is much better manufacturing practice than dependence upon a single vendor. As the data clearly shows, both vendors of the frames will produce good devices if the dimensions of K and L are changed. The lack of an O-ring will not produce bad devices.

Part Three

Multilevel Multivariable Experiments

16 Multilevel Experiments with Qualitative Variables

This chapter is concerned with experiments where the variables are qualitative and where at least one of the variables is at more than two levels. Typical examples of qualitative variables are different vendors, different grades of a material, the presence or absence of a component in a circuit, and different chemical additives to a formula.

Most of the principles and procedures described in the preceding chapters also apply to experiments involving qualitative variables. The experimenter must still choose values for α, β, and δ and must calculate a proper sample size (N). For experiments with qualitative variables, N is calculated with either of two formulas (the same formulas used in the preceding sections), depending upon whether σ^2 is known or only estimated:

- $N = 2(U_\alpha + U_\beta)^2 \dfrac{\sigma^2}{\delta^2}$ when σ^2 is known.
- $N = 2(t_\alpha + t_\beta)^2 \dfrac{S^2}{\delta^2}$ when σ^2 is estimated from previous data.

The designations of (+) and (−) cannot be used to signify multilevels of qualitative variables and the resultant treatment combinations because these designations, obviously, can be used only for two levels. Instead, a subscripted lowercase roman letter is used for each variable (e.g., $a_1, a_2, \ldots, a_i$; $b_1, b_2, \ldots, b_j$; etc.). If, for example, variable A is designated as type of polymer, and four polymers (Mylar, nylon, polyethylene, and acrylonitrile) are to be studied, Mylar could be designated as a_1, nylon as a_2, polyethylene as a_3, and acrylonitrile as a_4. Similar designations would be used for any other variables involved in the experiment. Thus, a treatment combination of $a_1b_2c_1d_3$ could designate, say, the first polymer, the second catalyst, the first mixing procedure, and the third posttreatment procedure.

The initial analysis of test data from experiments with qualitative variables requires the use of analysis of variance (ANOVA), as described below. The final analysis of the test data, however, involves the same procedures that have been described in previous chapters.

The next section will describe the techniques for using ANOVA. Following this will be a section describing Latin and Greco-Latin squares—useful techniques for many experiments concerned with qualitative variables.

Use of ANOVA

ANOVA (analysis of variance) is a mathematical procedure for testing hypotheses such as:

$$H_0: \mu_{a1} = \mu_{a2} = \mu_{a3} = \mu_{a4}$$
$$H_a: \mu_{ai} \neq \mu_{aj}$$

The null hypothesis states that all four levels of variable A have the same mean. The alternative hypothesis states that at least one of the four levels has a different mean. This is a very nice alternative hypothesis from a statistical viewpoint, but it is not very helpful from an engineering viewpoint; therefore, after testing these hypotheses with ANOVA, the experimenter will revert to the standard alternative hypotheses for final data analysis and for the engineering decisions:

$$H_{a1}: \mu_{a1} \neq \mu_{a2}$$
$$H_{a2}: \mu_{a2} \neq \mu_{a3}$$
$$H_{a3}: \mu_{a1} \neq \mu_{a3}$$
$$H_{a4}: \mu_{a1} \neq \mu_{a4}$$

etc.

Sample sizes will be chosen so that these latter hypotheses can be tested at the properly chosen α and β levels.

After the experiment has been properly designed, the test items fabricated and tested, and the data gathered, the data are analyzed by ANOVA according to the following sequence:

- Compute correction factor.
- Compute crude sum of squares.
- Compute sum of squares (SS).
- Compute total sum of squares (TSS).
- Compute sum of squares for two-factor interactions.
- Compute grand-total sum of squares ($GTSS$).
- Compute sum of squares for three-factor interaction.
- Prepare ANOVA table.

The following example describes the design and evaluation of an experiment involving three qualitative variables.

EXAMPLE

A film extruder is presently producing a product that is used as the substrate for a typewriter ribbon. The present material is polyethylene with C_2 additive and no posttreatment. The tensile strength of the present product is 8.1 pounds for a 12-mil thick ribbon. The sample standard deviation is 0.40 with $\phi = 10$. It is required that the tensile strength be increased, and several reasonable suggestions for doing this have been put forward:

- Change the base polymer to either Mylar or nylon.
- Posttreat the film by either calendering or orienting.
- Enhance the formula with one of four suggested new additives.

It is expected that there will be two-factor interactions, especially between the base polymer and the additives.

The number of treatment combinations required are all combinations of all levels of all variables. In this case, there are 45 treatment combinations (3 polymers × 5 additives × 3 posttreatment processes = 45). These treatment combinations are shown by the blocks in figure 16-1.

Posttreatment process ↓ / Additive →	Base polymer														
	Mylar (a_1)					Nylon (a_2)					Polyethylene (a_3)				
	c_1	c_2	c_3	c_4	c_5	c_1	c_2	c_3	c_4	c_5	c_1	c_2	c_3	c_4	c_5
Calendered (b_1)															
Oriented (b_2)															
Regular process (b_3)															

Fig. 16-1. Treatment combinations for film experiment.

A. State object of experiment.

$H_{01}: \mu_{\text{Mylar}} = \mu_{\text{nylon}} = \mu_{\text{polyethylene}}$
$H_{a1}: \mu_{\text{Mylar}} \neq \mu_{\text{nylon}} \neq \mu_{\text{polyethylene}}$
$H_{02}: \mu_{\text{calendered}} = \mu_{\text{oriented}} = \mu_{\text{regular process}}$
$H_{a2}: \mu_{\text{calendered}} \neq \mu_{\text{oriented}} \neq \mu_{\text{regular process}}$
$H_{03}: \mu_{c1} = \mu_{c2} = \mu_{c3} = \mu_{c4} = \mu_{c5}$
$H_{a3}: \mu_{c1} \neq \mu_{c2} \neq \mu_{c3} \neq \mu_{c4} \neq \mu_{c5}$

B. Choose α, β, and δ, and state S^2 and ϕ.

Let $\alpha = 0.10$, $\beta = 0.10$, $\delta = 0.6$, and $S^2 = 0.16$. $t_\alpha = 1.81$ from table 4. $t_\beta = 1.37$ from table 3. $\phi = 10$.

C. Compute $N_i = N_j = 2(t_\alpha + t_\beta)^2 \dfrac{S^2}{\delta^2}$

H_a will be doubled-sided. N_i is the number of samples required at any treatment level.

$$N_i = N_j = 2(1.81 + 1.37)^2 \frac{0.16}{0.36} = 8.99$$

D. Decide if replication is required. Count the number of blocks that have treatment c_1. The number of blocks that have c_2, c_3, etc., will be the same as the number of blocks that have c_1. The number of blocks that have a given c level will always be smaller than the number of blocks that have a given a or b level because there are five levels of variable C but only three levels each of variables A and B. Therefore, the variable with the largest number of levels is the deciding factor on the need for replication within blocks.

(1) If only one trial run is made at each treatment combination, the smallest number of trials at any level of a variable will be 9 at each level of additive, which is the correct sample size calculated above.

(2) For the other variables, there will be 15 trials at each level, which is more than the required sample size.

Decide on only one trial run at each treatment combination. The α and β values for decision on the additive will be exactly 0.10, but the α and β values for the decisions on the polymer and posttreatment will be less than 0.10 because the actual $N = 15$ at each level is larger than the required $N = 9$ calculated in (C), above.

E. Make one trial run at each treatment combination, and record the results in the design matrix (figure 16-2).

	a_1					a_2					a_3				
	c_1	c_2	c_3	c_4	c_5	c_1	c_2	c_3	c_4	c_5	c_1	c_2	c_3	c_4	c_5
b_1	9.2	8.7	9.1	12.4	10.5	8.2	7.7	11.4	8.1	9.5	9.2	13.4	9.7	9.1	8.5
b_2	8.2	8.7	8.8	11.5	10.6	7.2	7.7	10.5	7.8	9.6	8.4	12.5	9.1	9.1	8.9
b_3	8.4	8.3	8.7	8.5	8.8	7.4	7.8	7.3	7.7	7.1	8.2	8.5	8.8	8.1	8.3

Fig. 16-2. Results of experiment recorded in the design matrix.

F. Analyze the data.

The first step is to compute what is called the correction factor: sum all the results, square this sum, then divide by the number of results. In this case, the sum of all 45 results is 405.2; therefore,

$$\text{Correction} = \frac{(405.2)^2}{45} = 3{,}648.60.$$

The next step is to compute the crude sum of squares for each variable. To do this, the data are first reduced to three sets of two-way matrices: an AB matrix, an AC matrix, and a BC matrix. Figure 16-3 shows the AB matrix. The number in each block in the figure is obtained by adding together the five test results in the corresponding blocks in figure 16-2. Thus, the number 49.9 in block a_1b_1 is obtained by adding the results 9.2, 8.7, 9.1, 12.4, and 10.5 from the blocks in a_1b_1 in figure 16-2; i.e., sample results for $a_1b_1c_1 + a_1b_1c_2 + a_1b_1c_3 + a_1b_1c_4 + a_1b_1c_5$.

	a_1	a_2	a_3	$\sum b_i$	
b_1	49.9	44.9	49.9	144.7	
b_2	47.8	42.8	48.0	138.6	405.2
b_3	42.7	37.3	41.9	121.9	
$\sum a_i$	140.4	125.0	139.8		
	405.2				

Fig. 16-3. The AB matrix.

To obtain the crude sum of squares for variable A, each column total in figure 16-3 is squared, the squared results are added together, and the total is divided by the number of test results represented by the numbers in each column. For example, column a_1 in figure 16-3 represents 15 test results (5 results per block multiplied by three blocks). The crude sum of squares for variable A, therefore, is given by the relationship:

$$\text{Crude sum of squares A} = \frac{(140.4)^2 + (125.0)^2 + (139.8)^2}{15} = 3658.75.$$

The sum of squares for A (written SS_A) is given by the crude sum of squares for A minus the correction (computed above):

$$SS_A = \frac{(140.4)^2 + (125.0)^2 + (139.8)^2}{15} - \frac{(405.2)^2}{45}$$

$$= 3658.75 - 3648.60 = 10.15.$$

The sum of squares for B is obtained in the same manner by squaring each row total in figure 16-3:

$$SS_B = \frac{(144.7)^2 + (138.6)^2 + (121.9)^2}{15} - \frac{(405.2)^2}{45}$$

$$= 3667.18 - 3648.60 = 18.58.$$

A term called total sum of squares for AB (written TSS_{AB}) is next computed by squaring the number in each block of the AB matrix, dividing by five (the number of trials in each block), and then subtracting the correction:

$$TSS_{AB} = (49.9^2 + 47.8^2 + 42.7^2 + 44.9^2 + 42.8^2 + 37.3^2 + 49.9^2 + 48.0^2 + 41.9^2)/5 - 3648.60$$

$$= 28.78.$$

The sum of squares for the AB interaction is then calculated by difference:

$$SS_{AB} = TSS_{AB} - SS_A - SS_B = 28.78 - 10.15 - 18.58 = 0.05.$$

The values for SS_A, SS_B, and SS_{AB} will later be incorporated in the ANOVA table (described below).

	a_1	a_2	a_3	$\sum c_i$	
c_1	25.8	22.8	25.8	74.4	
c_2	25.7	23.2	34.4	83.3	
c_3	26.6	29.2	27.6	83.4	405.2
c_4	32.4	23.6	26.3	82.3	
c_5	29.9	26.2	25.7	81.8	
$\sum a_i$	140.4	125.0	139.8		
		405.2			

Fig. 16-4. The AC matrix.

The AC matrix is shown in figure 16-4. Values for SS_A, SS_C, TSS_{AC}, and SS_{AC} are calculated as described for the AB matrix:

- $SS_A = 10.15$, from above.
- $SS_C = (74.4^2 + 83.3^2 + 83.4^2 + 82.3^2 + 81.8^2)/9 - 3648.60 = 6.33.$
- $TSS_{AC} = (25.8^2 + 25.7^2 + 26.6^2 + \cdots + 25.7^2)/3 - 3648.60 = 49.57.$
- $SS_{AC} = 49.57 - 6.33 - 10.15 = 33.09.$

Note that the divisors are different for the AC matrix than for the AB matrix. By referring to figure 16-2, it may be seen that block a_1c_1 in figure 16-4 comprises three test values: $a_1b_1c_1$, $a_1b_2c_1$, and $a_1b_3c_1$; that is, 9.2, 8.2, and 8.4. Thus, each block in the AC matrix is the sum of three test values, while each block in the AB matrix was the sum of five test values.

The BC matrix is shown in figure 16-5. In the same manner as for the AB and AC matrices, the values for SS_B, SS_C, TSS_{BC}, and SS_{BC} are:

- $SS_B = 18.58$, from above.
- $SS_C = 6.33$, from above.
- $TSS_{BC} = (26.6^2 + 23.8^2 + 24.0^2 + 29.8^2 + \cdots + 24.2^2)/3 - 3648.60$ $= 28.05.$
- $SS_{BC} = TSS_{BC} - SS_B - SS_C = 28.05 - 18.58 - 6.33 = 3.14.$

	c_1	c_2	c_3	c_4	c_5	$\sum b_i$	
b_1	26.6	29.8	30.2	29.6	28.5	144.7	
b_2	23.8	29.9	28.4	28.4	29.1	138.6	405.2
b_3	24.0	24.6	24.8	24.3	24.2	121.9	
$\sum c_i$	74.4	83.3	83.4	82.3	81.8		
			405.2				

Fig. 16-5. The BC matrix.

The next term to be computed is the grand total sum of squares ($GTSS$). This is computed by squaring each result in the data matrix (figure 16-2), summing the resultant numbers, and subtracting the correction:

$$GTSS = (9.2^2 + 8.7^2 + 9.1^2 + \cdots + 8.3^2) - 3648.60 = 90.10.$$

The value for SS_{ABC} is then obtained by difference as follows:

$$SS_{ABC} = GTSS - SS_A - SS_B - SS_C - SS_{AB} - SS_{AC} - SS_{BC} = 18.76.$$

Having determined the sum of squares for each main effect and interaction, the experimenter is now ready to begin construction of an ANOVA table. This table has four columns:

- Variable or interaction estimated.
- Sum of squares for each effect or interaction.
- Degrees of freedom for each effect or interaction.
- Mean square for each effect or interaction.

The experimenter knows the effects and interactions and has computed the sum of squares for each of these. To complete the ANOVA table, the experimenter must now determine values for degrees of freedom and mean square for each variable or interaction.

The degrees of freedom (df) for a main effect in ANOVA of this type problem are always one less than the number of levels of the variable. Thus, there are three levels of variable A, so the number of degrees of freedom for A are $3 - 1 = 2$.

For interactions, the degrees of freedom are the product of the degrees of freedom of the individual variables that make up the interaction. In the present example, variable A, with three levels, has two degrees of freedom, and variable C, with five levels, has four degrees of freedom; therefore, the AC interaction has $2 \times 4 = 8$ degrees of freedom.

The mean square (MS) for each effect or interaction is obtained by dividing each sum of squares by the related degrees of freedom. Thus, for variable A, $SS_A = 10.15$ and df $= 2$; therefore, mean square $= 10.15/2 = 5.08$.

The ANOVA table can now be constructed as follows:

Effect or interaction	Sum of squares	Degrees of freedom	Mean square
A	10.15	2	5.08
B	18.58	2	9.29
C	6.33	4	1.58
AB	0.05	4	0.01
AC	33.09	8	4.14
BC	3.14	8	0.39
ABC	18.76	16	1.17
Total	90.10	44	

The information in the ANOVA table will be used in the evaluation of the test data, as will be described below.

G. Estimate variance.

In planning the experiment, an estimate of the standard deviation ($S = 0.4$) was available. If the above estimate was not available, and if the experimenter could assume that the ABC interaction did not exist, the mean square of this three-factor interaction could be used as an estimate of variance (S^2) with 16 degrees of freedom. However, in this case, this value ($S^2 = 1.17$) is substantially larger than the original estimate of $S^2 = 0.16$. Furthermore, inspection of the data matrix (figure 16-2) shows evidence of an ABC interaction. Therefore, we cannot conclude that there is no ABC interaction in the case. The specific things that lead to this conclusion will be discussed later.

H. Determine objective criterion.

Since there are more than two levels of the variables, the criterion used in two-level experiments cannot be used to test the significance of the main factors and interactions at more than two levels. However, the F distribution can be used, in conjunction with the estimate of the variance, by means of the following formula, to calculate the criterion:

$$MS_i^* = S^2 F_{(\phi_1, \phi_2, \alpha)},$$

where MS_i^* = criterion value for mean square; i = A, B, C, AB, AC, BC, ABC; F = value from table 10; ϕ_1 = degrees of freedom of main effect or interaction; and ϕ_2 = degrees of freedom of S^2.

From the ANOVA table, the experimenter determines that variables A and B each have 2 degrees of freedom, that variable C and interaction AB each have 4 degrees of freedom, that interactions AC and BC each have 8 degrees of freedom, and that ABC has 16 degrees of freedom. Thus, using the value for S^2 and the values of F obtained from table 10, the criteria for this problem are:

- $MS_A^* = MS_B^* = (0.4)^2[F_{(2,10,0.05)}] = (0.16)(4.10) = 0.66.$
- $MS_C^* = MS_{AB}^* = (0.4)^2[F_{(4,10,0.05)}] = (0.16)(3.48) = 0.56.$
- $MS_{AC}^* = MS_{BC}^* = (0.4)^2[F_{(8,10,0.05)}] = (0.16)(3.07) = 0.49.$
- $MS_{ABC}^* = (0.4)^2[F_{(16,10,0.05)}] = (0.16)(2.85) = 0.46.$

I. Make statistical decisions.

Since ($MS_A = 5.07$) > ($MS_A^* = 0.66$), one concludes with at least 95 percent confidence that the mean strength (μ) of at least one of the materials is different from the other two. This implies that maybe all three of the materials have different strengths.

The results for A, B, C, AC, and ABC are significant at greater than 95 percent confidence. Therefore, with at least 95 percent confidence, we accept the alternative hypothesis that at least one of the materials, at least one of the catalysts, and at least one of the posttreatments causes a difference in tensile strength.

J. Make engineering decisions.

The above statistical decisions are not very informative to engineering management. They want to know what material and process gives the best product. The engineer should therefore proceed as follows. (It should be noted that there are many statistical techniques for this stage, but only this one is given in this book.)

Inspection of the data matrix (figure 16-2), with due consideration of the statistical decisions above, indicates that the best choices for the product are:

- Mylar with additive c_4, either calendered or oriented.
- Nylon with additive c_3, either calendered or oriented.
- Polyethylene with additive c_2, either calendered or oriented.

The question of calendering vs orienting can be resolved by a simple t test:

$$t = \frac{\bar{X}_{\text{cal}} - \bar{X}_{\text{or}}}{S\sqrt{\dfrac{1}{N_{\text{cal}}} + \dfrac{1}{N_{\text{or}}}}} = \frac{(144.7/15) - (138.6/15)}{0.4\sqrt{\frac{1}{15} + \frac{1}{15}}} = \frac{0.407}{(0.4)(0.365)} = 2.79.$$

The term $\bar{X}_{\text{cal}}$ is the mean of the 15 results on samples produced by calendering; $\bar{X}_{\text{or}}$ is the mean of the 15 results on samples produced by orienting. From table 3, with $\phi = 10$, a t value of 2.76 is required for 99.0 percent confidence. Since the calculated t value is greater than 2.76, it is concluded, with greater than 99.0 percent confidence, that calendering is better than orienting; that is, $\mu_{\text{cal}} > \mu_{\text{or}}$.

The question could also be asked, Is calendered polyethylene with additive c_2 better than calendered Mylar with c_4 additive? Again using a t test:

$$t = \frac{13.4 - 12.4}{0.4\sqrt{(\frac{1}{1} + \frac{1}{1})}} = \frac{1.0}{(0.4)(1.414)} = 1.77.$$

From table 3, with $\phi = 10$, a t value of 1.81 is required for 95 percent confidence. Therefore, with approximately 95 percent confidence, it can be stated that calendered polyethylene with c_2 additive is the strongest of all the treatment combinations. Note that, in this case, there is only one sample ($N = 1$) made at the treatment combination (calendered, polyethylene, additive c_2) and only one sample at the treatment combination (calendered, nylon, additive c_4).

Latin and Greco-Latin Squares

With multilevel qualitative variables, it is sometimes possible to design Resolution IV or Resolution III experiments. Such designs are particularly applicable when one or more of the variables are nuisance variables. A nuisance variable is one that, if not controlled, can either bias the estimate of the variables of interest or increase the estimate of variance. For example, an automobile manufacturer might be interested in determining the gasoline mileage of one of the company's cars as compared to the mileage delivered by similar models made by competitors. The results of such a test will be influenced by the drivers, by the weather on the day a car is tested, by the time of day, etc. Only the cars are of interest to the experimenter; the other variables are nuisance variables.

Greco-Latin square designs are available for up to $p + 1$ variables each at p levels, where $p = 3, 4, 5, 7, 8$, or 9. The simplest Latin square is with $p = 3$, as shown in figure 16-6.

Note that, in the design shown in figure 16-6, each machine is tested by each operator and is tested on all three days. Thus, in comparing one machine with another, both the day effect and the operator effect should cancel out. In this manner, the nuisance variables are eliminated.

	Machine 1	Machine 2	Machine 3
Operator 1	Day 1	Day 2	Day 3
Operator 2	Day 3	Day 1	Day 2
Operator 3	Day 2	Day 3	Day 1

Fig. 16-6. Latin square for $p = 3$.

To illustrate the generation of a Greco-Latin square, assume that an experimenter will be dealing with a four-level experiment (i.e., $p = 4$). With $p = 4$, the experimenter will be able to handle $p + 1 = 5$ variables.

In this illustration, there are three of what are called mutually orthogonal Latin squares of size 4. Let X designate the first variable, with X_1 through X_4 being the four levels of X. Similarly, let Y_1 through Y_4 designate the four levels of the second variable, Y. The levels of the third variable are designated by the letters A, B, C, D. An initial Latin square can then be obtained as shown in figure 16-7.

	X_1	X_2	X_3	X_4
Y_1	A	B	C	D
Y_2	D	C	B	A
Y_3	B	A	D	C
Y_4	C	D	A	B

Fig. 16-7. Initial Latin square with $p = 4$.

A second orthogonal Latin square is obtained by writing down the first row and then rotating the other three rows:

A B C D → A B C D
D C B A ↘ C D A B ←
B A D C ↘ D C B A
C D A B ↘ B A D C

The uppercase Latin letters of this second Latin square are then converted to Greek letters to denote the levels of the fourth variable: A → α, B → β, C → γ, and D → δ. This second Latin square then becomes:

$\alpha\ \beta\ \gamma\ \delta$
$\gamma\ \delta\ \alpha\ \beta$
$\delta\ \gamma\ \beta\ \alpha$
$\beta\ \alpha\ \delta\ \gamma$

The third orthogonal set is obtained by repeating the above process with the second Latin square:

```
A B C D ——→ A B C D
C D A B      B A D C ←┐
D C B A  ↘   C D A B  │
B A D C  ↘   D C B A  │
   └──────────────────┘
```

This third Latin square is usually labeled with lowercase Latin letters to denote the fifth variable:

a b c d
b a d c
c d a b
d c b a

For all five variables, then, the complete experiment design is obtained by assembling all three Latin squares into one square as shown in figure 16-8. If there had been only four variables, only two of the above three steps would have been used. It should also be noted that the variable of interest and the variables of nuisance can be placed in any position.

	X_1	X_2	X_3	X_4
Y_1	Aαa	Bβb	Cγc	Dδd
Y_2	Dγb	Cδa	Bαd	Aβc
Y_3	Bδc	Aγd	Dβa	Cαb
Y_4	Cβd	Dαc	Aδb	Bγa

Fig. 16-8. Completed Greco-Latin square for 5-variable experiment.

To illustrate the use of this square, the treatment combination in the upper left-hand corner would be all five variables at their lowest levels; the treatment combination in the lower right-hand corner would be the highest level of the first two variables, the second level of the third variable, the third level of the fourth variable, and the lowest level of the fifth variable. The complete experiment would require 16 trials, one trial at each of the 16 treatment combinations.

Note that, in this design, all the degrees of freedom are used to estimate effects:

Variable 1	$\phi = 3$	(ϕ = Number of levels − 1)
2	$\phi = 3$	
3	$\phi = 3$	
4	$\phi = 3$	
5	$\phi = 3$	
Total	$\phi = 15$	

Since the total number of degrees of freedom is the total number of trials minus one, no degrees of freedom are available for an estimate of variance. Some prior estimate of the variance must be available to calculate an objective criterion to test the hypotheses. If no estimate of variance is available, this design cannot be used.

The following example illustrates the use of a Greco-Latin square to design the experiment and the use of ANOVA to analyze the data.

EXAMPLE

Four different keyboard configurations are proposed for a new typewriter. To evaluate the relative merits of each keyboard, one typewriter with each keyboard will be used to type four different types of documents that represent the range of work expected to be encountered in the office environment. The experimenter, being familiar with Greco-Latin squares, proposes that four typists should be used and that the test be conducted over a period of four days.

The responses of interest are the speed of typing and the error rate. From previous experiments of this type, it is known that the variance for typing speed is 2.5 words per minute (wpm) and that the variance for errors is 0.75 errors per 100 characters for the results of one day of typing.

A. State the object of the experiment.

Let variable A = keyboard configuration.

$H_{01}: \mu_{Ai} = \mu_{Aj}$ where H_1 = wpm.
$H_{a1}: \mu_{Ai} \neq \mu_{Aj}$

$H_{02}: \mu_{Ai} = \mu_{Aj}$ where H_2 = errors/100 characters.
$H_{a2}: \mu_{Ai} \neq \mu_{Aj}$

B. Choose α, β, and δ.

$\alpha = 0.05$
$\beta = 0.10$
$\delta_1 = 4$ wpm
$\delta_2 = 2$ errors per 100 characters

C. Calculate N = number of days of testing required.

$$N = 2(U_\alpha + U_\beta)^2 \frac{\sigma^2}{\delta^2}$$

U_α is obtained from table 2.
U_β is obtained from table 1.

$$N_{\text{wpm}} = 2(1.960 + 1.282)^2 \frac{2.5}{(4)^2} = 3.28$$

$$N_{\text{error}} = 2(1.960 + 1.282)^2 \frac{0.75}{(2)^2} = 3.94$$

(Note: in a 16-trial Greco-Latin square, there will be 4 trials on each machine configuration; hence, an unreplicated experiment is satisfactory.)

D. Determine the treatment combinations (figure 16-9).

E. Obtain the test data (figures 16-10 and 16-11).

	Keyboard A_1	Keyboard A_2	Keyboard A_3	Keyboard A_4
Test document 1	Day 1 Typist α	Day 2 Typist β	Day 3 Typist γ	Day 4 Typist δ
Test document 2	Day 4 Typist γ	Day 3 Typist δ	Day 2 Typist α	Day 1 Typist β
Test document 3	Day 2 Typist δ	Day 1 Typist γ	Day 4 Typist β	Day 3 Typist α
Test document 4	Day 3 Typist β	Day 4 Typist α	Day 1 Typist δ	Day 2 Typist γ

Fig. 16-9. Treatment combinations for keyboard experiment.

	Keyboard A_1	Keyboard A_2	Keyboard A_3	Keyboard A_4
Test document 1	39	52	69	73
Test document 2	65	64	44	54
Test document 3	60	63	51	44
Test document 4	48	38	68	73

Fig. 16-10. Test data—typing speed (wpm).

	Keyboard A_1	Keyboard A_2	Keyboard A_3	Keyboard A_4
Test document 1	3	1	0	4
Test document 2	0	3	2	1
Test document 3	3	0	1	5
Test document 4	1	4	2	2

Fig. 16-11. Test data—errors per 100 characters.

F. Analyze the typing-speed data using ANOVA.

- Correction = (sum of all data)^2/16
  = $(905)^2/16$
  = 819,025/16 = 51,189
- Sum of squares for keyboard configuration

  = (sum of each column)^2/4 − correction
  = $(39 + 65 + 60 + 48)^2 + (52 + 64 + 63 + 38)^2 + (69 + 44 + 51 + 68)^2 + (73 + 54 + 44 + 73)^2/4$ − correction
  = (44,944 + 47,089 + 53,824 + 59,536)/4 − 51,189
  = 205,393/4 − 51,189
  = 51,348 − 51,189 = 159
- Mean square for keyboard configuration = sum of squares/degrees of freedom = 159/3 = 53
- Since $\sigma_0^2 = 2.5$ is known, and the experimenter is only interested in making a decision on the keyboard configuration, compute:

  $$F = \frac{MS(\text{configuration})}{\sigma_0^2} = \frac{53}{2.5}$$
  $$= 21.2.$$

  Compare this value with the value of F (table 10) for $\phi_1 = 3$ and $\phi_2 = \infty$ degrees of freedom and $\alpha = 0.05$, which is 2.60.
- Therefore, accept the alternative hypothesis that at least one of the keyboard configurations has a different mean speed. The actual means are

  $\bar{X}_1 = 53.00$
  $\bar{X}_2 = 54.25$
  $\bar{X}_3 = 58.00$
  $\bar{X}_4 = 61.00$
- Test if keyboard A_4 is actually faster than keyboard A_3. (Remember σ^2 is known; therefore, the U test is correct.)

$$U = \frac{\bar{X}_4 - \bar{X}_3}{\sigma\sqrt{\frac{1}{4} + \frac{1}{4}}} = \frac{3}{\sqrt{2.5}\sqrt{0.5}}$$

$$= \frac{3}{(1.58)(0.71)} = 2.67$$

$U = 2.67$ corresponds to better than 95 percent confidence (table 1) that μ_4 is faster than μ_3. It can be calculated that μ_3 is faster than μ_2 and that $\mu_1 = \mu_2$ by using the same formula as above with $\bar{X}_3 - \bar{X}_2 = 3.75$ and $\bar{X}_2 - \bar{X}_1 = 1.25$.

- If the experimenter had been interested in determining the effect of document type on the typing speed, it would be necessary to calculate MS(document). It is suggested that the reader perform this calculation. (Document type 3 can be typed faster than the others.)

G. Analyze the error-rate data.

- Correction = (sum of all data)^2/16 $= (32)^2/16 = 64$
- Sum of squares (configuration) $= \sum(\text{sum of each column})^2/4 - \text{correction}$ $= [(7^2 + 8^2 + 5^2 + 12^2)/4] - 64$ $= 282/4 - 64 = 70.5 - 64 = 6.5$
- MS(configuration) $= SS/\text{df} = 6.5/3 = 2.17$
- $F = MS/\sigma_0^2 = 2.17/0.75 = 2.89$
- Since $F = 2.89$ is greater than the table 10 value for $F_{(3,\infty,0.05)} = 2.60$, accept the alternative hypothesis that at least one of the keyboards has an effect on error rate. The mean error rates are

$\bar{X}_1 = 1.75$
$\bar{X}_2 = 2.00$
$\bar{X}_3 = 1.25$
$\bar{X}_4 = 3.00$

The comparison of machine 3 with machine 4 is tested as follows:

$$U = \frac{\bar{X}_4 - \bar{X}_3}{\sigma\sqrt{\frac{1}{4} + \frac{1}{4}}} = \frac{3.0 - 1.25}{(0.87)\sqrt{0.5}} = 2.87$$

This value of U is compared with table 1 and indicates that machine 3 is better than machine 4, with better than 99.5 percent confidence. Using the same procedure, machine 1 is shown to be equal to machine 2.

H. Combine the decisions on speed and error rate.

- Keyboard configuration A_4 is fastest, whereas configuration A_3 has the fewest errors per 100 characters. Keyboard A_4 is poorest for errors. Therefore, keyboard A_3 is the best compromise.

If the variance had not been known in the above problem, the experimenter would calculate the total sum of squares, the sum of squares due to configuration, the sum of squares due to typist, the sum of squares due to days, and the sum of squares due to test document. Then:

$$TSS - SS(\text{configuration}) - SS(\text{typist}) - SS(\text{days}) - SS(\text{document}) = SS(\text{error}).$$

$$(\phi = 15) - (\phi = 3) - (\phi = 3) - (\phi = 3) - (\phi = 3) = (\phi = 3).$$

$$MS\ (\text{error}) = \frac{SS\ (\text{error})}{3} = S^2, \text{ with 3 degrees of freedom.}$$

This chapter has described the basic statistical procedure of ANOVA for multilevel, multivariable problems with qualitative variables. Two basic designs have been illustrated, but many variations of these designs to meet specific engineering limitations can be obtained. Such experiments are beyond the scope of this book. It is recommended that the reader consult a more advanced text before starting this type of experiment.

EXERCISES

1. A research team is attempting to use plastic bearings with steel shafts in a certain mechanism. There are four different plastic bearing materials, three different steel shaft materials, and two different lubricants that can be used.
   - Lay out the experimental plan to evaluate the effect of these variables on bearing wear.
   - What are the number of degrees of freedom associated with each main effect and each interaction?
2. An experiment has been conducted with six different rubber types, two different designs of a steel insert, and four different adhesives. The response of interest is the force required to remove the insert after curing the component. The results obtained are shown in the figure on page 235.

Analyze the data completely and make engineering decisions. Write a one-page report to management, giving them your conclusions.

	Adhesive 1		Adhesive 2		Adhesive 3		Adhesive 4	
	Design 1	Design 2	Design 1	Design 2	Design 1	Design 2	Design 1	Design 2
Rubber A	685	655	650	630	820	800	670	650
B	695	660	700	640	675	640	765	740
C	675	650	685	650	690	680	780	755
D	720	690	810	790	660	625	660	630
E	710	685	835	800	675	665	680	665
F	680	645	690	680	680	650	675	655

Results of experiment.

3. In the keyboard example in this chapter, calculate the mean square for documents, the mean square for days, and the mean square for typists for the speed data.
   a. Is there any difference in speed or error rate for different documents?
   b. Was there a day-to-day effect?
   c. Are the typists of equal ability?

REFERENCES

Box, G. E. P., Hunter, W. G., and Hunter, J. S. 1978. *Statistics for experimenters.* New York: John Wiley & Sons, Inc.

Davies, O. L. 1967. *Design and analysis of industrial experiments.* New York: Hafner Press.

John, P. W. M. 1971. *Statistical design and analysis of experiments.* New York: The Macmillan Co.

17 Multilevel Experiments with Quantitative Variables

If an experimenter is interested in determining the effects of temperature, time, and pressure on the strength of a molded part, all combinations of, perhaps, four temperatures, five times, and three pressures might be tested for a total of 60 trials. However, with a properly designed experiment, the same or better information can be obtained with as few as 20 trials. With more variables, the saving in samples and time that can be achieved with proper design is even more impressive. A 2-variable problem will be used to demonstrate the procedure.

Suppose the experimenter is interested in testing the effects of four temperatures and five times on the tensile strength of a molded item. Samples might be obtained by setting the temperature at 200° and making samples at 40 seconds, 45 seconds, 50 seconds, 55 seconds, and 60 seconds. Then the temperature could be changed to 220° and samples made at the five times, and so on for a total of 20 trials, with the test results shown in Figure 17-1.

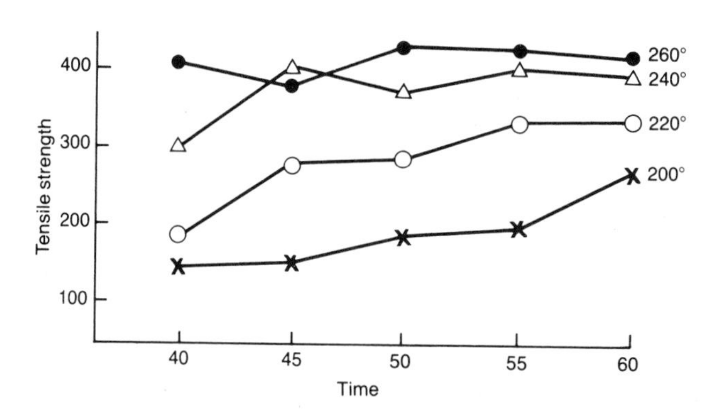

Fig. 17-1. Test results of 20 samples made at four temperatures and five times.

Almost certainly the lines shown in the figure do not correspond to the true lines that God defined for the effects of time and temperature on tensile strength. God's response lines are usually smooth and generally don't cross over in isolated spots. Remember, the purpose of an experiment is to estimate God's law.

Central Composite Rotatable Designs

A better procedure for this problem is the central composite rotatable design described in the following steps.

(1) Define the space of interest to the experimenter. In this case, the area is obviously 40 to 60 seconds and 200° to 260°.

(2) The general design of the trials to be conducted is expressed in coded terms below, and the terms are plotted as in figure 17-2.

Trial	A	B
1	-1	-1
2	$+1$	-1
3	-1	$+1$
4	$+1$	$+1$
5	$-\psi$	0
6	$+\psi$	0
7	0	$-\psi$
8	0	$+\psi$
9	0	0
10	0	0
11	0	0
12	0	0
13	0	0

The terms -1 and $+1$ have the same meaning here as in the Hadamard matrix, and the 0 level of a variable is the center value of the variable.

(3) Any value may be used for ψ, and there may be any number of replicated points in the center. However, the experiment design will have special properties if the value of ψ is set at 1.4142 and if the number of points in the center is exactly five, with two variables. Trials 5 through 8 are called

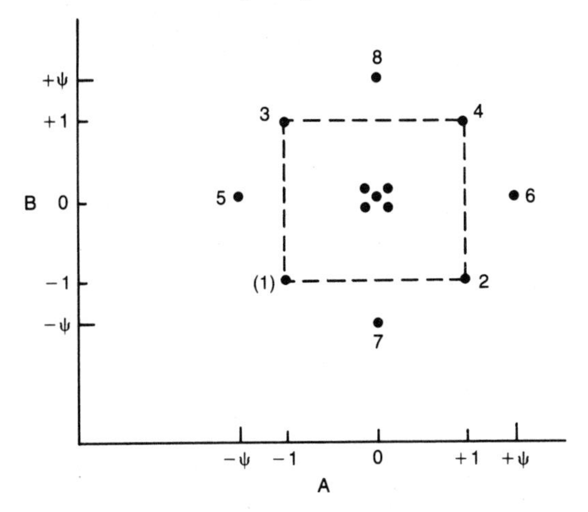

Fig. 17-2. Plot of trials in coded terms.

star points of the design, and trials 9 through 13 are called center points. If $\psi = 1.4142$ is set equal to half the ranges of the variables (e.g., time of 40 to 60 seconds), one can obtain the scaling factor for this specific experiment.

Scale of A: $1.4142A = (60 - 40)(1/2)$, $A = (20)(1/2)/1.4142$

$A = 7.07$ seconds/coded unit.

$$A_{\text{coded}} = \frac{(\text{time} - 50)}{7.07}, \quad \text{or} \quad \text{Real A} = (\text{A coded})(7.07) + 50.$$

For example, -1 on the coded scale $= (-1)(7.07) + 50$
$= 42.95$ seconds

$$B_{\text{coded}} = \frac{(\text{temp} - 230)}{21.21}, \quad \text{or} \quad \text{Real B} = (\text{B coded})(21.21) + 230.$$

For example, 0 on the coded scale $= (0)(21.21) + 230 = 230°$.

(4) In noncoded terms, the trials are:

Trial	Time	Temp
1	43	209
2	57	209
3	43	251
4	57	251
5	40	230
6	60	230
7	50	200
8	50	260
9 to 13	50	230

After the trials have been conducted and the test results have been obtained, the procedure for analysis is given in the following steps.

(1) The design matrix is expanded as follows, and the results are tabulated.

Trial	T	A	B	A^2	B^2	AB	Results, psi
1	1	−1	−1	1	1	+1	210
2	1	+1	−1	1	1	−1	280
3	1	−1	+1	1	1	−1	365
4	1	+1	+1	1	1	+1	420
5	1	−1.414	0	2	0	0	250
6	1	+1.414	0	2	0	0	380
7	1	0	−1.414	0	2	0	190
8	1	0	+1.414	0	2	0	420
9	1	0	0	0	0	0	330
10	1	0	0	0	0	0	335
11	1	0	0	0	0	0	340
12	1	0	0	0	0	0	335
13	1	0	0	0	0	0	335

Note that the A^2 column is obtained by squaring every term in the A column. The AB column is obtained by multiplying the term in the A column by the term in the B column.

(2) Multiply the term in each column by the result and sum the column; this is exactly the same procedure that is used with the Hadamard matrix experiments.

T	A	B	A^2	B^2	AB
(+1)(210)	(−1)(210)	(−1)(210)	(1)(210)	(1)(210)	(+1)(210)
(+1)(280)	(+1)(280)	(−1)(280)	(1)(280)	(1)(280)	(−1)(280)
(+1)(365)	(−1)(365)	(+1)(365)	(1)(365)	(1)(365)	(−1)(365)
(+1)(420)	(+1)(420)	(+1)(420)	(1)(420)	(1)(420)	(+1)(420)
(+1)(250)	(−1.414)(250)	(0)(250)	(2)(250)	0	0
(+1)(380)	(+1.414)(380)	(0)(380)	(2)(380)	0	0
(+1)(190)	(0)(190)	(−1.414)(190)	0	(2)(190)	0
(+1)(420)	(0)(420)	(+1.414)(420)	0	(2)(420)	0
(+1)(330)	(0)(330)	(0)(330)	0	0	0
(+1)(335)	(0)(335)	(0)(335)	0	0	0
(+1)(340)	(0)(340)	(0)(340)	0	0	0
(+1)(335)	(0)(335)	(0)(335)	0	0	0
(+1)(335)	(0)(335)	(0)(335)	0	0	0
4190	308.85	620.27	2535	2495	−15

(3) The above column totals ($T = 4190$, $A = 308.85$, $B = 620.27$, $A^2 = 2535$, $B^2 = 2495$, and $AB = -15$) are then used to compute the following terms:

$$b_0 = 0.2(T) - 0.1(A^2 + B^2)$$
$$b_1 = 0.125(A)$$
$$b_2 = 0.125(B)$$
$$b_{11} = 0.125(A^2) + 0.01875(A^2 + B^2) - 0.1(T)$$
$$b_{22} = 0.125(B^2) + 0.01875(A^2 + B^2) - 0.1(T)$$
$$b_{12} = 0.25(AB)$$

$$b_0 = (0.2)(4190) - 0.1(2535 + 2495) = 335$$
$$b_1 = 0.125(308.85) = 38.61$$
$$b_2 = 0.125(620.27) = 77.53$$
$$b_{11} = 0.125(2535) + 0.01875(2535 + 2495) - 0.1(4190) = -7.81$$
$$b_{22} = 0.125(2495) + 0.01875(2535 + 2495) - 0.1(4190) = -12.81$$
$$b_{12} = 0.25(-15) = -3.75$$

These terms, b_0, b_1, etc., are then substituted into what is called a regression equation. Therefore, tensile strength $= 335 + 38.61A + 77.53B - 3.75AB - 7.81A^2 - 12.81B^2$, where A can be any value of A in coded terms and B can be any value of B in coded terms.

(4) An estimate of the variance is obtained from the center points and the usual formula for S^2; that is,

$$S^2 = \frac{\sum(X_i - \bar{X})^2}{N_c - 1},$$

where N_c is the number of center points, which is 5; $S^2 = 12.5$ with 4 df; and $\phi = N_c - 1 = 4$. Therefore, $S = 3.54$.

(5) A test of significance can be made on each term in the regression equation using the following standard errors (SE) and the t test, where $t = b/SE$ of b.

$$SE(b_i) = S\sqrt{0.125} = 0.354S$$
$$= 1.28; \quad \text{this is the standard error of } b_1 \text{ and } b_2.$$

$$SE(b_{ii}) = S\sqrt{0.125 + 0.01875} = 0.379S$$
$$= 1.34; \quad \text{this is the standard error for } b_{11} \text{ and } b_{22}.$$

$$SE(b_{ij}) = S\sqrt{0.25} = 0.5S$$
$$= 1.77; \text{ this is the standard error for } b_{12}.$$

$t(b_1) = \dfrac{38.61}{1.28} = 30.89$; significant at 99+ percent confidence (30.89 is compared to the values of t with four degrees of freedom in table 3).

$t(b_2) = \dfrac{77.53}{1.28} = 62.02$; significant at 99+ percent confidence.

$t(b_{12}) = \dfrac{|-3.75|}{1.77} = 2.12$; significant at 95 percent confidence.

$t(b_{11}) = \dfrac{|-7.81|}{1.34} = 5.82$; significant at 99+ percent confidence.

$t(b_{22}) = \dfrac{|-12.81|}{1.34} = 9.56$; significant at 99+ percent confidence.

(6) If any term is not significant at $\alpha = 0.10$, it is eliminated from the response equation.

(7) The response equation can be used to compute the estimated response at any point in the system. It can also be used to generate a response-surface map, as shown in figure 17-3.

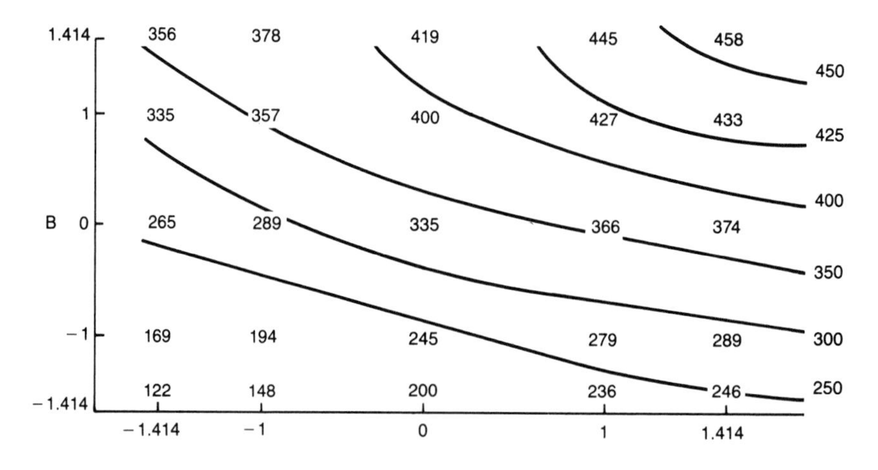

Fig. 17-3. Response-surface map.

The data of this experiment could have been analyzed with a regression analysis program using actual times and temperatures. Exactly the same figure as above would be obtained except that the labeling of A and B would be actual times and temperatures.

The reader should be cautioned that the data from the above experiment can be and should be subjected to advanced analytical techniques beyond the scope of this book (e.g., analysis of residuals or testing of the fit of the quadratic equation). In this particular case, for example, it would be shown that there is a big lack-of-fit term in an ANOVA table. However, it would also be seen by this same table that the only reason for the lack of fit (which is unimportant in this engineering situation) is that the variance of the experiment is so small.

Designs for more than two variables can be constructed in a manner similar to the above to obtain what are called uniform-precision, central-composite, rotatable designs. The parameters of such designs are given below:

No. of variables	No. of Hadamard matrix trials	No. of star trials	No. of center trials	ψ
2	4	4	5	1.4142
3	8	6	6	1.6820
4	16	8	7	2.0000
5	16	10	6	2.0000
6	32	12	9	2.3780
7	64	14	14	2.8280
8	128	16	20	3.3640

Thus, for example, if an experimenter has a 5-variable problem and must estimate all two-factor interactions and the curvature due to all variables, the design will consist of:

- A Resolution V Hadamard matrix design for five variables in 16 trials with I = ABCDE.
- Ten star points located at coded values of

A	B	C	D	E
−2	0	0	0	0
+2	0	0	0	0
0	−2	0	0	0
0	+2	0	0	0
0	0	−2	0	0
0	0	+2	0	0
0	0	0	−2	0
0	0	0	+2	0
0	0	0	0	−2
0	0	0	0	+2

- Six center points located at 00000.

The coefficients of the regression equations and the standard errors of the coefficients are computed as follows:

- Three Variables

  $b_0 = 0.1663402267(\mathrm{T}) - 0.05679210581\ (\mathrm{A}^2 + \mathrm{B}^2 + \mathrm{C}^2),$

  $b_i = 0.0732233047(i),$

  $b_{ii} = 0.062500(ii) + 0.00689003779(\mathrm{A}^2 + \mathrm{B}^2 + \mathrm{C}^2) - 0.05679210581(\mathrm{T})$, and

  $b_{ij} = 0.125000(ij),$

  where i = A, B, or C; $ii = \mathrm{A}^2, \mathrm{B}^2$, or C^2; T = total; ij = AB, BC, or AC.

  Standard error $b_i = 0.271S$

  Standard error $b_{ii} = 0.263S$

  Standard error $b_{ij} = 0.354S$

- Four Variables

  $b_0 = 0.1428571429(\mathrm{T}) - 0.03571428571(\mathrm{A}^2 + \mathrm{B}^2 + \mathrm{C}^2 + \mathrm{D}^2).$

  $b_i = 0.04166666667(i).$

  $b_{ii} = 0.031250(ii) + 0.003720238095(\mathrm{A}^2 + \mathrm{B}^2 + \mathrm{C}^2 + \mathrm{D}^2) - 0.03571428571(\mathrm{T}).$

  $b_{ij} = 0.0625(ij).$

  Standard error $b_i = 0.204S$

  Standard error $b_{ii} = 0.185S$

  Standard error $b_{iu} = 0.250S$

- Five Variables

  $b_0 = 0.1590909091(\mathrm{T}) - 0.03409090909(\mathrm{A}^2 + \mathrm{B}^2 + \mathrm{C}^2 + \mathrm{D}^2 + \mathrm{E}^2).$

  $b_i = 0.04166666667(i).$

  $b_{ii} = 0.031250(ii) + 0.002840909091(\mathrm{A}^2 + \mathrm{B}^2 + \mathrm{C}^2 + \mathrm{D}^2 + \mathrm{E}^2) - 0.03409090909(\mathrm{T}).$

  $b_{ij} = 0.0625(ij).$

  Standard error $b_i = 0.204S$

  Standard error $b_{ii} = 0.185S$

  Standard error $b_{ij} = 0.250S$

- Six Variables

  $b_0 = 0.1107488731(\mathrm{T}) - 0.01873806408(\sum ii),$
  where $\sum ii$ is $\mathrm{A}^2 + \mathrm{B}^2 + \mathrm{C}^2$, etc.

  $b_i = 0.02308737891(i).$

  $b_{ii} = 0.015625(ii) + 0.001217246271(\sum ii) - 0.01873806408(\mathrm{T}).$

  $b_{ij} = 0.03125(ij).$

  Standard error $b_i = 0.152S$

Standard error $b_{ii} = 0.130S$

Standard error $b_{ij} = 0.177S$

- Seven Variables

$b_0 = 0.0703125(\text{T}) - 0.009765625(\sum ii)$.

$b_i = 0.0125(i)$.

$b_{ii} = 0.0078125(ii) + 0.00048828125(\sum ii) - 0.009765625(\text{T})$.

$b_{ij} = 0.015625(ij)$.

Standard error $b_i = 0.112S$

Standard error $b_{ii} = 0.091S$

Standard error $b_{ij} = 0.125S$

- Eight Variables

$b_0 = 0.04505289317(\text{T}) - 0.005301719474(\sum ii)$.

$b_i = 0.006638897619(i)$.

$b_{ii} \doteq 0.00390635(ii) + 0.00023326899(\sum ii) - 0.005301719474(\text{T})$.

$b_{ij} = 0.0078125(ij)$.

Standard error $b_i = 0.0815S$

Standard error $b_{ii} = 0.064S$

Standard error $b_{ij} = 0.088S$

Designs for Experiments Where the Levels of Some of the Variables Are Different

The experimenter will frequently encounter situations where some of the variables are expected to have only linear effects and other variables are expected to have effects that are quadratic or even cubic. Designs for such experiments can be derived from the Hadamard matrices discussed in Chapter 6. These derived designs are called Addleman designs. The general procedure for deriving these designs will be demonstrated with the order-8 Hadamard matrix:

0	1	2	3	4	5	6	7
+	+	−	−	+	−	+	+
+	+	+	−	−	+	−	+
+	+	+	+	−	−	+	−
+	−	+	+	+	−	−	+
+	+	−	+	+	+	−	−
+	−	+	−	+	+	+	−
+	−	−	+	−	+	+	+
+	−	−	−	−	−	−	−

If an experimenter has a 2-variable problem, with variable A at two levels and variable B_1 at three levels, contrast column 1 can be labeled as A, but columns

2 and 3, normally labeled B and C, must be used to obtain three levels of variable B_1. The following transform rule is used:

Column 2	Column 3	B_1	
−	−	= −1	Lowest level of B_1, designated b^-.
−	+	= 0	Middle level of B_1, designated b^0.
+	−	= 0	Middle level of B_1, designated b^0.
+	+	= +1	Highest level of B_1, designated b^+.

Note that the symbol B_1 is used to identify the variable that will have three levels to prevent any possible confusion with the labeling of column 2 with the letter B in the two-level Hadamard matrix designs.

The treatment combinations can now be obtained from the following:

Trial	Total	A	B_1	Treatment combination
	0	1	2 & 3	
1	+	+	−1	ab^-
2	+	+	0	ab^0
3	+	+	+1	ab^+
4	+	−	+1	b^+
5	+	+	0	ab^0
6	+	−	0	b^0
7	+	−	0	b^0
8	+	−	−1	b^-

The conventional labeling of the contrast columns of the 8-trial Hadamard matrix is A, B, C, −AB, −BC, ABC, and −AC. This labeling must be modified before the matrix can be used for analysis of experimental data obtained from an Addleman experiment. Several modifications of the principles of the defining contrast (described in Chapter 6) must be made, but this procedure is then used to determine the labeling of the matrix. The following general rules are used:

- $A \times A = I$,

  $B \times B = I$, etc.,

  where A, B, etc., are two-level variables or two-level contrast columns used to obtain three levels of a variable.

  $A \times I = A$,

  $B \times I = B$, etc.,

  but,

  $B_1 \times B_1 = B_1^2$, etc.,

  where B_1 is the three-level variable obtained by the Addleman procedures.

- The identity elements are obtained by multiplying the equality by the initial label in the contrast column.

$$\begin{array}{r l}
\text{B} = & \text{B}_1 \\
\times\ \text{B} & \\
\hline
\text{I} = & \text{BB}_1
\end{array}$$

$$\begin{array}{r l}
\text{C} = & \text{B}_1 \\
\times\ \text{C} & \\
\hline
\text{I} = & \text{B}_1\text{C}
\end{array}$$

$$\text{I} = (\text{BB}_1)(\text{B}_1\text{C}) = \text{BCB}_1^2$$

- Therefore, the defining contrast is

  $\text{I} = \text{BB}_1 + \text{B}_1\text{C} + \text{BCB}_1^2.$

The labeling of the contrast columns is then obtained by multiplying each column label by the defining contrast:

$$\begin{aligned}
\text{A}(\text{I} &= \text{BB}_1 + \text{B}_1\text{C} + \text{BCB}_1^2) = (\text{A} = \text{ABB}_1 + \text{AB}_1\text{C} + \text{ABCB}_1^2)\\
\text{B}(\text{I} &= \text{BB}_1 + \text{B}_1\text{C} + \text{BCB}_1^2) = (\text{B} = \text{B}_1 + \text{BCB}_1 + \text{CB}_1^2)\\
\text{C}(\text{I} &= \text{BB}_1 + \text{B}_1\text{C} + \text{BCB}_1^2) = (\text{C} = \text{BCB}_1 + \text{B}_1 + \text{BB}_1^2)\\
-\text{AB}(\text{I} &= \text{BB}_1 + \text{B}_1\text{C} + \text{BCB}_1^2) = (-\text{AB} = -\text{AB}_1 - \text{ABCB}_1 - \text{ACB}_1^2)\\
-\text{BC}(\text{I} &= \text{BB}_1 + \text{B}_1\text{C} + \text{BCB}_1^2) = (-\text{BC} = -\text{CB}_1 - \text{BB}_1 - \text{B}_1^2)\\
\text{ABC}(\text{I} &= \text{BB}_1 + \text{B}_1\text{C} + \text{BCB}_1^2) = (\text{ABC} = \text{ACB}_1 + \text{ABB}_1 + \text{AB}_1^2)\\
-\text{AC}(\text{I} &= \text{BB}_1 + \text{B}_1\text{C} + \text{BCB}_1^2) = (-\text{AC} = -\text{ABCB}_1 - \text{AB}_1 - \text{ABB}_1^2)
\end{aligned}$$

All confounding terms that include B or C are crossed out, since there is neither a variable B nor a variable C in the experiment.

The correct labeling of the original Hadamard matrix is, therefore:

Trial			B_1	B_1	$-AB_1$	$-B_1^2$	AB_1^2	$-AB_1$	Treatment combination
		A	~~B~~	~~C~~	−~~AB~~	−~~BC~~	~~ABC~~	−~~AC~~	
1	+	+	−	−	+	−	+	+	ab^-
2	+	+	+	−	−	+	−	+	ab^0
3	+	+	+	+	−	−	+	−	ab^+
4	+	−	+	+	+	−	−	+	b^+
5	+	+	−	+	+	+	−	−	ab^0
6	+	−	+	−	+	+	+	−	b^0
7	+	−	−	+	−	+	+	+	b^0
8	+	−	−	−	−	−	−	−	b^-

Several variations must also be made in the use of the above matrix for the analysis of the data:

- Only the four results that do not include b^0 can be used to estimate the linear effect of A; that is

  $(\bar{X}_{\text{high}} - \bar{X}_{\text{low}})_{\text{A}} = (\text{ab}^- + \text{ab}^+ - \text{b}^+ - \text{b}^-)/2.$

  Therefore, the criterion for testing $H_0{:}\mu_{\text{high A}} = \mu_{\text{low A}}$ is

  $|\bar{X}_{\text{high A}} - \bar{X}_{\text{low A}}|^* = U_\alpha\sigma\sqrt{\tfrac{1}{2} + \tfrac{1}{2}}$ or $t_\alpha S\sqrt{\tfrac{1}{2} + \tfrac{1}{2}}.$

- Two contrast columns are labeled B_1. Each of these two columns is summed as usual, using all the data, and then this sum is divided by 4, as usual. The sum of these two averages is the linear effect of B_1. The criterion for testing $H_0: \mu_{\text{high } B_1} = \mu_{\text{low } B_1}$ is

$|\bar{X}_{\text{high } B_1} - \bar{X}_{\text{low } B_1}|^* = U_\alpha \sigma \sqrt{\frac{1}{2} + \frac{1}{2}}$ or $t_\alpha S \sqrt{\frac{1}{2} + \frac{1}{2}}$.

- Two contrast columns are labeled $-AB_1$. Each of these two columns is summed as usual, using all the data, and then this sum is divided by 4, as usual. The sum of these two averages is the linear, linear interaction of A and B_1; i.e., AB_1. The criterion for testing $H_0: \mu_{AB_1} = 0$ is

  $|\bar{X}_{\text{high } AB_1} - \bar{X}_{\text{low } AB_1}|^* = U_\alpha \sigma \sqrt{\frac{1}{2} + \frac{1}{2}}$ or $t_\alpha S \sqrt{\frac{1}{2} + \frac{1}{2}}$.
- Only one contrast column is labeled $-B_1^2$. This column is summed as usual, using all the data, and then this sum is divided by 4, as usual. The criterion is the same as above except that $N_{\text{high}} = N_{\text{low}} = 4$.
- Only one contrast column is labeled AB_1^2. This column is treated exactly the same as the $-B_1^2$ column above.
- An estimate of the variance with one degree of freedom can be obtained from the sample results of trials 2 and 5:

  $$S^2 = \frac{(\text{trial 2} - \text{trial 5})^2}{2}.$$

  A second estimate of the variance with one degree of freedom can be obtained from trials 6 and 7:

  $$S^2 = \frac{(\text{trial 6} - \text{trial 7})^2}{2}.$$

  These two S^2 values are then averaged to obtain S_{avg}^2 with two degrees of freedom.

EXAMPLE

A certain resin is used as an undercoat to promote adhesion in a magnetic tape. At the present time, an average force of 21 inch-pounds (μ_0) is required to break the bond. The variance (σ^2) is 1.0. It is proposed that either additive X or additive Y might enhance the adhesion. The concentration of the additive is probably critical, but it is expected that the correct amount should be between 1 percent and 2 percent. There probably will be a quadratic effect of concentration. An improvement of 4 inch-pounds (δ) would be very desirable.

A. Define the object of the experiment.

$H_{01}: \mu_X = \mu_Y$
$H_{a1}: \mu_X \neq \mu_Y$

$H_{02}: \mu_{2\%} = \mu_{1\%}$
$H_{a2}: \mu_{2\%} \neq \mu_{1\%}$

$H_{03}: \mu_{\text{linear } AB_1} = 0$
$H_{a3}: \mu_{\text{linear } AB_1} \neq 0$

$H_{04}: \mu_{\text{quad } B_1} = 0$
$H_{a4}: \mu_{\text{quad } B_1} \neq 0$

$H_{05}: \mu_{AB_1^2} = 0$
$H_{a5}: \mu_{AB_1^2} \neq 0$

Variable A = additive type
Variable B_1 = % additive

B. Choose α, β, and δ.

$\alpha = 0.10$
$\beta = 0.01$
$\delta = 4$

C. Compute N.

$$N = 2(U_\alpha + U_\beta)^2 \frac{\sigma^2}{\delta^2}$$

$U_{0.10} = 1.645$ from table 2 because the alternative hypotheses are double-sided.
$U_{0.01} = 2.326$ from table 1.
$N = 2(1.645 + 2.326)^2 \frac{1}{16} = 1.97$

D. Decide upon a proper experiment design, and designate levels of the variables.

The 8-trial Addleman design will measure all the effects with adequate precision.

Variables	Levels		
	−1	0	+1
A (additive)	Additive X		Additive Y
B_1 (concentration)	1%	1.5%	2%

E. Make and test the required eight samples.

Trial	Treatment combination	Results
1	ab^-	23
2	ab^0	30
3	ab^+	26
4	b^+	25
5	ab^0	32
6	b^0	24.5
7	b^0	22.5
8	b^-	22

F. Compute criterion.

$$|\bar{X}_{high} - \bar{X}_{low}|^* = U_\alpha \sigma \sqrt{\frac{1}{N_{high}} + \frac{1}{N_{low}}}$$

$|\bar{X}_{high} - \bar{X}_{low}|^* = (1.645)(1)\sqrt{\frac{1}{2} + \frac{1}{2}} = 1.645$
for effects of A, B_1, and AB_1.
$|\bar{X}_{high} - \bar{X}_{low}|^* = (1.645)(1)\sqrt{\frac{1}{4} + \frac{1}{4}} = 1.152$
for effects of B_1^2 and AB_1^2.

G. Compute mean effects.

	A	B_1	B_1	$-AB_1$	$-B_1^2$	AB_1^2	$-AB_1$
	+23	−23	−23	+23	−23	+23	+23
		+30	−30	−30	+30	−30	+30
	+26	+26	+26	−26	−26	+26	−26
	−25	+25	+25	+25	−25	−25	+25
		−32	+32	+32	+32	−32	−32
		+24.5	−24.5	+24.5	+24.5	+24.5	−24.5
		−22.5	+22.5	−22.5	−22.5	+22.5	+22.5
	−22	−22	−22	−22	−22	−22	−22
$\sum(X_{high} - X_{low}) =$	2	6	6	4	13	−13	−4
$(\bar{X}_{high} - \bar{X}_{low}) =$	1	1.5	1.5	1.0	3.25	−3.25	−1.0

H. Compare $(\bar{X}_{high} - \bar{X}_{low})$ with $|\bar{X}_{high} - \bar{X}_{low}|^*$ and make decisions.

Variable B_1 has a significant effect
$[(B_1 = 1.5 + 1.5) = (3.0)] > [(\bar{X}_{high} - \bar{X}_{low})^* = 1.645]$

There is also a quadratic effect, B_1^2 and a linear quadratic interaction, AB_1^2. There is no linear, linear interaction $[AB_1 = 1 + (-1) = 0]$.

I. Plot the results (figure 17-4).

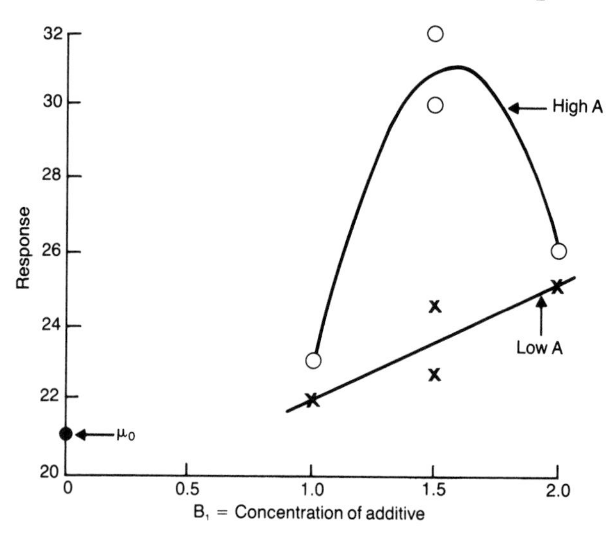

Fig. 17-4. Plot of test results.

Note: if the experiment had been conducted with only two levels of A and two levels of B_1 (namely $B_1 = 1.0$ and $B_1 = 2.0$), and these results connected with straight lines, the lines would be parallel. Therefore, we say there is no AB_1, that is, no linear, linear interaction. The AB_1^2 interaction is called a linear, quadratic interaction; that is, one of the lines has different curvature than the other. In this case, one line has no curvature and the other line substantial curvature. AB_1^2 would also be significant if one line was curved concave and the other was curved convex.

J. Make an engineering statement of the conclusions based on this experiment.

Additive X improves the adhesion; the amount of the improvement increases linearly as the concentration increases from 1 percent to 2 percent. Additive Y increases the adhesion as the concentration is increased from 1 percent to 1.5 percent, but a still higher concentration decreases the adhesion. Use additive Y at a concentration of 1.5 percent for maximum adhesion. If an adhesion of about 25 inch-pounds is satisfactory, and if additive X is cheaper, you should use additive X at 2 percent concentration. It is probable that a still higher concentration of additive X will give a result higher than 25 inch-pounds.

A Resolution III design for one variable at three levels and up to four variables at two levels may be obtained by labeling the transformed Hadamard matrix as follows:

Trial	Total	A	B_1	D	B_1^2	E	F	Treatment combinations
		1	2 & 3	4	5	6	7	
1	+	+	−1	+	−	+	+	$ab^{-1}def$
2	+	+	0	−	+	−	+	ab^0f
3	+	+	+1	−	−	+	−	ab^+e
4	+	−	+1	+	−	−	+	b^+df
5	+	+	0	+	+	−	−	ab^0d
6	+	−	0	+	+	+	−	b^0de
7	+	−	0	−	+	+	+	b^0ef
8	+	−	−1	−	−	−	−	b^-

Note: contrasts 2 and 3 were used to derive the levels of variable B_1; contrast 4 was then labeled D, etc. Contrast column 5 measures the quadratic effect of B_1: therefore, it could not be used for variable E.

It should also be noted that, in this Resolution III design, only four trials, those without b^0, are used to estimate the linear effect of variable B_1 and that the criterion is

$$|\bar{X}_{\text{high}} - \bar{X}_{\text{low}}|^* = t_\alpha S\sqrt{\tfrac{1}{2} + \tfrac{1}{2}} \quad \text{or} \quad U_\alpha \sigma\sqrt{\tfrac{1}{2} + \tfrac{1}{2}}.$$

B_1^2 and all other linear effects are estimated with all eight trials; therefore, the criterion is

$$|\bar{X}_{\text{high}} - \bar{X}_{\text{low}}|^* = t_\alpha S\sqrt{\tfrac{1}{4} + \tfrac{1}{4}} \quad \text{or} \quad U_\alpha \sigma\sqrt{\tfrac{1}{4} + \tfrac{1}{4}}.$$

16-Trial Experiment Designs with Variables at Two Levels and Three Levels

Two Variables at Two Levels and One Variable at Three Levels

The 16-trial Hadamard matrix can also be used to generate multilevel designs using the Addleman procedure. A Resolution V design for two variables at two levels and one variable at three levels can be obtained by using contrast columns 1 and 2 for the two-level variables and combining contrast columns 3 and 4 for the three-level variables. The identity elements will then be:

$$\begin{array}{r} C = C_1 \\ \times\ C \\ \hline I = CC_1 \end{array}$$

$$\begin{array}{r} D = C_1 \\ \times\ D \\ \hline I = DC_1 \end{array}$$

$$I = (CC_1)(DC_1) = CDC_1^2.$$

The defining contrast is then

$$I = CC_1 + DC_1 + CDC_1^2.$$

The various contrast column headings become:

1	2	3	4	5	6	7	8	9	10	11	12	13	14	15
A	B	C_1	C_1	AB	BC_1	C_1^2	ABC_1	AC_1	BC_1	ABC_1	BC_1^2	ABC_1^2	AC_1^2	AC_1

Values for A, B, and AB are estimated only with the eight trials that are not at c^0. Therefore, the criterion is

$$|\bar{X}_{\text{high}} - \bar{X}_{\text{low}}|^* = U_\alpha \sigma\sqrt{\tfrac{1}{4} + \tfrac{1}{4}} \quad \text{or} \quad t_\alpha S\sqrt{\tfrac{1}{4} + \tfrac{1}{4}}.$$

Values for C_1, AC_1, BC_1, and ABC_1 are each estimated from two columns. The algebraic sum of the mean effect of these two columns is compared to the criterion

$$|\bar{X}_{\text{high}} - \bar{X}_{\text{low}}|^* = U_\alpha \sigma\sqrt{\tfrac{1}{4} + \tfrac{1}{4}} \quad \text{or} \quad t_\alpha S\sqrt{\tfrac{1}{4} + \tfrac{1}{4}}.$$

Values for C_1^2, AC_1^2, BC_1^2, and ABC_1^2 are each measured by only one contrast column, and the criterion for each of these effects is

$$|\bar{X}_{\text{high}} - \bar{X}_{\text{low}}|^* = U_\alpha \sigma\sqrt{\tfrac{1}{8} + \tfrac{1}{8}} \quad \text{or} \quad t_\alpha S\sqrt{\tfrac{1}{8} + \tfrac{1}{8}}.$$

Three Variables at Two Levels and One Variable at Three Levels

A Resolution V design with three variables at two levels and one variable at three levels can be obtained from the 16-trial Hadamard matrix.

- Use contrast columns 1 and 2 and the Addleman principle to generate the three levels of variable A_1.
- Use contrast columns 3, 4, and 13 for variables C, D, and E at two levels. Note that contrast column 13 is ABCD, so, in effect, we are creating an identity I = ABCDE.
- Compute identities for the three-level variable:

$$\begin{array}{l} A = A_1 \\ \times\ A \\ \hline I = AA_1 \end{array}$$

$$\begin{array}{l} B = A_1 \\ \times\ B \\ \hline I = BA_1 \end{array}$$

$$I = (AA_1)(BA_1) = ABA_1^2.$$

- Each of the above identities must also be multiplied by I = ABCDE.

$$I = (AA_1)(ABCDE) = BCDEA_1$$
$$I = (BA_1)(ABCDE) = ACDEA_1$$
$$I = (ABA_1^2)(ABCDE) = CDEA_1^2$$

- The defining contrast is, therefore,

$$I = AA_1 + BA_1 + ABA_1^2 + ABCDE + BCDEA_1 + ACDEA_1 + CDEA_1^2.$$

- The labeling of the various contrast columns is

1	2	3	4	5	6	7	8	9	10	11	12	13	14	15
A_1	A_1	C	D	A_1^2	CA_1	EA_1^2	DA_1^2	CA_1	DA_1	CA_1^2	EA_1	E	EA_1	DA_1
						CD	CE			DE				

- The algebraic sum of the mean effects of the two contrasts for A_1, for CA_1, for DA_1, and for EA_1 are compared to the criterion with $N_{high} = N_{low} = 4$.
- A_1^2, $(CD + EA_1^2)$, $(CE + DA_1^2)$, and $(DE + CA_1^2)$ are measured only once with all the data, and the criterion uses $N_{high} = N_{low} = 8$.

If the computer program of Chapter 13 is used for the analysis of this design, type in the statement

16 FAN 5

The first set of output data will be labeled

Main Effects

A B C D E

The A and B columns should be relabeled A_1. The values for C, D, and E will be incorrect since they include the values at a^0. The linear effect of C, D, and E must be calculated by paper and pencil. The sum of the mean effects of the two A_1 columns is the linear effect of A.

The second set of output data will be labeled

Interactions

AB BC CD AC BD AD AE BE CE DE

These must be relabeled

A_1^2	CA_1	CD	A_1C	DA_1	A_1D	A_1E	A_1E	CE	DE
		EA_1^2						DA_1^2	CA_1^2

The A_1^2 column is the correct quadratic effect of A_1. The sum of the two contrasts for A_1C, A_1D, and A_1E is the correct value of these interactions.

The $(CD + EA_1^2)$, $(CE + DA_1^2)$, and $(DE + CA_1^2)$ columns are correct estimates. EA_1^2, DA_1^2, and CA_1^2 are assumed to be zero.

Two Variables at Two Levels and Two Variables at Three Levels

A Resolution V design can be obtained from the 16-trial Hadamard matrix by using contrast columns A and B for variable A_1, contrast columns C and D for B_1, column $-AC$ for E, and column $-BD$ for F. Then the primary identity elements will be

$$I = AA_1 = BA_1 = CB_1 = DB_1 = ABA_1^2 = CDB_1^2 = ACA_1B_1$$
$$= BDA_1B_1 = BCA_1B_1 = ADA_1B_1$$

and

$$I = -ACE = -BDF = ABCDEF.$$

Of course other identity elements can be obtained by multiplying the elements in the first group with elements in the second group. For example,

$$I = (AA_1)(-ACE) = -CEA_1$$
$$I = (BA_1)(-BDF) = -DFB_1$$
$$I = (CB_1)(-ACE) = -AEB_1.$$

Thus, the various contrast columns are labeled:

1	2	3	4	5	6	7	8	9	10	11	12	13	14	15
A_1	A_1	B_1	B_1	$-A_1^2$		$-B_1^2$		E	F					
$-EB_1$	$-FB_1$	$-EA_1$	$-FA_1$		$-A_1B_1$		$-FA_1$	$-A_1B_1$	$-A_1B_1$	$-EA_1$	$-FB_1$	EF	$-EB_1$	$-A_1B_1$

- The effect of $A_{1/2}$ is measured with either column 1 or 2, using only the results that are not at a^0. If column 1 is used, the value of $-EB_1$ from column 14 must be first subtracted from the result. The criterion uses $N_{high} = N_{low} = 4$.
- A_1^2, B_1^2, and EF are obtained from columns 5, 7, and 13, respectively, using all the data. $N_{high} = N_{low} = 8$ for the criterion.
- FA_1 and EA_1 are obtained from their respective columns using only the results that are not at a^0. EB_1 and FB_1 use only the results that are not at b^0. $N_{high} = N_{low} = 4$ is used in the criterion.
- A_1B_1 is obtained from the four results that do not contain either a^0 or b^0. $N_{high} = N_{low} = 2$ for the criterion.
- E and F are obtained from their respective columns using all the results. The mean effect of A_1B_1 must be added to the above calculated values. $N_{high} = N_{low} = 8$ for the criterion.

(Note: The computer program of Chapter 13 *cannot* be used for design or analysis of any part of this experiment.)

Four Variables at Two Levels and One Variable at Three Levels

The first four contrast columns are used for the two-level variables A, B, C, and D. Contrast column 8 (ABD) and contrast column 11 (ABC) are used for the three-level variable E_1.

The defining contrast is

$$I = ABDE_1 + ABCE_1 + CDE_1^2.$$

Therefore, CD will be confounded with the quadratic effect E_1^2, and the experimenter will have to select the variables to be called C and D so that there is no possibility of an interaction between them.

The contrast columns of the Hadamard matrix, therefore, measure the following main effect and interactions:

1	2	3	4	5	6	7	8	9	10	11	12	13	14	15
A	B	C DE_1^2	D CE_1^2	$-DE_1$ $-CE_1$ $-AB$	$-AE_1$ $-BC$	$-E_1^2$ $-CD$	E_1	$-BE_1$ $-AC$	$-AE_1$ $-BD$	E_1	BE_1^2	CE_1 DE_1	AE_1^2	$-BE_1$ $-AD$

- Values for A, B, C, and D are measured by columns 1, 2, 3, and 4, respectively, using all the data. $N_{\text{high}} = N_{\text{low}} = 8$ for the criterion. (Note: DE_1^2 and CE_1^2 must be assumed to be zero.)
- E_1 is measured by either contrast column 8 or 11, using only the data not at e^0. $N_{\text{high}} = N_{\text{low}} = 4$ for the criterion.
- E_1^2 is measured by contrast column 7. $N_{\text{high}} = N_{\text{low}} = 8$ for the criterion.
- The linear quadratic interactions BE_1^2 and AE_1^2 can be estimated from columns 12 and 14. $N_{\text{high}} = N_{\text{low}} = 8$ for the criterion.
- The mean effects of contrast columns 5, 6, 9, 10, 13, and 15 measure groups of interactions. The criterion for these contrasts uses $N_{\text{high}} = N_{\text{low}} = 8$.

(Note: The computer program of Chapter 13 *cannot* be used for design or analysis of any part of this experiment.)

32-Trial Experiment Designs with Variables at Two Levels and Three Levels

Some very useful experiment designs can be obtained from the 32-trial Hadamard matrix using the Addleman principle. A Resolution V design can be obtained with one variable at three levels and four variables at two levels. Contrast columns 1 and 2 are combined to give the three levels of variable A_1 and contrast column 16 (ABCDE) is labeled F. The defining contrast is then

$$I = AA_1 + BA_1 + ABA_1^2 + ABCDEF + BCDEFA_1 + ACDEFA_1 + CDEFA_1^2.$$

The mean linear effect of A_1 is then the sum of contrast columns 1 and 2. Contrast column 19 (AB) is the mean quadratic effect of A_1 (i.e., A_1^2). Contrast column 16 is a measure of the average effect of variable F.

The computer program in Chapter 13 can be used for the design and analysis of this experiment. The computer instructions must be

32 DES 6 and 32 FAN 6.

The computer program will correctly identify all variables and interactions except A_1, A_1^2, A_1C, A_1D, A_1E, and A_1F. The experimenter must cross out AB and replace it with A_1^2, cross out AC and replace it with A_1C, cross out BC and replace it with A_1C, etc. The true effect of A_1C is the sum of the two columns headed by A_1C; likewise for A_1D, etc.

The criterion for A_1^2, C, D, E, F, and all the interactions of C, D, E, and F is

$$U_\alpha \sigma \sqrt{\tfrac{1}{16} + \tfrac{1}{16}} \quad \text{or} \quad t_\alpha S \sqrt{\tfrac{1}{16} + \tfrac{1}{16}}.$$

The criterion for A_1 and all interactions of A_1 is

$$U_\alpha \sigma \sqrt{\tfrac{1}{8} + \tfrac{1}{8}} \quad \text{or} \quad t_\alpha S \sqrt{\tfrac{1}{8} + \tfrac{1}{8}}.$$

Resolution IV designs can be obtained for 1 variable at three levels and 5 to 14 variables at two levels. The variable with three levels is obtained by combining contrasts 1 and 2 and labeling the variable A_1. Variables C, D, and E occupy their regular positions in the matrix. Variables F and G are put into contrast columns 9 (ACD) and 10 (BDE), respectively, to obtain the design for five variables at two levels. The contrast columns where each main effect and interaction appears is given in figure 17-5. (The computer program in Chapter 13 can be used to obtain all the designs in figures 17-5, 17-6, and 17-7.) For the experiment with one variable at three levels and five variables at two levels, the design is obtained with the command

32 DES 7

The first two columns of the design are converted into three levels, using the Addleman principle, and labeled A_1. The analysis of the data is performed by the computer using the statement

32 FAN 7

	Number of variables at two levels									
Contrast	5	6	7	8	9	10	11	12	13	14
1	A_1									
2	A_1									
3	C									
4	D									
5	E									
6	$DF+A_1C$	A_1H			GK		EM	JN	IO	LP
7	A_1D+EG	FH	CI				KM	LN	A_1O	JP
8	CE		GI	DJ		HL	A_1M	FN	KO	A_1P
9	F									
10	G									
11	A_1E			FJ	IK	A_1L	CM	DN	GO	HP
12		H								
13			I							
14				J						
15	EF	GH		A_1J	A_1K	IL	DM	CN		OP
16					K					
17	A_1G			HJ	CK	DL	IM	A_1N	EO	FP
18						L				
19	A_1^2	CH	FI		JK	EL		GN	DO	MP
20	A_1C	A_1H	DI	GJ			LM	KN	FO	EP
21	$CD+A_1F$		A_1I	EJ		KL		MN	HO	GP
22	$DE+A_1G$			CJ	HK		FM	A_1N	LO	IP
23							M			
24	A_1F	DH	A_1I		EK	JL	GM		CO	NP
25	CG		EI	A_1J	A_1K	FL		HN	MO	DP
26								N		
27	FG	EH			DK	CL	A_1M	IN	JO	A_1P
28									O	
29										P
30	$CF+A_1D$		HI			GL	JM	EN	A_1O	KP
31	$DG+A_1E$			IJ	FK	A_1L	HM		NO	CP

Fig. 17-5. One variable at three levels for 32-trial designs.

The computer will print out the following main effects:

A B C D E F G.

The labels A and B are crossed out and replaced with A_1. The computer will then print out the number of the interaction columns:

6 7 8 11 15 17 19, etc.

Figure 15-5 gives the actual interactions that should be written in under each number. For example, contrast 6 measures DF + A_1C, contrast 8 measures CE, contrast 19 measures A_1^2, etc., for five variables at two levels.

The criterion for the sum of the mean effects of the two A_1 columns is

$$|\bar{X}_{\text{high}} - \bar{X}_{\text{low}}|^* = U_\alpha \sigma \sqrt{\tfrac{1}{8} + \tfrac{1}{8}} \quad \text{or} \quad t_\alpha S \sqrt{\tfrac{1}{8} + \tfrac{1}{8}}.$$

The criterion for all other columns is

$$U_\alpha \sigma \sqrt{\tfrac{1}{16} + \tfrac{1}{16}} \quad \text{or} \quad t_\alpha S \sqrt{\tfrac{1}{16} + \tfrac{1}{16}}.$$

A design with one variable at three levels and six variables at two levels is obtained when one additional three-factor interaction (ABC) is labeled H. The experiment design is obtained from the computer program with the command

32 DES 8

The three levels of variable A_1 are obtained exactly as described above, and the analysis of the data and relabeling of the output, etc., proceeds exactly as described in the previous paragraphs. The labeling of the analysis output is given in figure 17-5. (Note: Since there are six variables at two levels, the interaction column numbers measure all the interactions listed under five variables as well as those listed under six variables. For example, contrast 15 measures EF + GH, and contrast 19 measures (A_1^2 + CH.)

It should be pointed out that A_1^2 is confounded with CH. Therefore, the experimenter should apply the C and the H labels to variables that are not expected to interact with each other.

Additional variables at two levels can be added to the design as follows:

BCD	I*	(Note: I* = variable I, not identity.)
CDE	J	
ABCDE	K	
ABE	L	
ACE	M	
ADE	N	
ABD	O	
BCE	P	

All the procedures for these designs are exactly the same as described above. The labeling of the analysis output is given in figure 17-5.

Resolution IV designs with 2 variables at three levels and 3 to 12 variables at two levels can be obtained by exactly the same procedure as described above for one variable at three levels. The labeling of the contrast columns is shown in figure 17-6. The computer program of Chapter 13 can be used exactly as described above for all these designs. For two variables at three levels and four variables at two levels, use the command

32 DES 8

Resolution IV designs for 3 variables at three levels and 1 to 10 variables at two levels can be obtained exactly as described above. The labeling of the contrast columns is given in figure 17-7.

With two or three variables at three levels, $N_{\text{high}} = N_{\text{low}}$ for the criterion is exactly the same as with one variable at three levels. With three variables at three levels and two variables at two levels, the computer command for the design is

32 DES 8

The general rule for these designs is to count each three-level variable as the equivalent of two two-level variables.

Contrast	Number of variables at two levels									
	3	4	5	6	7	8	9	10	11	12
1	A_1									
2	A_1									
3	B_1									
4	B_1									
5	E									
6	B_1F	A_1H			GK		EM	JN	IO	LP
7	EG	FH	B_1I				KM	LN	A_1O	JP
8	B_1E		GI	B_1J		HL	A_1M	FN	KO	A_1P
9	F									
10	G									
11	A_1E			FJ	IK	A_1L	B_1M	B_1N	GO	HP
12		H								
13			I							
14				J						
15	EF	GH		A_1J	A_1K	IL	B_1M	B_1N		OP
16					K					
17	A_1G			HJ	B_1K	B_1L	IM	A_1N	EO	FP
18						L				
19	A_1^2	B_1H	FI		JK	EL		GN	B_1O	MP
20		A_1H	B_1I	GJ			LM	KN	FO	EP
21	$B_1^2 + A_1F$		A_1I	EJ		KL		MN	HO	GP
22	$A_1G + B_1E$			B_1J	HK		FM	A_1N	LO	IP
23							M			
24	A_1F	B_1H	A_1I		EK	JL	GM		B_1O	NP
25			EI	A_1J	A_1K	FL		HN	MO	B_1P
26								N		
27	FG	EH			B_1K	B_1L	A_1M	IN	JO	A_1P
28									O	
29										P
30	B_1F		HI			GL	JM	EN	A_1O	KP
31	$B_1G + A_1E$			IJ	FK	A_1L	HM		NO	B_1P

Fig. 17-6. Two variables at three levels for 32-trial designs.

Contrast	Number of variables at two levels									
	1	2	3	4	5	6	7	8	9	10
1	A_1									
2	A_1									
3	B_1									
4	B_1									
5	C_1									
6	B_1C_1	A_1H			GK		C_1M	JN	IO	LP
7	$A_1B_1 + B_1C_1 + C_1G$	C_1H	B_1I				KM	LN	A_1O	JP
8	B_1C_1		GI	B_1J		HL	A_1M	C_1N	KO	A_1P
9	C_1									
10	G									
11	A_1C_1			C_1J	IK	A_1L	B_1M	B_1N	GO	HP
12		H								
13			I							
14				J						
15	C_1^2	GH		A_1J	A_1K	IL	B_1M	B_1N		OP
16					K					
17	A_1G			HJ	B_1K	B_1L	IM	A_1N	C_1O	C_1P
18						L				
19	A_1^2	B_1H	C_1I		JK	C_1L		GN	B_1O	MP
20	A_1B_1	A_1H	B_1I	GJ			LM	KN	C_1O	C_1P
21	$B_1^2 + A_1C_1$		A_1I	C_1J		KL		MN	HO	GP
22	$B_1C_1 + B_1G + A_1G$			B_1J	HK		C_1M	A_1N	LO	IP
23							M			
24	A_1C_1	B_1H	A_1I		C_1K	JL	GM		B_1O	NP
25	B_1G		C_1I	A_1J	A_1K	C_1L		HN	MO	B_1P
26								N		
27		C_1H			B_1K	B_1L	A_1M	IN	JO	A_1P
28									O	
29										P
30	$A_1B_1 + B_1C_1$		HI			GL	JM	C_1N	A_1O	KP
31	$A_1C_1 + B_1G$			IJ	C_1K	A_1L	HM		NO	B_1P

Fig. 17-7. Three variables at three levels for 32-trial designs.

64-Trial Experiment Designs with Variables at Two Levels and Three Levels

The same procedure that was used with 32-trial experiments is used to obtain these designs. Resolution V designs are obtained by using the first six contrast columns of the Hadamard matrix, contrast column 20 (ABCD), and contrast column 24 (ACEF).

With the above contrast columns, the experimenter can obtain Resolution V designs as follows:

- One variable at three levels, six variables at two levels.
- Two variables at three levels, four variables at two levels.
- Three variables at three levels, two variables at two levels.

The computer program in Chapter 13 can be used to obtain all of these designs with the statement

64 DES 8

and modification of the output using the Addleman procedure.

Resolution IV designs can be obtained for 3 variables at three levels and from 3 to 26 variables at two levels in 64 trials. These designs can be obtained from the

Contrast			Variables at two levels 3	4	5	6	7	8
1	A_1							
2	A_1							
3	B_1							
4	B_1							
5	C_1							
6	C_1							
7	$A_1^2 + C_1G$	$+C_1H$				B_1L		
8	A_1B_1		$+C_1I$			A_1L	B_1M	
9	B_1^2						A_1M	C_1N
10	B_1C_1			GJ				B_1N
11	C_1^2	$+GH$						
12	G							
13	A_1B_1		$+GI$			A_1L		
14	A_1B_1				GK		B_1M	
15	B_1C_1					HL		B_1N
16	B_1C_1			HJ	A_1K		IM	
17		H						
18			I					
19					IK	B_1L	A_1M	HN
20							C_1M	A_1N
21						JL		C_1N
22				J				
23			$+HI$				JM	KN
24					K			
25	A_1C_1	$+A_1H$						
26	$A_1C_1 + A_1G$		$+B_1I$					
27						L		
28							M	
29								N
30								
31	C_1G	$+C_1H$		B_1J				
32								
33	A_1B_1				C_1K		LM	
34	A_1C_1	$+A_1H$			JK			MN
35	B_1C_1		$+A_1I$			GL		
36								
37								
38								
39		B_1H		C_1J				LN
40			C_1I					
41								
42				IJ			HM	A_1N
43				A_1J	HK			IN
44								
45					B_1K		GM	
46								
47								
48	B_1G		$+A_1I$			C_1L	KM	JN
49								
50								
51								
52	B_1G			C_1J	A_1K			
53								
54								
55		B_1H				C_1L		
56			B_1I			KL	C_1M	
57								
58								
59				B_1J				GN
60								
61				A_1J	C_1K			
62								
63	$A_1C_1 + A_1G$				B_1K	IL		

Fig. 17-8. Three variables at three levels for 64-trial designs.

computer program by using the statement

64 DES NV

NV is the effective number of variables; that is, each three-level variable counts as two and each two-level variable counts as one. Therefore, with three variables at three levels and five variables at two levels, NV = 11.

The data from these experiments can be analyzed with the computer program in Chapter 13. The command statement will be

64 FAN NV

where NV is the same number computed above. All the rules previously described for this type of design must be used. The main effects and interactions measured by each contrast column are given in figure 17-8 for designs with up to eight two-level variables.

Resolution IV designs for 2 variables at three levels and from 5 to 28 variables at two levels can be attained by using the computer program as described above. The main effects and interactions measured by each contrast are given in figure 17-9 for up to eight variables at two levels.

Other Multilevel Experiment Designs

Sidney Addleman, in *Technometrics* (Vol. 4, No. 1, 1962), presents a series of matrices that can be used to generate designs where the variables are at a variety of different levels. The 8-, 16-, and 32-trial designs are Hadamard matrices; the other designs, which require 9, 18, 25, and 27 trials, are not Hadamard matrices, but they are particularly useful when two or more of the variables have either three or five levels. The 9-trial design plan for four variables at three levels is:

Trial	A	B	C	D
1	−1	−1	−1	−1
2	−1	0	0	+1
3	−1	+1	+1	0
4	0	−1	0	0
5	0	0	+1	−1
6	0	+1	−1	+1
7	+1	−1	+1	+1
8	+1	0	−1	0
9	+1	+1	0	−1

Note that this is a Resolution III design; the linear and quadratic contrasts may be estimated, but interactions may not.

Usually, it doesn't make much sense to use four levels of a quantitative variable. However, if it is necessary to have four levels for a particular experiment, the procedure with the Hadamard matrices is similar to that described above for three levels. For example, with one variable at four levels and one variable at two levels in an 8-trial Hadamard matrix, columns 2 and 3 are transformed as follows:

Column 2	Column 3	B_1	
−	−	= −3	Lowest level of variable.
−	+	= −1	Next-to-lowest level of variable.
+	−	= +1	Next-to-highest level of variable.
+	+	= +3	Highest level of variable.

Contrast	Variables at 2 levels					5	6	7	8
1	A_1								
2	A_1								
3	B_1								
4	B_1								
5	E								
6	F								
7	A_1^2	+	FG	+	EH				B_1L
8	A_1B_1	+				FI			A_1L
9	B_1^2								
10	B_1E						GJ		
11	EF	+			GH				
12			G						
13	A_1B_1	+				GI			A_1L
14	A_1B_1							GK	
15	B_1E								HL
16	B_1F						HJ	A_1K	
17					H				
18						I			
19								IK	B_1L
20									
21									JL
22							J		
23						HI			
24								K	
25	A_1E	+			A_1H				
26	A_1F	+	A_1G	+		B_1I			
27									L
28									
29									
30									
31			EG	+	FH		B_1J		
32									
33	A_1B_1							FK	
34	A_1E	+			A_1H			JK	
35	B_1F	+				A_1I			GL
36									
37									
38									
39					B_1H		FJ		
40						EI			
41									
42							IJ		
43							A_1J	HK	
44									
45								CK	
46									
47									
48			B_1G	+		A_1I			FL
49									
50									
51									
52			B_1G				EJ	A_1K	
53									
54									
55					B_1H				EL
56						B_1I			KL
57									
58									
59							B_1J		
60									
61							A_1J	EK	
62									
63	A_1F	+	A_1G					B_1K	IL

Fig. 17-9. Two variables at three levels for 64-trial designs.

Note in particular that the interval between levels is the same; in this case, two coded units. For example, if the variable is temperature, the actual levels of the variable might be:

Coded level	Actual level
−3	60°
−1	80°
+1	100°
+3	120°

The treatment combination for this design will, therefore, be:

ab^{-3}
ab^{+1}
ab^{+3}
b^{+3}
ab^{-1}
b^{+1}
b^{-1}
b^{-3}

The analysis of this design is beyond the scope of this book. Therefore, if such a design is used, the reader should consult a statistician for the analysis of the data.

EXERCISES

1. You are in charge of a project to develop a fuse for an electronic device operating at a relatively high temperature. The fuse is not supposed to blow at temperatures from 200° to 300° and at voltages from 100 to 150. Wire diameters of 0.020 and 0.030 are available at this time. The response to be measured is time in seconds at an operating condition before blowout, which, in the experimental space, must be greater than 4000 seconds at all temperatures below 280° and at all voltages below 140. Above 150 volts, the fuse should blow within 3500 seconds at any temperature above 300°. The desired $\alpha = 0.05$.
   - Design the experiment. Let A = temperature and B = voltage.
   - Plot the design points on graph paper. Indicate both coded and real values of the variables.
2. The following results are obtained on the above project with 0.02 and 0.03 wire:

0.02 Wire A	B	Seconds	0.03 Wire A	B	Seconds
+	+	520	+	+	1400
−	+	2680	−	+	3500
+	−	3910	+	−	5800
−	−	4900	−	−	6200
$+\psi$	0	850	$+\psi$	0	1275
$-\psi$	0	3680	$-\psi$	0	5875
0	$+\psi$	1075	0	$+\psi$	2050
0	$-\psi$	4620	0	$-\psi$	6300
0	0	2690	0	0	4290
0	0	2810	0	0	4680
0	0	2440	0	0	4310
0	0	2680	0	0	4590
0	0	2510	0	0	4440

Analyze the data; make an engineering decision; and propose wire thickness and tolerance for a satisfactory fuse.

3. Design a 16-trial experiment for one variable (A_1) at three levels and three variables (C, D, and E) at two levels, using the Addleman procedure.

- Label the trials from 1 to 16.
- Locate each trial in the following block diagram of the experiment:

		1		e	
		1	d	1	d
a^{-1}	1	Trial 16			
	c				
a^0	1				
	c				
a^{+1}	1				
	c				

Block diagram of experiment.

4. Suppose the following results are obtained in the above experiment:

Treatment combination	Results
a^0e	13
a^+	8
a^+ce	7
a^+cd	9
a^0cde	14
a^0cde	14
a^0d	14
a^0c	16
a^+de	7
a^0c	15
a^-cd	6
a^0d	15
a^0e	12
a^-ce	4
a^-de	4
a^-	7

Analyze these results, and make an engineering decision (assume that a high result is desirable).

REFERENCE

Myers, R. H. 1971. *Response surface methodology*. Boston: Allyn & Bacon.

18 Experiment Designs for Chemical-Composition Experiments

In previous chapters, we have discussed experiment designs for discrete variables and for continuous variables where the levels of one variable were completely independent of the levels of all other variables. For example, in a chemical process, setting the temperature at some value did not prevent setting the time at any desired value. Such is not the case when the chemist conducts an experiment to determine the effect of varying the chemical composition on some property of a product. If the chemist wishes to increase the percentage of ingredient A, the percentage of at least one of the other ingredients must be decreased by an equal amount; the sum of all the ingredients must equal 100 percent. Obviously, one cannot conduct a factorial experiment with variables such as percentage of ingredient A, percentage of ingredient B, etc. However, experiments can be designed that have most of the nice properties of matrix designs; these designs are called extreme-vertices designs and were discovered by McLean and Anderson.

Extreme-Vertices Designs

The extreme-vertices design and its comparison to the Hadamard matrix design is best portrayed by a graphic example with three components. Suppose that there is a product with a content of ingredient A that cannot be less than 10 percent or more than 30 percent. The content of ingredient B must be at least 25 percent and could be as high as 60 percent, and the content of ingredient C could be as low as 20 percent and as high as 55 percent. The total space of interest is the cross hatched space shown in figure 18-1. Note that there are some points outside the space of interest that meet two of the required levels but that fail to meet the requirement of the third component. The basic treatment combinations are the six points (or vertices) of the cross hatched area in the figure. Details of the analysis will be given later, but, at this time, it should be observed that the results of (1) − (2) and (5) − (6) measure the effect of decreasing C and increasing

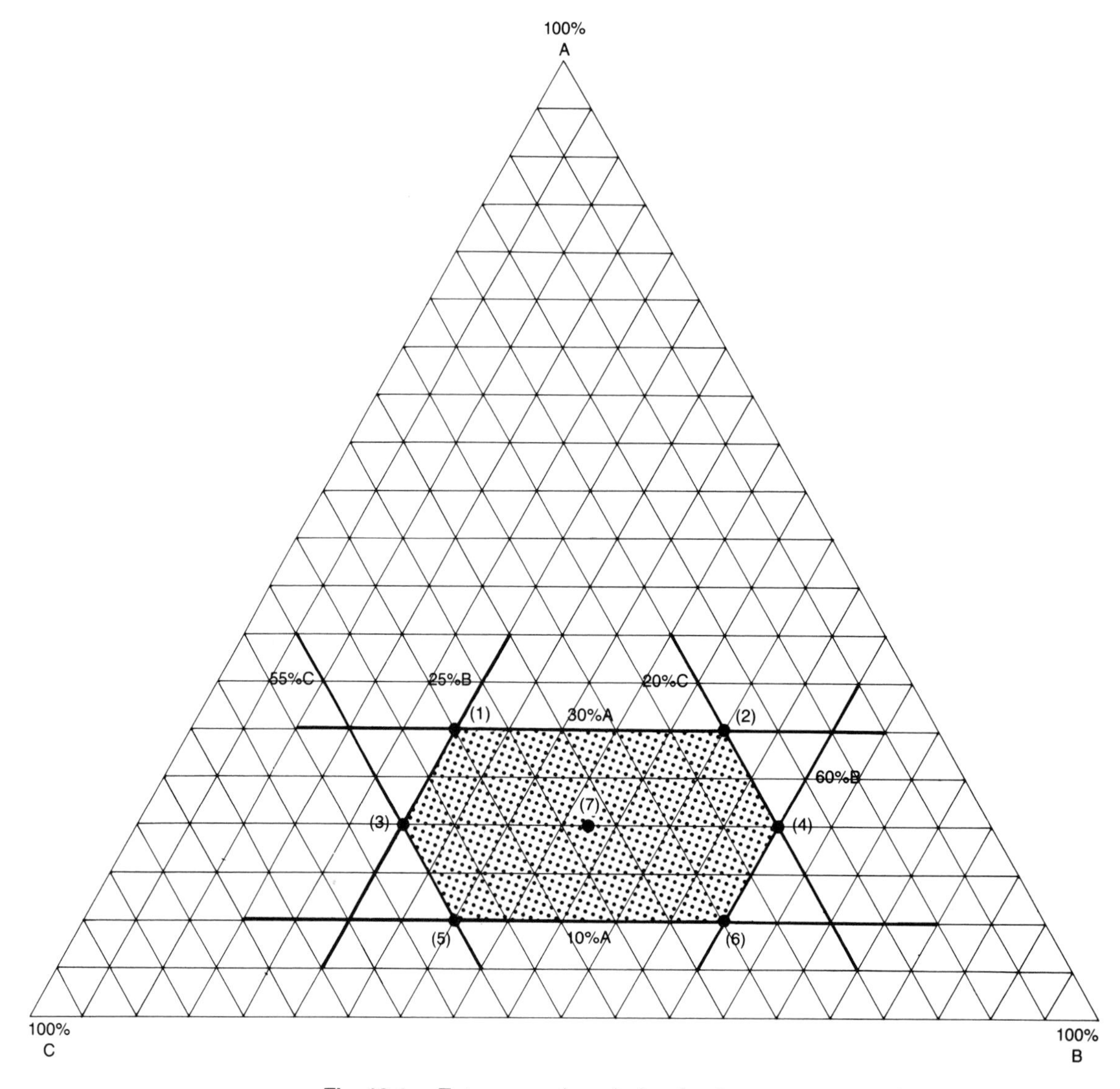

Fig. 18-1. Extreme-vertices design for three-component mixture.

B while keeping A constant at either 30 percent or 10 percent. Likewise (1) – (3) measures the effect of decreasing A while increasing C and holding B constant at 25 percent. Many other direct comparisons should be apparent to the reader.

The addition of point (7), called the centroid, improved the design, since an estimate of curvature can now be obtained from the data.

EXAMPLE

The present formula for a typewriter-ribbon ink formulation consists of 30 percent carbon black, 20 percent dye, 25 percent Carbowax, and 25 percent binder in what is called part A of the formulation. Part A is combined with part

B, which consists of a number of additives and a dispersing agent, and the total formula is coated on a film and dried. The formula is satisfactory in all respects except for the sharpness of the print work, which is definitely a function of part A only. The sharpness is measured by an optical instrument that rates the sharpness from 0 (very poor) to 10 (very good). A rating of 8 or better is required for the product to meet market requirements.

A. Define the limits of the variables in part A.

A =	carbon black	20%–40%
B =	dye	15%–25%
C =	Carbowax	15%–35%
D =	binder	20%–30%

B. Determine the treatment combinations.

All combinations of the first three variables are specified. Those combinations that can add up to exactly 100 percent by adding some value of the fourth variable, within the limits of the fourth variable, are valid treatment combinations and vertices of the *n*-dimensional space.

A	B	C	D	Trial No.
20	15	15	—	
40	15	15	30	(1)
20	25	15	—	
40	25	15	20	(2)
20	15	35	30	(3)
40	15	35	—	
20	25	35	20	(4)
40	25	35	—	

The procedure is then repeated with the first, second, and fourth variables defined; repeated again with the first, third, and fourth variables defined; and finally, repeated again with the second, third, and fourth variables defined.

A	B	C	D	Trial No.	
20	15	—	20		
40	15	25	20	(5)	
20	25	35	20	(4)	Same as 4 above.
40	25	15	20	(2)	Same as 2 above.
20	15	35	30	(3)	Same as 3 above.
40	15	15	30	(1)	Same as 1 above.
20	25	25	30	(6)	
40	25	—	30		

A	B	C	D	Trial No.
20	—	15	20	
40	25	15	20	(2)
20	—	25	20	
40	15	25	20	(5)
20	—	15	30	
40	15	15	30	(1)
20	25	25	30	(6)
40	—	25	30	

A	B	C	D	Trial No.
—	15	15	20	
40	25	15	20	(2)
30	15	35	20	(7)
20	25	35	20	(4)
40	15	15	30	(1)
30	25	15	30	(8)
20	15	35	30	(3)
—	25	35	30	

(Note: There are only eight distinct treatment combinations for this experiment.)

A	B	C	D	Trial No.
40	15	15	30	(1)
40	25	15	20	(2)
20	15	35	30	(3)
20	25	35	20	(4)
40	15	25	20	(5)
20	25	25	30	(6)
30	15	35	20	(7)
30	25	15	30	(8)

C. Determine the centroid of the polyhedron. Take the average of all the A values, all the B values, etc., of the vertices.

A	B	C	D	Trial No.
30	20	25	25	(9)

D. In addition, one can determine the centroids of each face of the polyhedron. If a variable has at least three points at a given level, those points define a surface. The centroid of the surface is obtained by taking the average of the levels of the other variables in that surface.

Trial	A	B	C	D	Treatment combinations in surface
10	40.0	18.3	18.3	23.3	(1), (2), (5)
11	20.0	21.7	31.7	26.7	(3), (4), (6)
12	32.5	15.0	27.5	25.0	(1), (3), (5), (7)
13	27.5	25.0	22.5	25.0	(2), (4), (6), (8)
14	36.7	21.7	15.0	26.7	(1), (2), (8)
15	23.3	18.3	35.0	23.3	(3), (4), (7)
16	32.5	20.0	27.5	20.0	(2), (4), (5), (7)
17	27.5	20.0	22.5	30.0	(1), (3), (6), (8)

In this experiment, it was decided to make the eight trials of the main design plus the centroid of the design and to conduct only those face centroids that appeared promising after analysis of the data for the first nine trials.

E. Make the nine test ribbons according to the treatment combinations determined in B, above.

Trial No.	Sharpness
(1)	2.6
(2)	5.0
(3)	3.0
(4)	6.5
(5)	5.2
(6)	2.8
(7)	8.0
(8)	2.9
(9)	3.2

F. Analyze the data.

The data could be analyzed using regression analysis. In this case, however, the conclusions are obvious from an inspection of the data. Treatment combination 7 (i.e., 30 percent A, 15 percent B, 35 percent C, and 20 percent D) is obviously superior.

G. Draw a figure of the experiment design and the results.

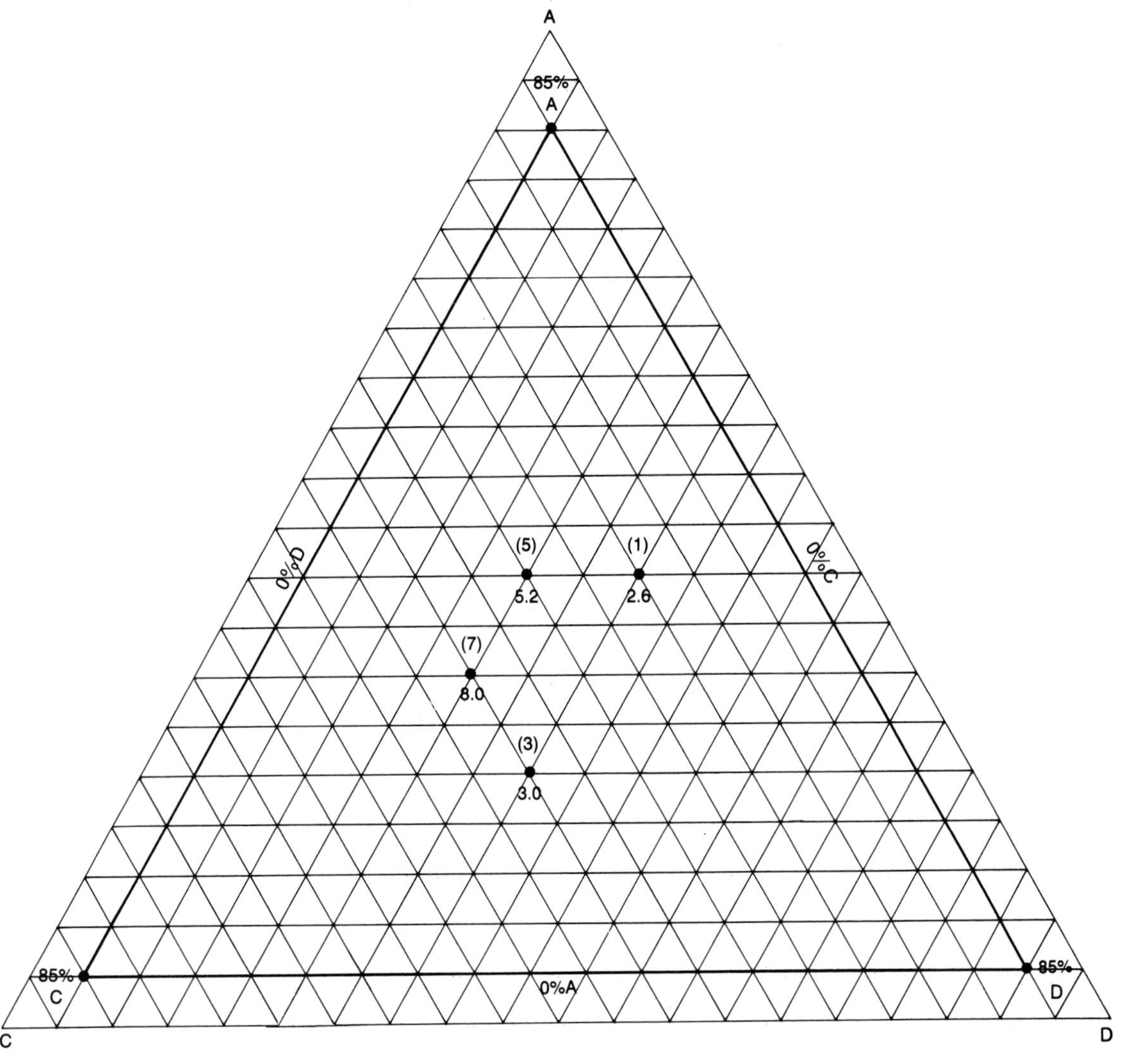

Fig. 18-2. Graph of experiment design with B = 15 percent.

(a) Choose as the first variable, if possible, one that has only two levels. (b) Label the vertices of the triangle with the letters of the remaining three variables. Make up one chart for the low level of the first variable and one chart for the high level of the first variable.

Since variable B has only two levels (except for the centroid), let variable B be the first variable. Make up a triangular graph with B = 15 percent (figure 18-2). Therefore, the sum of variables A, C, and D must equal 100% − 15% = 85%.

The 0 level of A is located at 1/3 of 15 percent, and the 85 percent level of A is located at 100 percent − (2/3 of 15 percent). Likewise for C and D.

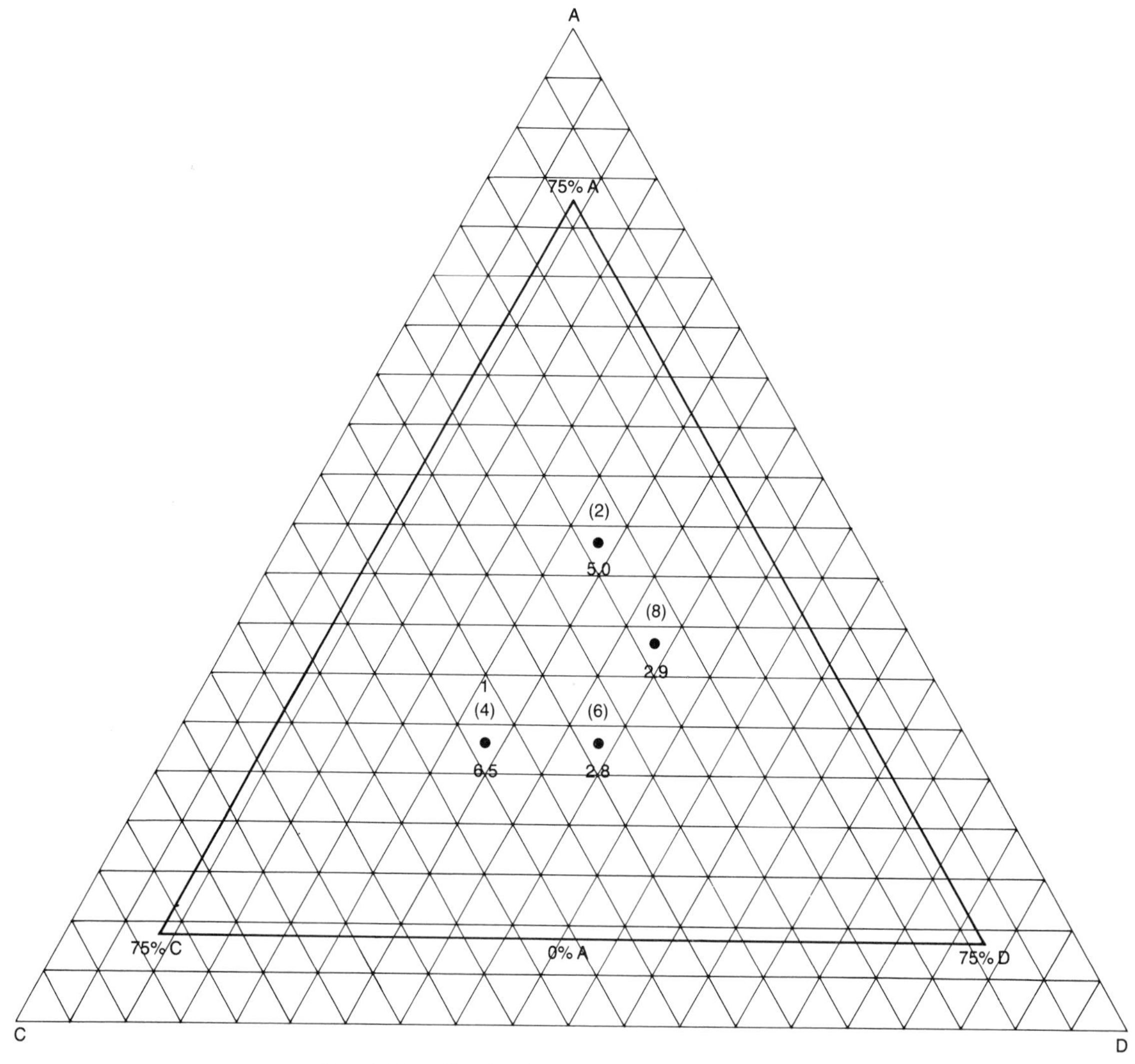

Fig. 18-3. Graph of experiment design with B = 25 percent.

(c) Draw a smaller triangle with vertices at (100% − low level of the first variable); that is, a slice through a pyramid at the low level of the first variable.

(d) Draw a still smaller triangle with vertices at (100% − high level of the first variable).

Treatment combinations (1), (3), (5), and (7) are plotted on this plane together with their results (figure 18-2). A similar procedure is used for figure 18-3, where B = 25 percent.

Note: It is helpful to plot these figures on thin transparent plastic sheets and insert the sheets into a frame so they are parallel to each other. This is particularly necessary if all the variables are at more than two levels.

H. Make engineering decisions.

(a) A composition of 20 percent D or lower is an absolute requirement for high sharpness.
(b) A composition of 15 percent B and 30 percent A, or lower B and A, is an absolute requirement for high sharpness.
(c) A composition of 35 percent C or higher is an absolute requirement for high sharpness.
Since a response of 8 on the sharpness scale was adequate for the product, no further experiments were conducted. However, several formulations that should give even better results should be obvious to the reader.

Simplex Designs

Simplex designs were discovered by Scheffé in 1958. They were specifically intended for problems where there are no restrictions on the limits of the per-

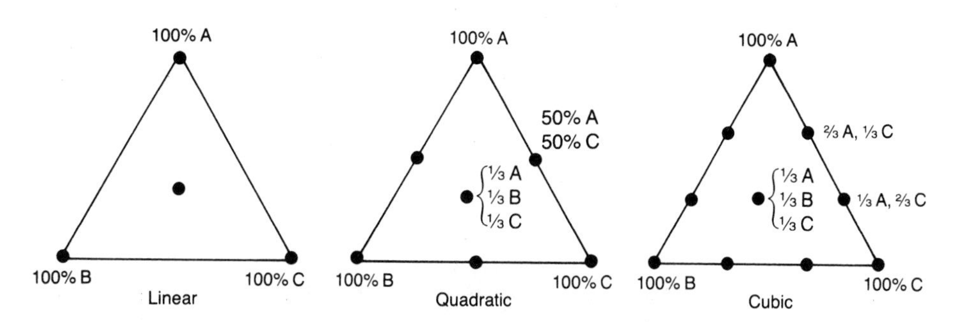

Fig. 18-4. Simplex designs available for three variables.

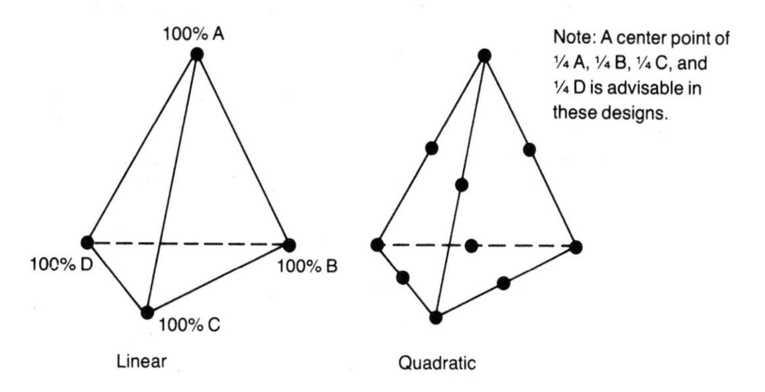

Fig. 18-5. Simplex designs available for four variables.

centage of any component in the composition (but, of course, the total must be exactly 100 percent). Thus, with three variables, the designs shown in figure 18-4 are available.

For four variables, the treatment combinations are points on a pyramid, as shown in figure 18-5.

EXAMPLE

In the synthesis of a certain chemical, a contaminating by-product is produced that must be removed by a solvent rinse. Three solvents are proposed—methanol, acetone, and trichloroethylene. Any combination of these could be most effective, and it is expected that some combination of the solvents will be required. The quadratic three-component simplex is obviously the best initial design.

A. Determine the treatment combinations from the above quadratic figure for three components.

	A Methanol	B Acetone	C Trichloroethylene
(1)	100%	0	0
(2)	0	100%	0
(3)	0	0	100%
(4)	50%	50%	0
(5)	50%	0	50%
(6)	0	50%	50%
(7)	33-1/3%	33-1/3%	33-1/3%

B. Make one pilot run of the process, and divide the resultant product into seven parts. Treat each part with one of the solvent combinations, and measure the percentage of by-product remaining after the treatment.

Part	By-product remaining, %
(1)	6.2
(2)	8.4
(3)	3.9
(4)	7.4
(5)	2.8
(6)	6.1
(7)	2.2

C. Plot the results as shown in figure 18-6.

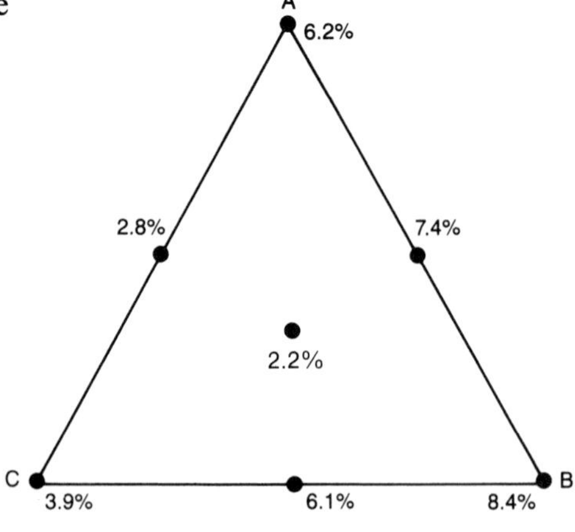

Fig. 18-6. Plot of test results.

D. Analyze the results.

Procedures exist for analyzing the data and reporting the results as a regression equation (*Technometrics*, Vol. 4, No. 4). However, in most cases such as this example, the conclusions are obvious. The best result is at 33-1/3 percent methanol, 33-1/3 percent acetone, and 33-1/3 percent trichloroethylene. There is obviously some curvature to the response-variable relationship, and a still better composition is possible in the vicinity of the center point.

E. Make engineering decision.

The by-product content of 2.2 percent does not result in a satisfactory product. It is expected that a better composition can be found, possibly by increasing C and/or decreasing B, and/or increasing A. An extreme-vertices design is indicated.

F. Design an extreme-vertices experiment.

The best approach to such a design at this stage is graphically shown in figure 18-7. If the lower level of each solvent is chosen as 25 percent and the high level is chosen as 50 percent, the design will be an equilateral triangle with only three trials.

	A	B	C
(8)	50%	25%	25%
(9)	25%	50%	25%
(10)	25%	25%	50%

Because point 5 is the second-best result in the first experiment, it is advisable to add an extra point in that direction:

(11) 37.5% 25% 37.5%

It is also advisable to repeat the treatment combination that was the best in the first experiment and is now the centroid of the follow-up experiment:

(12) 33.7% 33.3% 33.3%

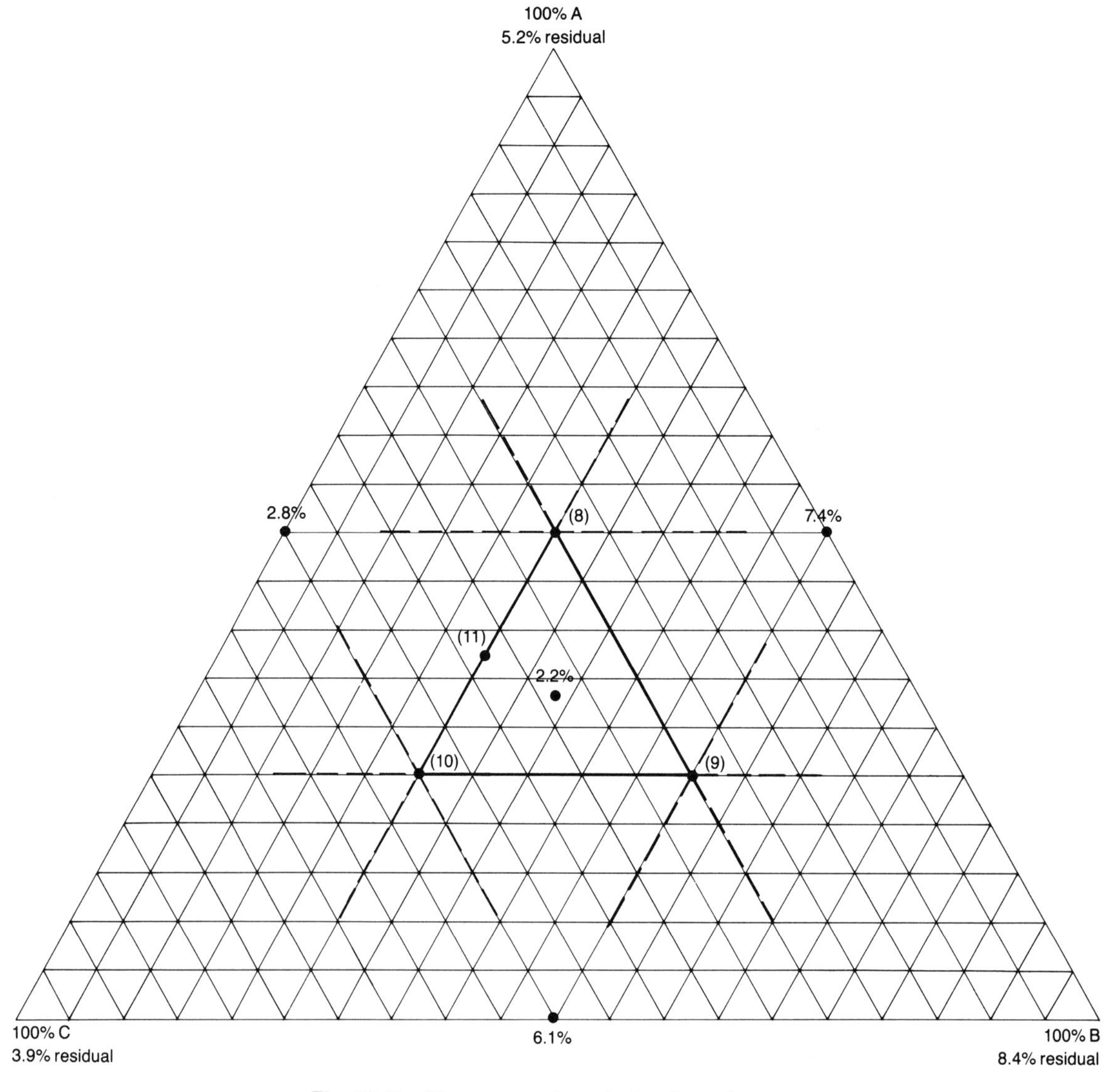

Fig. 18-7. Extreme-vertices design for solvent experiment.

G. Make a pilot run of the process, and divide the product into five parts. Treat each part with one of the five solvent combinations of this experiment, and measure the percentage of by-product remaining after the solvent treatment.

Residual	Content, %
(8)	3.3
(9)	4.8
(10)	1.4
(11)	1.2
(12)	2.4

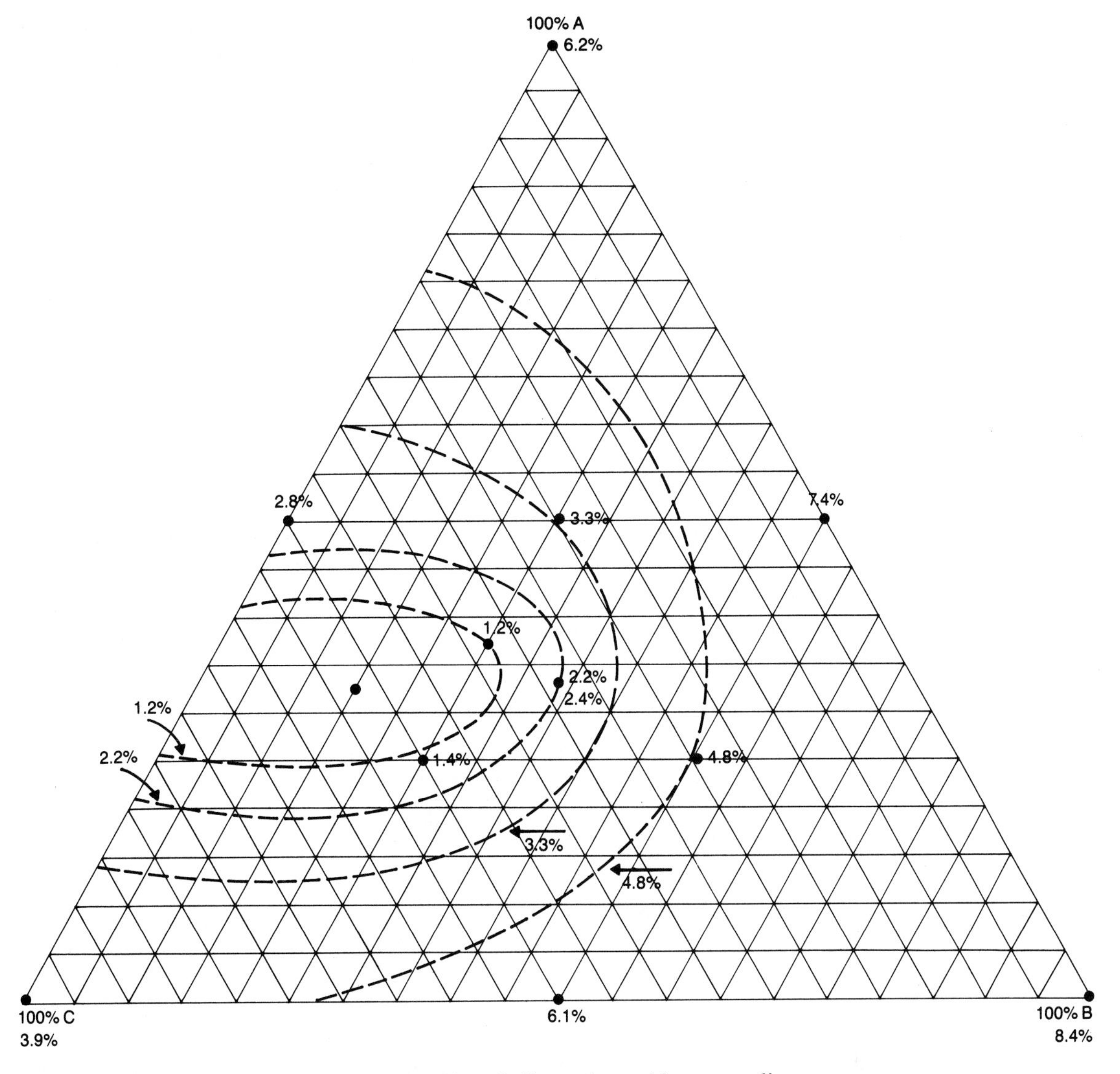

Fig. 18-8. Plot of all test data with contour lines.

H. Plot all the data of this and the previous experiment, and draw approximate contour lines of equal response.	See figure 18-8.
I. Make engineering decisions.	From all the data, it appears that there should be a minimum at about 33 percent A, 15 percent B, and 52 percent C.

Make one run of the process, and extract the product with this solvent combination.

J. Make and test the process with the indicated solvent combinations.

(13) 0.45% residual content

This is a completely satisfactory product. Therefore, even though some additional work might find a still better combination, the project should be closed down and the treatment combination of trial 13 written into the process specification.

EXERCISES

1. A chemist is formulating an insecticide that contains three active ingredients and wishes to test the formula with two different concentrations of total active ingredients. All possible combinations from 0 to 100 percent of each ingredient are feasible. Design the experiment to determine both linear and quadratic effects of the ingredients and the linear effects of concentrations from 1 percent to 3 percent.

2. The following results are obtained from an extreme-vertices experiment on a paint formulation:

Trial	A	B	C	Result
1	65%	10%	25%	Smoothness index = 36
2	65%	15%	20%	Smoothness index = 39
3	50%	30%	20%	Smoothness index = 17
4	50%	15%	35%	Smoothness index = 24
5	55%	10%	35%	Smoothness index = 11
6	57%	16%	27%	Smoothness index = 45

The higher the smoothness index, the better the paint. Propose an experiment to possibly find a still better formulation than formula (6) above.

REFERENCE

Anderson, V. L. and McLean, R. A. 1974. *Design of experiments.* New York: Marcel Dekker

19 Random-Strategy Experiments

The principles and procedure of random strategy will be discussed in this chapter. Let us look at basic probability theory. Suppose that a dart board is 10 inches × 10 inches and that somewhere on the dart board is a space, 4 inches × 4 inches, which is labeled "good" (figure 19-1).

The fraction of good space is $\frac{4 \times 4}{10 \times 10} = 0.16$; that is, 16 percent of the total space is good space. If one were to throw 10 darts at this board, and, in every case, the dart would land within the 10-inch × 10-inch board, but exactly where it landed was pure chance, the probability that at least one of the darts would be in the good space is

$$P = 1 - (1 - \gamma)^N,$$

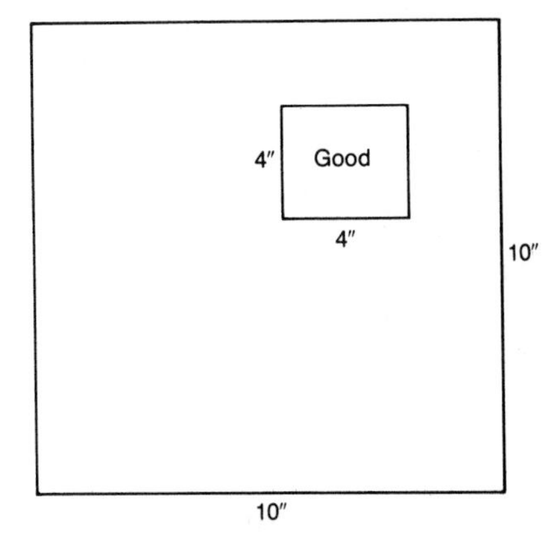

Fig. 19-1. Location of good space.

where N is the number of darts thrown, and γ is the fraction of the space that is good. For this example,

$$P = 1 - (1 - 0.16)^{10} = (1 - 0.17) = 0.83.$$

Needless to say, this probability is the correct probability, even if the dart thrower doesn't know either where the good space is located or even if it really exists.

In an experimental project, the same law of probability applies. Suppose there are two variables that influence some response of interest to the experimenter, who knows the limits of the variables but has no idea of the levels of the variables that produce a good result. Assume the situation shown in figure 19-2.

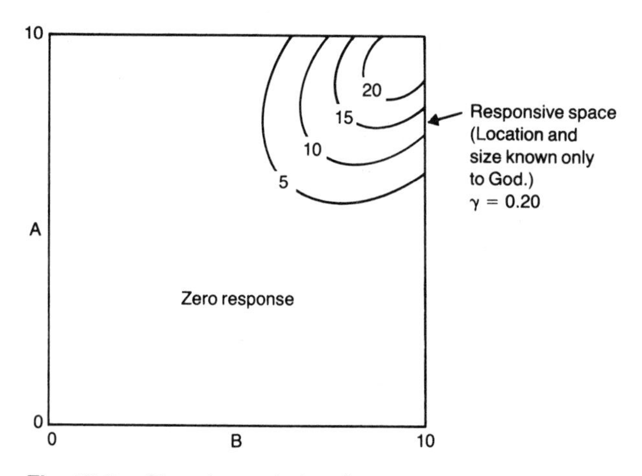

Fig. 19-2. Experimental situation.

If the experimenter randomly chooses 10 values of A with 10 values of B from a population with uniform distribution from 0 to 10, then produces and measures some property of the 10 samples, what is the probability that at least one of the randomly chosen trials will be within the responsive space? The answer is computed, as shown above,

$$P = 1 - (1 - 0.20)^{10} = 0.89.$$

Note that the same probability of success is assured if there are more than two variables but the γ is 0.20 for the system.

If it is possible to perform the experiment sequentially, it is advisable to do so, using the techniques of sequential random strategy that will be discussed later in this chapter. If this is not possible, the following procedure is used.

EXAMPLE

A machine for butt welding two shafts of unequal diameter has been purchased. Six settings on the machine are available to influence the strength of the weld.

A. Define all the variables and the theoretical limits of the variables.

Variables		Low limit	High limit
Preheat temp	= A	200°	700°
Dial setting for pressure	= B	0	10
Time (preheat)	= C	0	1 min
Time (welding)	= D	0	30 sec
Welding current	= E	1 amp	2 amps
Welding voltage	= F	100 volts	150 volts

B. Specify the object of the project.

The object of the project is a butt weld with a life of 400 hours when subjected to a repeated stress of 125 foot-pounds.

C. Select the number of random trials to be conducted. $N = 20$ is a good average number to use. If the trials are easy and/or cheap, use a higher N. If the trials are difficult or expensive, use a lower N.

$$P = 1 - (1 - \gamma)^N$$

A pseudopower curve can be drawn as a function of γ (figure 19-3).

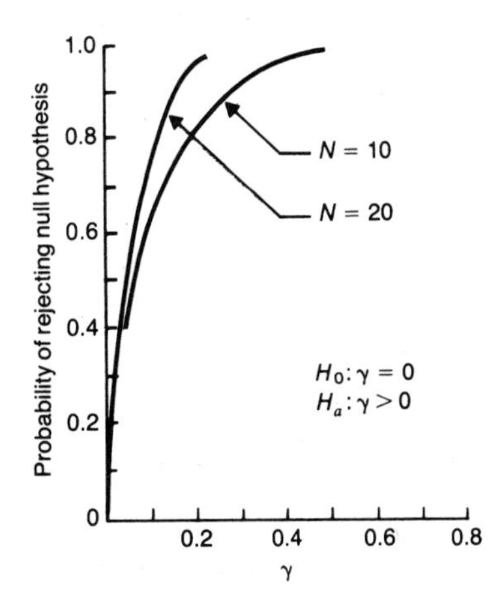

Fig. 19-3. Pseudopower curve as a function of γ.

Thus, if one supposes that γ will be at least 0.2 and desires about 90 percent confidence of detecting the good space, $N \doteq 20$.

D. Select 20 random levels of each variable from a uniform distribution.

Trial	A	B	C	D	E	F
1	354	1.2	0.64	4	1.54	110
2	275	9.7	0.81	19	1.79	117
3	620	9.2	0.22	17	1.02	141
4	540	4.8	0.35	28	1.35	136
↓	↓	↓	↓	↓	↓	↓
20	461	1.8	0.49	26	1.58	128

E. Make and test the 20 samples specified.

Several alternative results are possible: (a) Suppose trial 4 has a life of 400+ hours and all other samples failed prior to 400 hours. Pick the winner—specify that the preheat temperature should be 540°, the pressure dial set at 4.8, preheat time set at 0.35 minutes, welding time set at 28 seconds, welding current at 1.35 amps, and welding voltage at 136. (b) None of the results exceeds the required specification; however, one of the results is best. Use this best result as the starting point for a designed experiment. Accept $H_a: \gamma > 0$. (c) None of the results exceeds the required specification, and all the results are approximately the same. In particular, none of the results is greater than the criterion $|\bar{X} + t_\alpha S|^*$. Conclude that no combination of the variables will produce a satisfactory part. Accept $H_0: \gamma = 0$.

If it is possible to perform an experiment sequentially, the following procedure is used.

EXAMPLE

A. Define all the variables and the theoretical limits of the variables.

Let the variables and limits be the same as in the previous problem.

B. Specify the object of the project.

The object of the project is a butt weld with an impact strength of 85 foot-pounds.

C. Select a random level of each variable, produce a part (or parts) at that condition, and test. If a product is produced that exceeds the specification, stop. If the product obtained does not meet the specification, make another part at a second set of random levels. Rule: if $X_i > 85$, stop. If $X_i < 85$, make another part.

A = 258; B = 3.5; C = 0.68; D = 14; E = 1.64; F = 140. Impact strength = 45.0 = X_1.
$X_1 < 85$. Therefore, make a second sample at randomly chosen levels of A, B, C, D, E, and F. Impact strength = 55.1 = X_2.
$X_2 < 85$. Therefore, make a third sample at randomly chosen levels. Impact strength = 51.0 = X_3.

D. After the third sample, it is possible to compute a criterion D_i^* for deciding objectively when to stop.

$$D_i^* = \bar{X}_i + 1.5S_i$$
$$i = 3, 4, 5, \ldots, N$$

where $S_i = \sqrt{\dfrac{\sum(X_i - \bar{X})^2}{i - 1}}$

If any X_i is greater than D_i^*, stop the random-strategy experiment.
If no X_i is greater than D_i^*, make another sample.

$S_3 = 5.03 \quad \bar{X}_3 = 50.37$
$D_3^* = 50.37 + (1.5)(5.03) = 57.9$

Since no experimental result obtained to this point is greater than 57.9, make sample 4 at randomly chosen levels. Impact strength = 46.9 = X_4, $S_4 = 4.4$, $\bar{X}_4 = 49.5$, $D_4^* = 56.2$.
No result to this point is greater than D_4^*. Therefore, make sample 5 at randomly chosen levels. Impact strength = 46.0 = X_5, $S_5 = 4.20$, $\bar{X}_5 = 48.8$, $D_5^* = 55.09$. The tensile strength of the second trial (55.1) is greater than $D_5^* = 55.09$. Therefore, stop the random-strategy experiment, and use the levels of the variables in sample 2 as the starting point for a Hadamard matrix experiment.

The stopping rules for the sequential random strategy are:

a. A sample result is obtained that exceeds the requirements.
b. A sample result is obtained that exceeds the criterion $D_i^* = \bar{X}_i + 1.5S_i$.
c. Twenty trials have been conducted without invoking conditions (a) or (b), above.

If the full 20 (or N) trials are conducted without invoking one of the stopping criteria, pick the winner. Use this best trial combination as the starting point for a matrix experiment.

EXERCISE

A sequential random-strategy experiment is conducted on a tape-feeding mechanism. The response of interest is the noise level, which is supposed to be less than 5 decibels. There are 20 variables in the experiment. If the following sequence of results were obtained, at what trial would you stop the random-strategy experiment?

Trial	Result, decibels
1	22
2	14
3	18
4	17
5	24
6	21
7	19
8	23
9	23
10	19
11	13
12	15

Part Four

Related Topics

20 Blocking an Experiment

In many experiments, there is a possibility of inadvertently confounding the results. This confounding can result from the use of two or more operators or two or more pieces of process equipment or from conducting the trials on two or more days, etc. For example, suppose that an experimenter is going to make the following trials for a Resolution V experiment:

Trial	Treatment combination
1	a
2	ab
3	abc
4	bc
5	ac
6	b
7	c
8	(1)

Only one oven is available for processing the samples, which require a full day in the oven, and only four samples will fit in the oven at one time. Which samples should be processed the first day, and which the second day? Suppose trials 1, 2, 3, and 5 are processed the first day, and trials 4, 6, 7, and 8 the second day. It can be seen immediately that the four trials processed the first day are at high A, and the four trials processed the second day are at low A. Thus, if the temperature of the oven is different (unknown to the experimenter), or if the humidity is different on the second day, or if any number of things having some influence on the response are different on the second day, the experimenter will erroneously attribute the observed effects to variable A.

The solution to this problem is called blocking and is actually nothing more than a different aspect of the confounding of a higher-order interaction technique used previously. Look at the Hadamard matrix of order 8:

	A	B	C			ABC		Treatment combination	Block
+	+	−	−	+	−	+	+	a	1
+	+	+	−	−	+	−	+	ab	2
+	+	+	+	−	−	+	−	abc	1
+	−	+	+	+	−	−	+	bc	2
+	+	−	+	+	+	−	−	ac	2
+	−	+	−	+	+	+	−	b	1
+	−	−	+	−	+	+	+	c	1
+	−	−	−	−	−	−	−	(1)	2

Simply use the signs in the ABC column to choose which day to make each sample; (+) is day one, (−) is day two. Therefore, make trials a, abc, b, and c on day one, and make trials ab, bc, ac, and (1) on day two. If something in the experimental procedure changes overnight, it will have no influence on the estimate of the effect of A, B, C, or the two-factor interactions.

Suppose that, in a 3-variable problem, the true effect of A = 5, B = 2, and C = 1; and, for simplicity, assume that the variance is zero and there are no interactions. The correct response is read off a dial the first day, but overnight the dial is accidentally shifted so that it reads 20 units higher, and the experimenter is unaware of this. Therefore, on the second day, the results will all be 20 units higher than they should be, and the following results will be obtained:

Trial		A	B	C	−AB	−BC	ABC	−AC	Measured response
1	+	+	−	−	+	−	+	+	5
2	+	+	+	−	−	+	−	+	7 + 20 = 27
3	+	+	+	+	−	−	+	−	8
4	+	−	+	+	+	−	−	+	3 + 20 = 23
5	+	+	−	+	+	+	−	−	6 + 20 = 26
6	+	−	+	−	+	+	+	−	2
7	+	−	−	+	−	+	+	+	1
8	+	−	−	−	−	−	−	−	0 + 20 = 20

Obviously, the results on trials 2, 4, 5, and 8 are absolutely wrong because the dial was shifted; however, the computed effect of A will be absolutely correct:

$$\text{Effect of A} = (5 + 27 + 8 - 23 + 26 - 2 - 1 - 20)/4 = 5.$$

Likewise, all of the other contrasts will be correct; B = 2, C = 1, −AB = 0, etc.

The computed value of the ABC interaction will be −20. The experimenter, of course, wouldn't know for sure that something happened overnight. It might well be that there is a significant ABC interaction, though it would be a strange one. However, the experimenter should at least be alerted to the possibility that something may have happened overnight.

If there are more than two possible levels of confounding—e.g., four different ovens or four days—block on the next-to-highest order interactions. For example:

−AB	−BC	
+	−	Oven 1
−	+	Oven 2
−	−	Oven 3
+	+	Oven 4

(Note that a small matrix and four blocks is not very good protection against bias. This example is only to show the principle for use in larger matrices.)

Trial		A	B	C	↓ −AB	↓ −BC	ABC	−AC	Oven
1	+	+	−	−	+	−	+	+	1
2	+	+	+	−	−	+	−	+	2
3	+	+	+	+	−	−	+	−	3
4	+	−	+	+	+	−	−	+	1
5	+	+	−	+	+	+	−	−	4
6	+	−	+	−	+	+	+	−	4
7	+	−	−	+	−	+	+	+	2
8	+	−	−	−	−	−	−	−	3

It should be noted that in this small experiment, with four blocks, only the main factors can be estimated. With larger designs this is not a problem.

Three blocks could be obtained in the above experiment as follows:

−AB	−BC	
+	−	Oven 1
−	+	Oven 1
−	−	Oven 2
+	+	Oven 3

Blocking can also be used in a central composite rotatable design (CCRD), which was discussed in a previous chapter. For example, with the 3-variable design, block as follows:

	Block I			Block II			Block III	
A	B	C	A	B	C	A	B	C
−1	−1	+1	−1	−1	−1	−1.633	0	0
+1	−1	−1	+1	−1	+1	+1.633	0	0
−1	+1	−1	−1	+1	+1	0	−1.633	0
+1	+1	+1	+1	+1	−1	0	+1.633	0
0	0	0	0	0	0	0	0	−1.633
0	0	0	0	0	0	0	0	+1.633
						0	0	0
						0	0	0

Block I consists of one-half of the Hadamard matrix (a Resolution III part) plus two center points. Block II consists of the other half of the Hadamard matrix and two center points. Block III consists of the six star points and two center points. Note that, when blocked, the value of ψ is different than in the unblocked design. Likewise, the computation of the coefficients is different.

$$b_0 = 0.165385(\mathrm{T}) - 0.057692(\mathrm{A}^2 + \mathrm{B}^2 + \mathrm{C}^2).$$

$$b_i = 0.075(i).$$

$$b_{ii} = 0.070312(ii) + 0.005409(\mathrm{A}^2 + \mathrm{B}^2 + \mathrm{C}^2) - 0.057692(\mathrm{T}).$$

$$b_{ij} = 0.125(ij).$$

i = A, B, or C; if i = A, j = B or C; if i = B, j = A or C; if i = C, j = A or C.

Thus, if a CCRD with three variables is to be conducted on three different days, and there is a possibility that temperature or humidity or the amount of pollen in the air, etc., could influence the response, the proper procedure would be to make the six samples of block I on the first day; the six samples of block II on the second day; and the eight samples of block III on the third day. In this particular example, the two center points each day furnish a direct comparison between days. If the mean of the center points is significantly different from one day to the next, it is direct evidence of a day-to-day effect.

The blocking of these designs can also serve an entirely different purpose; namely, as a sequential procedure. For example, the experimenter might conduct an experiment with just block I. If the results were such that all the necessary information was obtained, blocks II and III would not be made. On the other hand, it might be decided that block II is needed, and, therefore, that block would be completed. Then the experimenter would make a decision regarding the need for block III.

Note that, after completion of the first two blocks, the experimenter has a Resolution V+ design with four center points. The analysis of the experimental data in regard to main factors and interactions is exactly the same as if there were no center points. The four center points furnish an estimate of σ^2 with three degrees of freedom in the usual manner. In addition, another useful estimate can be obtained from the data. $(\bar{Y}_C - \bar{Y}_H)$ is an estimate of the sum of the quadratic effects; $\bar{Y}_C$ is the mean of the center points, and $\bar{Y}_H$ is the mean of the Hadamard points.

EXAMPLE

Suppose the following results are obtained on the first two blocks of a 3-variable CCRD.

Treatment combination	Result
a	5.3
ab	6.9
abc	7.1
bc	2.4
ac	4.7
b	2.1
c	0.1
(1)	0.2
0,0,0	4.4
0,0,0	4.8
0,0,0	4.7
0,0,0	4.6

$$\bar{Y}_H = (5.3 + 6.9 + 7.1 + 2.4 + 4.7 + 2.1 + 0.1 + 0.2)/8 = 4.09$$

$$\bar{Y}_C = (4.4 + 4.8 + 4.7 + 4.6)/4 = 4.625$$

$$S^2 = 0.02892 \text{ with 3 df from the 4 center points}$$

$$S = 0.17$$

$$\bar{Y}_C - \bar{Y}_H = 4.625 - 4.09 = 0.535$$

The criterion for testing the significance of this contrast is

$$|\bar{Y}_C - \bar{Y}_H|^* = t_\alpha S \sqrt{\frac{1}{N_C} + \frac{1}{N_H}}.$$

For 95 percent confidence,

$$|\bar{Y}_C - \bar{Y}_H|^* = (2.35)(0.17)\sqrt{\tfrac{1}{4} + \tfrac{1}{8}} = 0.245.$$

Since $[(\bar{Y}_C - \bar{Y}_H) = 0.535] > [|\bar{Y}_C - \bar{Y}_F|^* = 0.245]$, we know that the response is not a linear effect of A, B, and C. The experimenter should, therefore, complete block III to determine the quadratic terms in the regression equation.

EXERCISES

1. A 16-trial experiment with four variables is to be conducted. Four different sets of experimental equipment are to be used. Which treatment combinations should be assigned to each set of experimental equipment?
2. Suppose a 16-trial experiment with five variables is to be conducted and two different technicians will conduct the trials. Which treatment combinations should be assigned to each technician?
3. The following Resolution IV Hadamard matrix experiment was conducted.

Day 1					Day 2				
A	B	C	D	Result	A	B	C	D	Result
+	−	−	+	20	+	−	+	−	13
+	+	−	−	14	−	+	−	+	10
+	+	+	+	21	−	−	+	+	10
−	+	+	−	8	−	−	−	−	9
0	0	0	0	11	0	0	0	0	12
0	0	0	0	13	0	0	0	0	14
0	0	0	0	12	0	0	0	0	11

Do a complete analysis of the data. What is the value of S^2? Is there a day-to-day effect? Is there curvature? At what levels of the variable could you expect a result of 30?

REFERENCES

Hicks, C. R. 1973. *Fundamental concepts in the design of experiments.* New York: Holt, Rinehart & Winston.

Ott, L. 1977. *Introduction to statistical methods and data analysis.* North Scituate: Duxbury Press.

21 Validation of Test Methods

Many projects are hindered for months or years by the failure of the experimenter to validate testing procedures in the earliest stages of the project. As a result, much misinformation can be generated. Furthermore, the greater the variance of the test method, the greater the number of samples required for a valid experiment. It is, therefore, good economy in the early stages of a project to reduce the experimental variance as much as possible. To intelligently reduce the variance, the experimenter must determine the source of the variance.

The characteristics of a good test method are:

- The test discriminates; i.e., a sample that is good gives a good test result, and a sample that is bad gives a bad test result.
- The test is accurate; i.e., the recorded value of some measurement is, on the average, close to the true value, which is known only to God.
- The test is precise; i.e., repeated readings on the same item are reasonably close to each other.

EXAMPLE

A certain chemical is purchased from a vendor and is mixed with other materials. It is then coated on Mylar to form a photoconductor. Samples produced to evaluate the incoming material are 4 feet long. One foot of this material is adequate for a test of the dark decay, with a specification of 30 volts/second or less. Results have been very erratic on this measurement, and an experiment must be conducted to determine if this variation is due to the raw material, the fabrication, or the testing procedure.

The experiment design consists of the following:

- Five batches of the vendor's material.
- Three panels of photoconductor made from each batch.
- Four tests made on each panel.
- All other materials and the processing to be held rigidly constant.

	Panel 1				Panel 2				Panel 3			
	Test 1	2	3	4	1	2	3	4	1	2	3	4
Batch 1	37.5	30.7	28.4	35.6	33.2	35.8	35.2	39.4	30.1	38.2	33.6	31.1
2	28.5	30.9	35.4	32.1	34.2	26.5	29.8	25.0	30.7	26.2	31.4	29.1
3	25.2	24.7	30.1	28.8	26.5	32.4	28.2	27.5	24.1	26.8	26.2	27.1
4	31.6	34.8	29.3	31.2	28.5	36.4	32.5	31.7	32.1	25.5	29.7	30.5
5	28.5	25.2	27.5	30.2	30.7	28.1	29.5	34.0	31.5	27.2	35.1	29.2

Fig. 21-1. Experimental results on photoconductor material.

The experimental results obtained are shown in figure 21-1.

The testing error is determined from the replicate test within each panel within each batch. Compute the testing sum of squares for each panel:

$$SS_0 = \frac{\sum X_i^2}{N_1} - (\sum X_i)^2/N_2.$$

In this case, and in all subsequent uses of this formula, the value of N_1 and N_2 in the denominators will always be the number of sample results that are summed in the numerators. In each panel, there will be four test results; therefore, for SS_0, $N_2 = 4$. In this case, $N_1 = 1$. The number of degrees of freedom on each panel = 4 tests − 1 = 3 df.

$$(37.5)^2 + (30.7)^2 + (28.4)^2 + (35.6)^2 - \frac{(37.5 + 30.7 + 28.4 + 35.6)^2}{4}$$

$$= 53.4 = SS_0 \text{ with } 3\,\text{df}.$$

$$(28.5)^2 + (30.9)^2 + (35.4)^2 + (32.1)^2 - \frac{(28.5 + 30.9 + 35.4 + 32.1)^2}{4}$$

$$= 24.7 = SS_0 \text{ with } 3\,\text{df}.$$

. .

. .

. .

$$(31.5)^2 + (27.2)^2 + (35.1)^2 + (29.2)^2 - \frac{(31.5 + 27.2 + 35.1 + 29.2)^2}{4}$$

$$= 34.5 = SS_0 \text{ with } 3\,\text{df}.$$

The complete computation results are shown in figure 19-2.

$$\sum SS_0 = 387.7 \text{ with 45 degrees of freedom.}$$

$$MS_0 = \frac{\sum SS_0}{\text{df}} = \frac{387.7}{45} = 8.6 \text{ with 45 degrees of freedom.}$$

	Panel 1	Panel 2	Panel 3
Batch 1	SS_0 = 53.4	20.0	39.2
2	= 24.7	49.9	16.0
3	= 21.2	20.2	5.5
4	15.6	31.6	23.8
5	13.1	19.0	34.5

Fig. 21-2. Computation results on photoconductor experiment.

	Panel 1	Panel 2	Panel 3	
Batch 1	132.2	143.6	133.0	408.8
2	126.9	115.5	117.4	359.8
3	108.8	114.6	104.2	327.6
4	126.9	129.1	117.8	373.8
5	111.4	122.3	123.0	356.7
	606.2	625.1	595.4	1826.70

Fig. 21-3. Two-way chart of batch vs panel.

The variations between panels within batches and the variations between batches are obtained by making up a two-way chart of batch vs panel (figure 21-3).

On each batch, the sum of squares due to panels is computed by the same formula as above. Each X_i, however, in this stage, is the sum of the four test results on a given panel within a given batch. Therefore, $N_1 = 4$, $N_2 = 3 \times 4 = 12$; df is the number of panels minus one $= 3 - 1 = 2$. SS_1 for batch 1 is

$$\frac{132.2^2 + 143.6^2 + 133.0^2}{4} - \frac{408.8^2}{3 \times 4} = 20.24 = SS_1 \text{ with 2 df.}$$

$$\begin{aligned}
\text{Batch 2} &= 18.65 = SS_1 \text{ with 2 df.}\\
\text{Batch 3} &= 13.58 = SS_1 \text{ with 2 df.}\\
\text{Batch 4} &= 17.95 = SS_1 \text{ with 2 df.}\\
\text{Batch 5} &= 21.16 = SS_1 \text{ with 2 df.}\\
&\overline{91.57} = \sum SS_1 \text{ with 10 df.}
\end{aligned}$$

$$MS_1 = \frac{\sum SS_1}{\text{df}} = \frac{91.57}{10} = 9.16$$

The sum of squares between batches is computed by the same formula given above. Each X_i, however, is the sum of the four test results on the three panels made with a batch. Therefore, $N_1 = 12$ and $N_2 = 60$; df is the number of batches minus one $= 5 - 1 = 4$.

$$SS_2 = (408.8^2 + 359.8^2 + 327.6^2 + 373.8^2 + 356.7^2)/12 - \frac{1826.70^2}{60}$$

$$= 55904.71 - 55613.89 = 290.83 \text{ with 4 df.}$$

$$MS_2 = \frac{290.83}{4} = 72.71$$

A variance component table is now constructed from the data:

Source of variation	df	Sum of squares	Mean square	Variances estimated by mean square
Between batches	4	290.83	72.71	$\sigma_0^2 + 4\sigma_1^2 + 12\sigma_2^2$
Between panels within batches	10	91.57	9.16	$\sigma_0^2 + 4\sigma_1^2$
Between tests within panels	45	387.10	8.60	σ_0^2

Since $S_0^2 + 4S_1^2 = 9.16$ and $S_0^2 = 8.60$,

$$S_1^2 = \frac{9.16 - 8.60}{4} = 0.14,$$

and

$$S_0^2 + 4S_1^2 + 12S_2^2 = 72.71.$$

$$S_2^2 = \frac{72.71 - 8.60 - 4(.14)}{12} = 5.30$$

The engineering conclusions from this experiment are:

1. The testing procedure is subject to large variations ($S_0^2 = 8.6$) and is the major source of the different results that are obtained. If the test procedure is not improved, the best precision that one can hope for on a single test of the photoconductor is:

   $\pm t_\alpha S = \pm(2.02)(2.93) = \pm 5.92$ volts/second with 95 percent confidence.

   If a precision of ± 2 with 95 percent confidence is desired for a given treatment combination (one panel), then the number of tests on that panel must be 9.

   $$\frac{t_\alpha S}{\sqrt{N}} = 2$$

   $$\sqrt{N} = \frac{t_\alpha S}{2}$$

   $$N = \left[\frac{t_\alpha S}{2}\right]^2 = \left[\frac{(2.02)(2.93)}{2}\right]^2 = 8.76$$

2. The procedure for making the panels is extremely reproducible ($S_1^2 = 0.14$). In testing any treatment combination, therefore, it is only necessary to make one panel.

3. The vendor of the material has a quality problem ($S_2^2 = 5.30$). The average of all the material supplied is 29.72, but at least two of the batch sample means are above the maximum limit of the specification. Only one of the batches is below 30 volts/second with high confidence.

$$(1 - \alpha) \text{ batch confidence} = \pm t_\alpha \left(\frac{S_1^2}{n} + \frac{S_0^2}{kn} \right)^{1/2} \qquad \begin{array}{l} n = \text{number of panels} \\ k = \text{number of tests} \end{array}$$

The 95 percent batch confidence $= \pm 2.02 \left(\frac{0.14}{3} + \frac{8.60}{3 \times 4} \right)^{1/2} = \pm 1.76$ volts/second, if three panels are tested and four tests are made on each panel. Thus, batch 3 with a mean of 27.3 has a 95 percent confidence interval of 27.3 ± 1.76 or $25.54 \leq \mu \leq 29.06$, but batch 5 with a mean of 29.725 has a 95 percent confidence interval of $27.97 \leq \mu \leq 31.49$.

Because of the large testing variance, it would be pure nonsense to proceed with experimentation to discover the effect of film thickness or curing temperature on the dark decay of this photoconductor if, for example, one were interested in δ values of 1 or 2 volts/second. It would also be very unwise to use different batches in any one experiment because of the high variation between batches.

In most cases, the experimenter cannot do much about improving the uniformity of the vendor's product other than to bring the matter to the attention of the vendor. In the case of the testing procedure being used, however, an attempt should be made to reduce the variance if it is excessively large.

The sources of variance in every test are different and, in most cases, too numerous to attempt to enumerate. However, there are certain common causes of poor tests that can be eliminated at almost no cost or effort:

- Failure to write down the testing procedure.
- Failure to specify the test procedure in detail; e.g., "stir the mixture." For how long? How vigorously? With what?
- Failure to frequently calibrate equipment.
- Failure to follow the test procedure in minute detail.
- Use of multiple pieces of the "same" equipment or multiple operators without checking them for bias.
- Inadequate equipment.
- Failure of technicians to appreciate the importance of their job, with a resultant careless attitude.
- Failure to mechanize as much as possible of a test procedure.

With the experiment design discussed above, the degrees of freedom are highest at the end of the process and lowest at the beginning of the process. This means that the test variance is estimated with the most degrees of freedom and, in this sense, is estimated the best. The vendor or batch variance has the fewest degrees of freedom and is, therefore, estimated the worst. There are many other designs that are more efficient than this (see references). For example, the design described above can be represented as shown in figure 21-4. Another design with considerable merit is shown in figure 21-5.

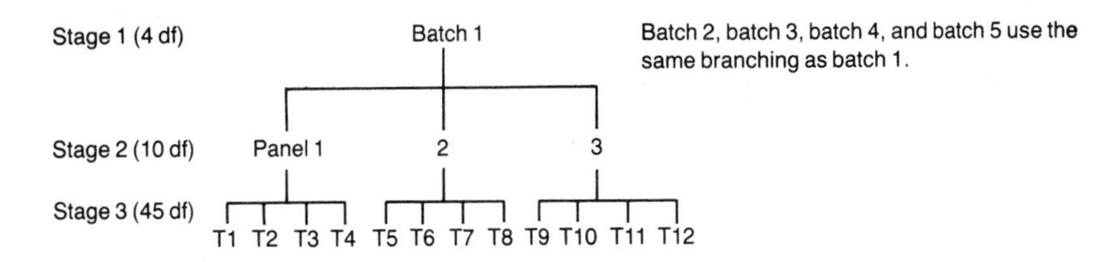

Fig. 21-4. Experiment design of example.

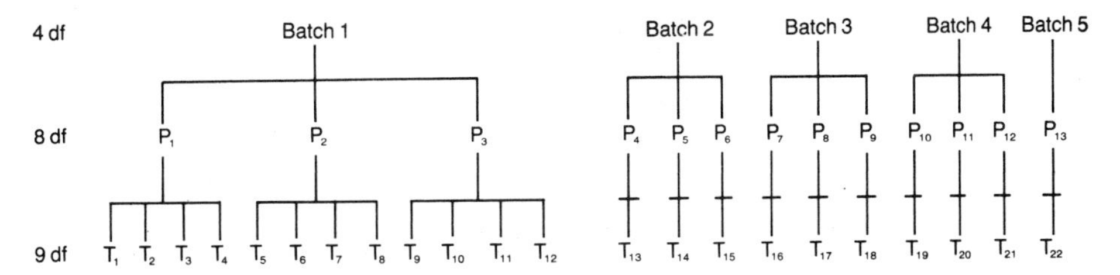

Fig. 21-5. Another alternative experiment design.

With batch 1, three panels are made, and each panel is tested four times. With batches 2, 3, and 4, three panels are made, and each panel is tested only once. With batch 5, only one panel is made, and it is tested only once. In this design, the number of degrees of freedom for stage 1 is the same as in the first design. The degrees of freedom for stage 2 are approximately the same as in the first design. The degrees of freedom for stage 3 are considerably less than the excessive 45 in the first design. Only 22 tests are required; this is considerably fewer than the 60 tests of the first design.

Many other variations of this type of design are available. In all of them except the symmetrical design first described, the variance component analysis is not straightforward, as described in the example. Furthermore, the variances estimated by the mean squares are not easily determined and, in most cases, will not be nice and neat whole numbers but, rather, something like the following: $MS_2 = \sigma_0^2 + 3.72\sigma_1^2 + 14.15\sigma_2^2$. This should not dissuade the reader from using such designs, however, as they are well described in several advanced texts, and the savings in time and money can be substantial. Certainly, a statistician should be consulted.

In addition to variance problems with testing procedures, it is also possible that bias problems can exist. The most frequent cases are operator bias or instrument bias. If several instruments and/or several operators are an absolute necessity on a project, they must be checked out for bias. This can be done at the same time as a variance component experiment. Such a design could consist of the branching shown in figure 21-6. Again, analysis of this experiment is somewhat different than in the original example, and an advanced text or a statistician should be consulted.

In the previous example, which is called a nested, random-model experiment, the batches, the panels within batches, and the test within panels were all randomly selected from an infinity of batches, panels within batches, and within panels.

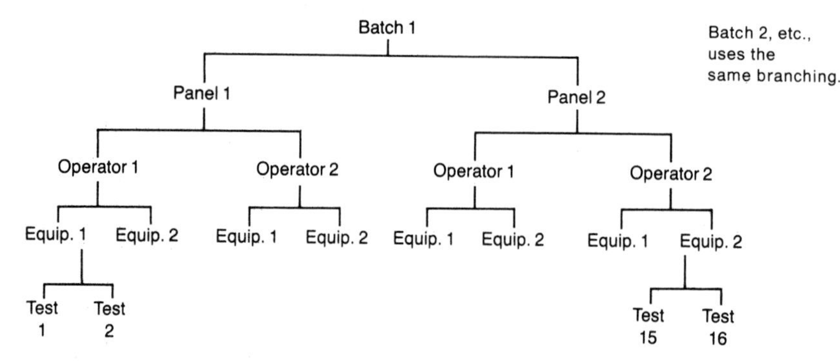

Fig. 21-6. Experiment design for mixed model.

In some cases in test validation, however, we encounter a mixed model; i.e., some of the variables are random, whereas others are fixed. For example, samples and tests might be random within a batch, but there might be two fixed levels of testing equipment and/or three fixed levels of test equipment operators.

EXAMPLE

A new test procedure that requires much less time than the old method has been proposed for measuring the mean particle size of toner to be used in a photocopier machine. The test will be used to accept or reject 100-pound lots of material in production. Three different operators and two different pieces of test equipment will be employed in production testing. Two different batches of material, one at the high limit and one at the low limit of the specification, are available to validate the new test method.

		Low batch				High batch			
		Equip. 1		Equip. 2		Equip. 1		Equip. 2	
	Sample	Test 1	Test 2	Test 1	Test 2	Test 1	Test 2	Test 1	Test 2
Operator A	1	6.81	6.74	7.46	7.54	8.82	8.84	9.34	9.39
	2	6.62	6.58	7.34	7.35	8.95	8.97	9.40	9.39
	3	6.73	6.79	7.40	7.46	8.75	8.81	9.28	9.32
	4	6.72	6.80	7.39	7.41	8.87	8.93	9.45	9.38
Operator B	1	6.64	6.77	7.54	7.68	8.79	8.95	9.22	9.36
	2	6.71	6.54	7.21	7.39	8.71	8.95	9.38	9.51
	3	6.71	6.84	7.37	7.52	8.65	8.91	9.15	9.34
	4	6.95	6.80	7.30	7.74	8.67	8.81	9.55	9.60
Operator C	1	6.72	6.79	7.50	7.51	8.79	8.83	9.30	9.35
	2	6.64	6.55	7.30	7.36	8.97	9.00	9.42	9.45
	3	6.75	6.81	7.45	7.51	8.73	8.82	9.22	9.27
	4	6.73	6.79	7.40	7.45	8.91	8.85	9.50	9.42

Fig. 21-7. Experiment design for validation of new test method.

A. The experiment design and average particle-size results, in microns (μ), are shown in figure 21-7.

B. The testing error is determined from the replicate test within each sample for a given batch, given operator, and given equipment. The formula is exactly the same as for the previous example:

$$SS_0 = \frac{\sum X_i^2}{N_1} - \frac{(\sum X_i)^2}{N_2}, \qquad \text{where } N_1 = 1 \text{ and } N_2 = 2 \text{ in this example.}$$

Note that, in the case where $N_2 = 2$, the calculation can be simplified:

$$SS_0^2 = \frac{(X_1 - X_2)^2}{2}.$$

Computation results are shown in figure 21-8.

Several things are immediately apparent from figure 21-8:

(1) The testing variance of operator B is completely out of line with the other two operators. A test of significance can be made with the F-test.

$$F = \frac{\text{Avg } SS_{\text{B}}}{(\text{Avg } SS_{\text{A}} + \text{Avg } SS_{\text{C}})/2} = \frac{0.01871}{0.001265} = 14.79$$

$$F^*(16, 32, 0.01) \doteq 2.67$$

Therefore, with greater than 99 percent confidence, operator B is doing something wrong and causing erratic results. Note, however, there is no bias, since the mean of all three operators is the same.

(2) Testing equipment 2 is giving higher mean results than testing equipment 1.

$$\text{With low batch, } (\bar{X}_2 - \bar{X}_1) = (14.88 - 13.46)/2 = .71$$
$$\text{With high batch, } (\bar{X}_2 - \bar{X}_1) = (18.75 - 17.69)/2 = .53$$
$$\text{Avg } (\bar{X}_2 - \bar{X}_1) = 0.62$$

After the variance of the experiment is calculated, these statistics can be tested with the t test, but it is fairly obvious that it will be significant with at least 99 percent confidence.

(3) The experiment proves that the test does discriminate adequately.

$$\bar{X}_{\text{high}} = (17.69 + 18.75)/4 = 9.11$$
$$\bar{X}_{\text{low}} = (13.46 + 14.88)/4 = 7.08$$

C. The variation between samples within a batch with the same operator and equipment is given by the formula

$$SS_{\text{sample}} = \frac{\sum X_i^2}{N_1} - \frac{(\sum X_i)^2}{N_2},$$

where X_i is the sum of the two test results from figure 21-8.

	Low batch				High batch				
	Equip. 1		Equip. 2		Equip. 1		Equip. 2		
Op A	$\sum =$ 13.55	$SS_0 =$ 0.00245	15.00	0.00320	17.66	0.00020	18.73	0.00125	$\sum X_i =$ 259.03
	13.20	0.00080	14.69	0.00005	17.92	0.00020	18.79	0.00005	$\bar{X}_A =$ 16.19
	13.52	0.00045	14.86	0.00180	17.56	0.00180	18.60	0.00080	Avg $SS_0 =$ 0.00126
	13.52	0.00320	14.80	0.00020	17.80	0.00125	18.83	0.00245	$\phi =$ 16
Op B	13.41	0.00845	15.22	0.00962	17.74	0.01280	18.58	0.00980	$\sum X_i =$ 259.26
	13.25	0.01445	14.60	0.01620	17.66	0.02880	18.89	0.00845	$\bar{X}_B =$ 16.20
	13.55	0.00845	14.89	0.01128	17.56	0.03380	18.49	0.01805	Avg $SS_0 =$ 0.01871
	13.75	0.01125	15.04	0.09680*	17.48	0.00980	19.15	0.00125	$\phi =$ 16
Op C	13.51	0.00245	15.01	0.00005	17.62	0.00080	18.65	0.00125	$\sum X_i =$ 259.09
	13.19	0.00405	14.66	0.00180	17.97	0.00045	18.87	0.00045	$\bar{X}_C =$ 16.19
	13.56	0.00180	14.96	0.00180	17.55	0.00125	18.49	0.00125	Avg $SS_0 =$ 0.00127
	13.52	0.00180	14.85	0.00125	17.76	0.00180	18.92	0.00320	$\phi =$ 16
	$\sum X_i =$ 161.53, $\bar{X}_1 =$ 13.46	Avg $SS_0 =$ 0.00497, $\phi =$ 12	$\sum X_i =$ 178.58, $\bar{X}_2 =$ 14.88	Avg $SS_0 =$ 0.01200, $\phi =$ 12	$\sum X_i =$ 212.28, $\bar{X}_1 =$ 17.69	Avg $SS_0 =$ 0.00775, $\phi =$ 12	$\sum X_i =$ 224.99, $\bar{X}_2 =$ 18.75	Avg $SS_0 =$ 0.00402, $\phi =$ 12	
	$\sum X_i =$ 340.11, $\bar{X} =$ 14.17	Avg $SS_0 =$ 0.00625, $\phi =$ 24			$\sum X_i =$ 437.27, $\bar{X} =$ 18.22	Avg $SS_0 =$ 0.00589, $\phi =$ 24			

* This datum looks incorrect; at least it doesn't seem to fit with all the other data. The possibility that this datum might be false should, at least, be kept in mind throughout all subsequent analyses.

Fig. 21-8. Computation results for $SS_{testing}$.

	Low batch		High batch		
	Equip. 1	Equip. 2	Equip. 1	Equip. 2	Average
Op A	0.0411	0.0250	0.0374	0.0151	0.0297
Op B	0.0676	0.1032	0.0194	0.1365	0.0816
Op C	0.0441	0.0361	0.0515	0.0598	0.0479

Fig. 21-9. Values for SS_{sample}.

For example, with operator A, equipment 1, and low batch,

$$SS_{\text{sample}} = \frac{(13.55)^2 + (13.20)^2 + (13.52)^2 + (13.52)^2}{2} - \frac{(13.55 + 13.20 + 13.52 + 13.52)^2}{8}$$

$$= \frac{723.4233}{2} - \frac{(53.79)^2}{8} = 361.711 - 361.670 = 0.0411$$

The values for SS_{sample} for the other blocks are given in figure 21-9.

Since SS_0 was larger for operator B than for operators A and C, the SS_{sample} data must also be segregated by operators.

Avg $SS_{\text{sample}}(\text{Op A} + \text{C}) = 0.0388$;

$MS_{\text{sample}}(\text{Op A} + \text{C}) = \dfrac{0.0388}{3} = 0.0129.$

$\phi = 24$

Avg $SS_{\text{sample}}(\text{Op B}) = 0.0816$; $MS_{\text{sample}}(\text{Op B}) = \dfrac{0.0816}{3} = 0.0272.$

Degrees of freedom = 12.

Source of variation	MS	ϕ	Variance estimated by MS
Between samples within lot (operators A and C) and within equipment.	0.0129	24	$\sigma_0^2 + 2\sigma_s^2$
Between tests within samples (operators A and C) from figure 19-8.	0.00126	32	σ_0^2
Between samples within lot (operator B) and within equipment.	0.0272	12	$\sigma_0^2 + 2\sigma_s^2$
Between tests within samples (operator B).	0.01871	16	σ_0^2

$S^2_{\text{samples}} = (0.0129 - 0.00126)/2 = 0.00582$ for operators A and C.

$S^2_{\text{samples}} = (0.0272 - 0.01871)/2 = 0.00425$ for operator B.

$S^2_{\text{samples}} = 0.0053$(weighted avg).

Therefore, the estimated variance for the estimate of the μ of a lot is:

$$S^2_{\text{lot}} = \frac{S^2_{\text{samples}}}{n} + \frac{S_0^2}{nk} = \frac{0.00530}{n} + \frac{0.00126}{nk} \quad \text{for operators A and B,}$$

where k is number of tests, and n is the number of samples.

The variance for various test plans can now be calculated:

- With one sample and 10 tests on each sample,

$$S^2_{\text{lot }\mu} = \frac{0.0053}{1} + \frac{0.00126}{(1)(10)} = 0.005426.$$

- With two samples and five tests on each sample,

$$S^2_{\text{lot }\mu} = \frac{0.0053}{2} + \frac{0.00126}{(2)(5)} = 0.002776.$$

- With five samples and two tests on each sample,

$$S^2_{\text{lot }\mu} = \frac{0.0053}{5} + \frac{0.00126}{(5)(2)} = 0.001186.$$

- With 10 samples and one test on each sample,

$$S^2_{\text{lot }\mu} = \frac{0.0053}{10} + \frac{0.00126}{(10)(1)} = 0.000656.$$

The latter procedure is obviously the best (lowest variance) for a total of 10 tests. The confidence interval for μ_{lot} for this test is $\bar{X} \pm t_\alpha S_{\text{lot }\mu} = \bar{X} \pm (2)(0.026) = \bar{X} \pm 0.052$ for 95 percent confidence.

The following conclusions are reached from the experiment:

1. The testing variance is small (0.00126) for operators A and C; therefore, the test is approved for precision.
2. Operator B is not following procedures precisely, and this is showing up as a very large testing variance. However, there is no bias between operators.
3. The sampling variance (0.0053) is large compared to the testing variance. This indicates either lack of homogeneity within a batch of material or variation in the sample-making procedure. Perhaps this should be investigated further by product engineering.
4. The test discriminates very well between material with high particle size and material with low particle size. The confidence interval of the mean of a batch is $\pm 0.05\ \mu$ for a sample of 10 with one test per sample.
5. The accuracy of the test equipment is not satisfactory. One or both of the test devices is inaccurate; at the least, one is biased 0.62 μ from the other. This, of course, is much greater than any sampling or testing error. Product engineering must determine the cause of the difference and validate the accuracy of one or the other instrument by using a standard material.

EXERCISES

1. In the final development stage of a project, an engineer is attempting to set up a testing procedure for lots of an electron circuit where the important characteristic is rise time in milliseconds. The specification calls for a rise time of 1.2 to 1.7 msec. A second objective of the experiment is to evaluate the reproducibility of the vendor's product. Four different lots are sampled, and three circuits are chosen from each lot. Each selected circuit is tested twice.

The following results (msec) are obtained:

Lot 1	Sample A	1.65	1.60
	B	1.75	1.73
	C	1.45	1.48
Lot 2	Sample D	1.18	1.22
	E	1.42	1.38
	F	1.27	1.33
Lot 3	Sample G	1.32	1.36
	H	1.58	1.54
	I	1.72	1.65
Lot 4	Sample J	1.59	1.53
	K	1.49	1.54
	L	1.42	1.42

Lay out the data in a block diagram, then analyze the results, and set up the test plan for production quality control. Are you satisfied with the vendor's product? What will be the batch confidence with your test plan? What will be the acceptance limits of rise time on your sample for 95 percent confidence?

2. Repeat the analysis of the toner particle size example, using the following data in place of the questionable data of figure 19-8.

Operator B	Test 1 = 7.40
Low batch	Test 2 = 7.49
Equipment 2	

REFERENCES

Anderson, V. L. and McLean, R. A. 1974. *Design of experiments.* New York: Marcel Dekker.

Ott, L. 1977. *An introduction to statistical methods and data analysis.* North Scituate: Duxbury Press.

Anderson, R. L. and Bancroft, T. A. 1952. *Statistical theory in research.* New York: McGraw-Hill Book Co.

22 Concepts for a Complete Project Strategy

In this chapter, we shall integrate all the previous discussion on experiment strategy into a general concept for a strategy of conducting a complete project. It was pointed out in the first three chapters that there is a mathematical relationship between every response of interest to an experimenter and a variety of variables, at least some of which are known to the experimenter. The general equation for all projects can be written:

$$\text{Response} = f(\text{A}) + f(\text{BC}) + f(\text{DE}) + f(\text{FGH}) + f(\text{IJ}), \quad \text{etc.}$$

This equation implies that either a high or a low value of response can be obtained if we can experimentally determine an estimate of, for example, $f(\text{A})$ or $f(\text{BC})$. Of course, to determine an estimate of $f(\text{BC})$ experimentally, we must first establish experimentally that variables B and C interact.

The general equation above can be pictured as a set of additive response surfaces, as shown in figure 22-1.

Of course, it is also possible to have four-factor (or higher) interactions in a project. However, we can't draw the response surfaces of such interactions; furthermore, there is no practical way of detecting them experimentally in the usual engineering project with many variables. It is most important to note that engineering projects do not consist of an N-dimensional space, such as $Y = f(\text{A, B, C} \ldots\ldots)$, but rather the sum of many two- or three- or maybe even four-dimensional spaces.

The response surfaces shown in figure 22-1 are just a few of the many types of response surfaces available. The important characteristics of response surfaces, insofar as potential for a successful project by the experimenter, are: (a) the percentage of the total space that is responsive (i.e., γ); (b) the steepness of the slopes; and (c) the number of additive functions that exist in a project and can be identified by the experimenter.

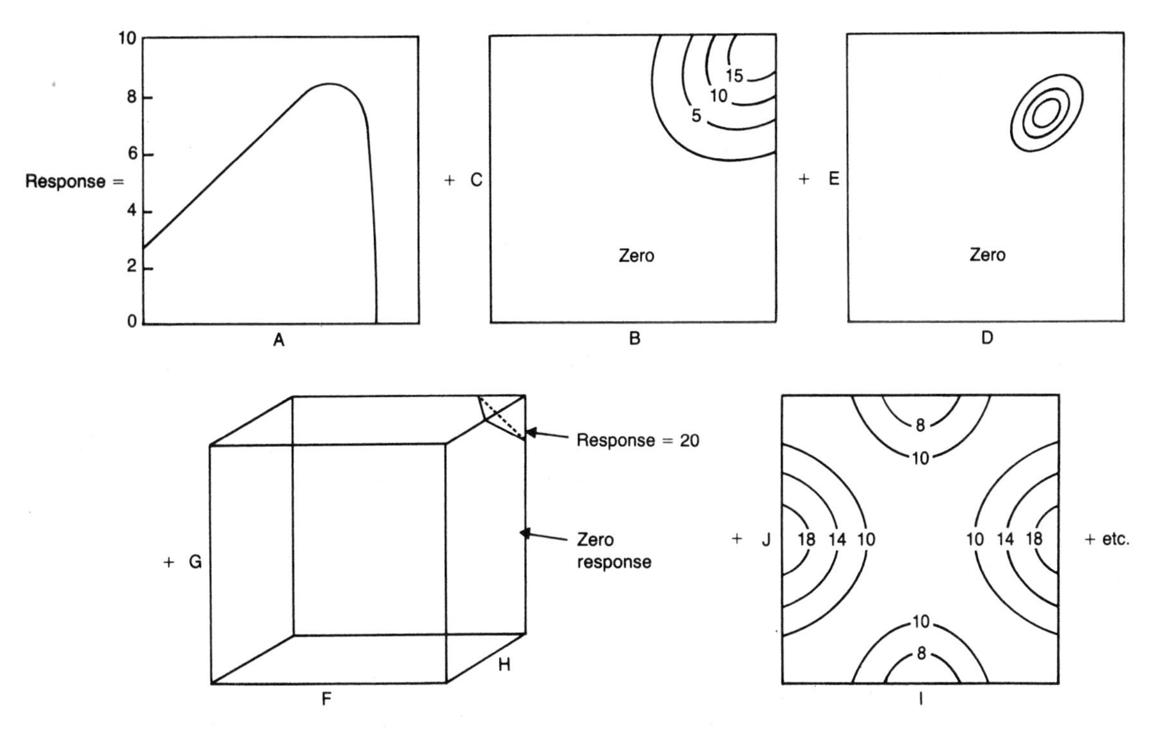

Fig. 22-1. Example of a set of additive response surfaces.

The following procedure is suggested for conducting an engineering project:

- Define the objective of the project.
- Define all the variables that can influence the responses of interest.
- Define the range of all the variables.
- Apply all known engineering theory; set all variables to their best level where the functions are known.
- Conduct experiments on variables whose functions are unknown.

When conducting experiments, the experimenter should use the following procedure:

- Find a best starting point with random strategy.
- Find a good direction from the starting point with a Resolution III or IV experiment. That is, identify the most important variables and the direction for improving the response due to these variables.
- Pursue this direction to the fullest extent with Resolution IV or V experiments or even one-factor-at-a-time trials. Establish the best level for the most important variables.
- Repeat the first three steps as required, with a reduced number of variables in each iteration as the variables are identified and fixed at their estimated best values.

The general formula for the probability of success, i.e., finding *a* trial with better-than-average response with the random strategy, is:

$$p = 1 - \left[\prod_{i=1}^{j} (1 - \gamma_i)\right]^N,$$

where γ_i is the γ of the ith additive group, j is the number of additive groups of variables in the project, and N is the number of random trials. Note carefully, the above is *not* the probability of achieving the objective of the project.

Consider the response function (known only to God) in figure 22-2.

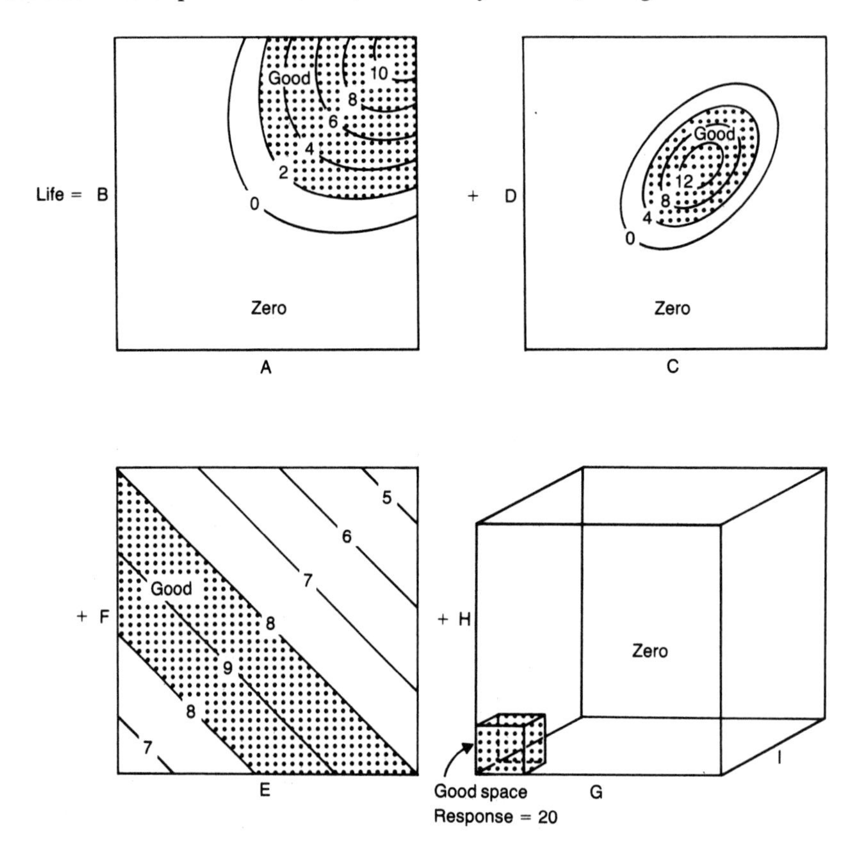

Fig. 22-2. Response function.

For variables A and B, $j = 1$, and $\gamma_{(1)}$ (1 = A and B group) is about 0.25. For variables C and D, $j = 2$, and $\gamma_{(2)}$ (2 = the C and D group) is about 0.15. For variables E and F, the entire space is responsive, but good space can be defined as, for example, a value greater than 8. Then, $\gamma_{(3)}$ is about 0.30. For variables G, H, and I, $j = 4$, and $\gamma_{(4)}$ equals $0.2 \times 0.2 \times 0.2 = 0.008$.

If the experimenter selects random levels of A, B, C, D, E, F, G, H, and I, and conducts one trial at the selected levels, the probability that this trial will have an unusually high result (i.e., greater than 8) will be

$$p = 1 - [(1 - 0.25)(1 - 0.15)(1 - 0.30)(1 - 0.008)] = 0.44.$$

For example, if the randomly chosen levels of A and B are such that they are on the 10 line, if F and E are on the 7 line, and if D, C, H, G, and I are all in zero space, the experimental result will be approximately 17, which is a good response. If 20 random samples are made in this system, the probability that at least one of the responses will be good is given by

$$\begin{aligned} p &= 1 - [(1 - 0.25)(1 - 0.15)(1 - 0.30)(1 - 0.008)]^{20} \\ &= 1 - 0.443^{20} = 1. \end{aligned}$$

Of course, the experimenter does not know that it is almost a certainty that one outstandingly better result will be obtained if a random-strategy experiment is conducted. Only God knows the values of γ_i and can compute the value of p. After the experiment is complete, and an outstandingly good result has been obtained, the experimenter suspects that at least one of the γ_i must be large and steep. In this example, the true but unknown probabilities of success are:

- For 10 trials, $p = 1 - 0.443^{10} = 1$.
- For 5 trials, $p = 1 - 0.443^{5} = 0.98$.
- For 3 trials, $p = 1 - 0.443^{3} = 0.91$.

Thus, if the random strategy is performed sequentially, it is likely that, by the fifth trial, one of the stopping criteria would occur. It is almost certain that the trial with the best response is due to a productive point in the AB subsystem and/or in the EF subsystem. Of course, the experimenter has no way of knowing that A and B or C and D are parts of interactive subsystems. A might just as well be interactive with G.

However, having obtained a best point, and proven by the criterion D_i^* that continued random trials will not be productive, the experimenter should conduct a matrix experiment; e.g., a Resolution IV design with nine variables in 24 trials, or a Resolution III design with nine variables in 12 trials. Since the experimenter has no idea of the important variables, or of the correct direction to move each of the important variables, the purpose of this matrix experiment will be to identify the important variables and the proper directions to move them. The initial direction chosen is unimportant; if it turns out that, by increasing variable A, for instance, the response is decreased, the experimenter will have achieved the objective: the proper direction for A is not up, but down.

The empirical rules for selecting the levels of the variables are:

- Use the levels of the best random trials as the low levels of the variables for the matrix experiment.
- Increase or decrease (whichever seems wisest) the low level of each variable by *15 percent of the total range of the variable* to obtain the high level for the matrix experiment.

Suppose, for example, this first matrix experiment (a Resolution IV design) is conducted, and it indicates that high A, high B, low E, and low F are significantly better than the alternate levels and that certain groups of interactions are significant. The procedure is to follow up on the variables that are indicated to be significant. In this case, with four variables significant, it is possible to conduct a Resolution V design with 12 trials or a Resolution IV design with only 8 trials.

For the second matrix experiment, the low level of variable A will be the same as the high level in the previous experiment, and likewise for variable B.

The new high level of A and B will be determined by the average effect of A and B in the first matrix experiment. Thus, if the effect of A was 4.0 for a change from -1 to $+1$ and a δ of 2.0 is desired in the next experiment, then high A should be $+2$.

For variables E and F, the low level of these variables should be the low level of the previous experiment and the new high level should be a still lower value. Again, the new level will be determined by the average effect of E and F in the first experiment. If the effect of E was -2.0 for a change from -1 to $+1$ and δ of 2.0 is desired in the next experiment, then low A $= -1$ and high A $= -3$.

If the Resolution V design is used, the interactions should be identified immediately. If the Resolution IV design is used, the interactions should be identified by inference; the significant set of interactions in experiment 1 and the significant set of interactions in experiment 2 should have only AB and EF in common.

Experiments should continue with variables A, B, E, and F until the experimenter has obtained a maximum response due to these variables.

If the "maximum" response obtained by a sequence of experiments on variables A, B, E, and F is less than the objective of the project, the experimenter should conduct a second iteration of the above procedure:

- Set A, B, E, and F at their best levels, as determined in the first iteration.
- Randomly vary C, D, G, H, and I.
- Stop the random strategy when one of the stopping rules applies.
- Conduct a matrix experiment on variables C, D, G, H, and I.
- Conduct follow-up matrix experiments on variables identified as important in the fourth step.

The probability of success for the second random strategy can be computed by God as:

$$P = 1 - [(1 - 0.15)(1 - 0.008)]^{20}.$$
$$P = 1 - [(0.85)(0.992)]^{20} = 1 - 0.033 = 0.967.$$

It is obvious that, in less than 20 random trials, a significantly better result than any obtained on the first iteration will be achieved. This will probably be due to obtaining a trial in the responsive spaces of variables C and D. After maximizing the response due to C and D with matrix experiments, a final random strategy would be conducted with variables G, H, and I. The probability of obtaining a point in the responsive space of this subsystem is

$$P = 1\ (0.992)^{20} = 0.15.$$

There is very little probability that the experimenter will detect these three variables. However, if 40 trials are conducted, the probability of detection is increased:

$$P = 1 - (0.992)^{40} = 0.275.$$

Thus, it is recommended that, as variables are removed from the system by successive iterations, the number of random trials be increased.

The fact that the variables G, H, and I will escape detection should not upset

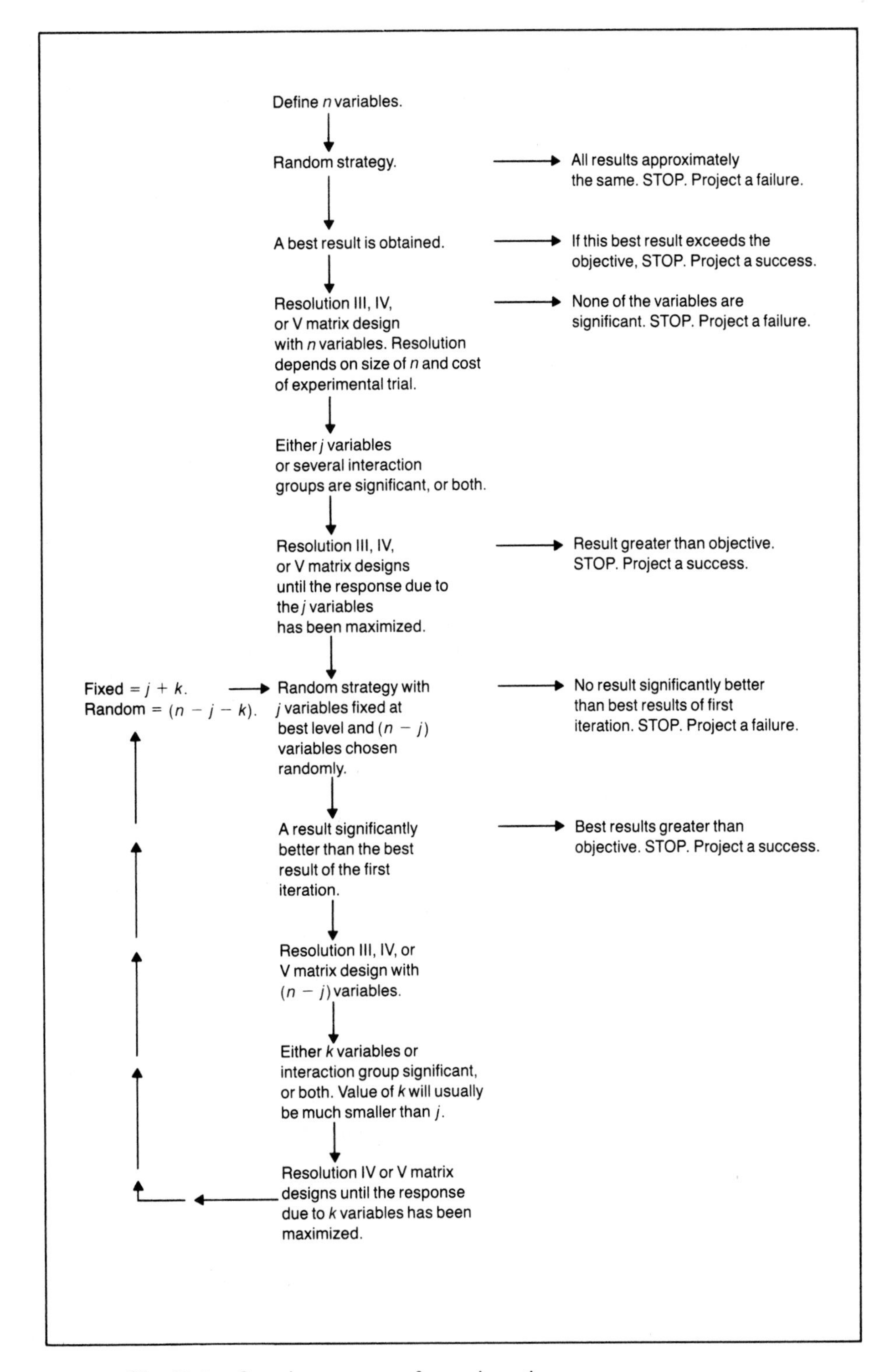

Fig. 22-3. Iterative strategy of experimenting.

the reader. When the γ of an additive group is very small (in this case 0.008), no strategy of experimenting has a better chance of detecting it than the random strategy discussed in this text. In any project, the experimenter will miss some variables because the γ is too small.

Figure 22-3 outlines the sequence of experiments and the various decision points in the iterative strategy of experimenting.

EXERCISE

Suppose a sequential random-strategy experiment has been conducted with the following results.

A	B	C	D	E	F	G	H	I	J	K	L	M	N	O	Response, %
(0–10)	(125–200)	(0.1–0.5)	(1–2)	(600–800)	(5–7)	(10–20)	(1–3)	(9–12)	(1–1.5)	(0–100)	(5–6)	(90–100)	(25–125)	(2000–4000)	
2.2	175	0.32	1.64	690	5.2	17.1	1.04	11.1	1.15	94	5.50	91.2	78	3642	178
5.9	185	0.44	1.79	625	5.1	12.2	1.05	11.9	1.25	80	5.90	90.2	120	2725	155
6.8	130	0.47	1.50	680	6.9	19.0	2.05	10.4	1.45	35	5.85	98.5	121	3218	165
1.2	150	0.15	1.52	710	6.8	14.3	1.62	9.1	1.10	39	5.02	94.6	84	2764	158
9.5	170	0.22	1.31	705	6.1	16.9	2.55	9.3	1.21	42	5.41	91.2	89	2981	140
8.4	135	0.37	1.19	655	5.4	10.1	1.72	10.9	1.33	61	5.36	95.0	92	2165	149

Verify that the experimenter stopped properly after six trials. Design the next experiment if the objective of the project is a part with an elongation of at least 200 percent.

23 Project Engineer's Game

It will have been noted that, contrary to the practice of textbooks, there is not a large number of problems in previous chapters to be worked out by the student. The reasons for this are:

- The only problems that are easy to formulate are problems for analysis of the data, but the analysis of another set of data would be no different than the analyses of the examples that are in each chapter. The reader who can follow the examples can analyze any similar set of numbers.
- The prime purpose of this book is to teach the reader how to design experiments, but problems in the design of experiments are almost impossible to formulate. First, it would take pages of description to define a problem that would be reasonably complete and would reasonably match a typical experimental problem. Second, there is no absolutely correct answer to experiment design problems. With exactly the same information, two experimenters might come up with entirely different experiments; in most cases, it would be hard to say which was the correct answer.

Despite the above reasons, it is certainly desirable for the reader to practice these design methods in a controlled situation. Therefore, this author has devised a computer simulation game that can be used by the reader to obtain data for any experiment design. With these data, and the analysis of the data, the reader can then design a second experiment, etc. Thus, the reader can gain experience in both design and analysis in what appears to be a real-life project. Likewise, the reader can experience both success and failure, or, at least, apparent failure.

To play the game, the reader must have a fellow worker enter the computer program, given at the end of this chapter, into the reader's APL work space. *Caution—do not look at the appendix to this chapter.* The appendix contains the equation that computes the response for each variable, and, obviously, there is no problem if the reader sees the true equation. It should be noted, however, that it is relatively easy for a fellow worker with experience in APL to change the given equation to obtain a new problem.

It is likewise quite easy to write these programs in Basic or Fortran.

Project Assignment

The research laboratory has discovered a new type of widget, which has a number of applications in various products of the company. All properties of the widget meet the requirements of the products, except that the life is too short. A life of 55 days is required if the product is to be commercially feasible. The best result, to date, has been 10 days. An accelerated test is available—1 day on the accelerated test is equivalent to 60 days in real life. All results obtained from the computer are in equivalent days. Two months are available for the project. Four trials can be conducted in one day.

The research lab has established that the only process variables that can influence the response are A, C, D, I, L, and R. The material variables that are thought to be important are B, E, F, K, N, and Q. The machine variables that are thought to be important, either on their own or as an interaction with material and/or process variables, are G, H, J, M, O, and P.

You cannot test the effects of material or process variables except on machines. Therefore, every trial must specify a level for all variables.

The research laboratory has established the following limits on the variables. (Note, if you go outside these limits, your laboratory will blow up, and you lose the game.)

A	0–3.5
B	0–3.0
C	0–2.0
D	0–5.0
E	0–2.5
F	0–2.0
G	0–2.5
H	0–2.0
I	0.4–2.0
J	0–3.0
K	0–1.2
L	0–2.0
M	0–2.0
N	0–1.5
O	0.4–3.5
P	0–2.0
Q	0–10
R	0–18

Use of the Computer Program

As previously indicated, the game must have been entered into the computer work space by a fellow worker so that the game player does not know the true equation, the variance, the interactions, etc. To obtain the simulated experimental result of a trial, the experimenter enters the word ONE and presses carriage return. The computer will then type

```
ENTER VALUES.
[ ]:
```

The experimenter then enters the values of variables A, B, C, D, etc. After the last number is entered, press carriage return. The computer will then print the response; e.g., 32.61.

To obtain a second trial result at different levels of the variables, again type ONE and enter the data (levels of the variables) as above.

Playing the Game

If the game is properly entered into the computer work space, the experimenter can obtain simulated test results from the system over any period of real time. The procedure is to design a first experiment; e.g., a random-strategy experiment. This program features a subprogram to generate a random trial within the experiment design space and to test these results against the D_i^* criterion.

Type RANDOM M

where M is the maximum number of trials.

The program then requests

ENTER MIN

□:

The experimenter types in the list of minimum values for the range of each variable.

The program then requests

ENTER MAX

□:

The experimenter types in the list of maximum values for the range of each variable. If random trials are to be conducted on fewer variables than the number designated by the game, type in the same number for both min and max of the variables that are not to be included as variables.

If the experimenter then wishes to conduct a Hadamard matrix experiment, the treatment combinations are obtained using the computer program in Chapter 13. These treatment combinations, with real values for each variable, are then entered into this program one at a time to obtain the responses to the treatment combinations. The results are then put back into the computer program of Chapter 13 for analysis.

A skilled programmer, of course, can easily work out a combination of these two programs so that there is less data entry required. In general, however, it is felt that the learning process is more realistic if each step is separate and some effort on the part of the experimenter is required to obtain the simulated experimental results.

After thorough study of the analysis and the data, the experimenter must then design a second, third, etc., experiment until the required objective is obtained or until it is decided that the objective cannot be obtained within the given experimental space or within the given time.

After the project is completed, the experimenter can go to the appendix of this chapter and compare the estimate of the project system with the true values. It is suggested that the experimenter also plot the response surfaces of the project system and use these to explain the results of individual experiments.

It should be noted that this project is extremely difficult, probably more difficult than any project that is encountered in real life.

Other Project Engineer Games

Two additional projects are included in the computer program. The friend or associate who sets up the program can choose the project for the experimenter.

- Game 2

Variables	Range
A	0–3.5
B	0–3.0
C	0–2.0
D	0–2.0
E	0–2.5
F	0–2.0
G	0.4–2.0
H	0–1.2
I	0–2.0
J	0–1.5
K	0.4–3.5
L	0–2.0

Objective of project = 43.

- Game 3

Variables	Range
A	0–3
B	0–2
C	0–3
D	0–2
E	0–3
F	0–3
G	0–3
H	0–2
I	0–2
J	0–2
K	0–2
L	0–3

Objective of project = 50.

Chapter 23 Appendix

Computer Program and Response Equation for Project Engineer's Game

```
      ∇DIR[⎕]∇
    ∇ DIR
[1]    'GAME NO 1 REQUIRES 14 VARIABLES'
[2]    'GAME NO 2 REQUIRES 10 VARIABLES'
[3]    'GAME NO 3 REQUIRES 12 VARIABLES'
[4]    'IN ALL GAMES DUMMY VARIABLES CAN BE ADDED'
[5]    'ENTER  GAME NO'
[6]    GAME←⎕
[7]   A1:' ENTER MIN'
[8]    NV←NUV←⍴UN←,UN←⎕
[9]    'ENTER MAX'
[10]   →(NUV=⍴UX←,UX←⎕)/A2
[11]   NUV,⍴UX;'REENTER UN UX'
[12]   →A1
[13]  A2:'ENTER SIGMA'
[14]   SIG←⎕
[15]   G2← 1 2 3 6 7 8 9 11 12 14
[16]   'THE CONTROL VARIABLES HAVE THE FOLLOWING VALUES'
[17]   'DR←0      UA←0         SPC←1.5       ZZ←0        '
[18]   'THESE CAN BE CHANGED'
[19]   DR←ZZ←0
[20]   UA←0
[21]   SPC←1.5
[22]   Q←3
[23]   'THIS WORKSPACE SHOULD BE SAVED.  USE THE )SAVE COMMAND'
    ∇
```

Note: This computer program is not an IBM product and is not guaranteed or maintained by IBM.

Instruction to readers: do not look at or read this appendix until you have read the preceding pages of this chapter.

```
      ∇EQUA[⎕]∇
    ∇ EQUA GAME
[1]   V←14⍴0
[2]   →(GAME=⍳4)/A1,A2,A3,A4
[3]   →0
[4]  A1:V←VV[⍳14]
[5]   →A6
[6]  A2:V[G2]←VV[⍳10]
[7]  A6:R←7⍴0
[8]   R[1]←(|15×(V[1]×2-V[1])+V[6]×2-V[6])*0.5
[9]   R[2]←15×(V[2]*2)××0.64-(V[2]*2)+10×(V[2]-V[7])*2
[10]  R[3]←50×V[3]××0.04-(V[3]*2)+10×((6×V[3])-V[8])*2
[11]  R[4]←V[4]×V[5]÷2
[12]  R[5]←5×|V[9]-V[11]
[13]  R[6]←(V[10]*2)+(V[13]*2)-V[10]×V[13]
[14]  R[7]←10××-100×((¯1.3+0.72×V[12])*2)+(¯1.3+V[14])*2
[15]  →A7
[16] A3:V←VV
[17]  R←6⍴0
[18]  R[1]←(|15×2-((V[1]-1)*2)+(V[8]-1)*2)*0.5
[19]  R[2]←50×V[2]××0.04-(V[2]*2)+10×((6×V[2])-V[4])*2
[20]  R[3]←4×1+((V[3]-1.5)*2)-((V[11]-1)*2)+0.2×(V[3]-1.5)×V[11]-1
[21]  R[4]←10×(V[5]*2)××1-(V[5]*2)+10×(V[5]-V[12])*2
[22]  R[5]←9-3×|¯3+V[6]+V[7]
[23]  R[6]←¯1+(V[10]*2)+V[9]*2
[24] A4:→A7
[25] A7:R←R,SIG×¯6++/((?1001×12⍴1)-1)÷1000
[26]  RES←+/R
[27]  →(DR=0)/0
[28]  8 PO R
    ∇
```

```
      ∇RANDOM[⎕]∇
    ∇ RANDOM M
[1]   →(ZZ=0)/A3
[2]   ⎕RL←(+/⎕TS)*2
[3]  A3:'ENTER MIN'
[4]   TEST
[5]   NV←⍴MIN←TT
[6]   'ENTER MAX'
[7]   TEST
[8]   MAX←TT
[9]   NRS←RES←⍳I←BEST←0
[10] A1:→(M<I←I+1)/0
[11]  VV←MIN+(MAX-MIN)×((?1001×NV⍴1)-1)÷1000
[12]  →(Q=1)/A2
[13]  EQUA GAME
[14]  →(BEST>RES)/A2
[15]  BEST←RES
```

```
[16]   BV←VV
[17] A2:7 PO VV,RES
[18]   →(Q=1)/A1
[19]   →(BEST<NMN+SPC×((+/(NRS-NMN←(+/NRS←NRS,RES)÷I)*2)÷I-1)*0.5)/A1
[20]   ' '
[21]   →(UA=0)/0
[22]   PS←PS-2×UA×(MAX-MIN)×MAX<PS←BV+UA×MAX-MIN
[23]   VDIF←⍳I←0
[24] A4:→(NUV<I←I+1)/A5
[25]   →(BV[I]=PS[I])/A4
[26]   →A4,VDIF←VDIF,I
[27] A5:(8×4-(2×8≥NV)+4≥NV) POB NV←⍴VDIF
     ∇
```

```
       ∇TEST[⎕]∇
     ∇ TEST
[1]   A1:→(NUV=⍴TT←,TT←⎕)/A2
[2]    ⍴TT;' REENTER'
[3]    →A1
[4]   A2:→(Q=1)/0
[5]    →(0=+/TT<UN)/A3
[6]    'MIN';UN;'REENTER'
[7]    →A1
[8]   A3:→(0=+/TT>UX)/0
[9]    'MAX';UX;' REENTER'
[10]   →A1
     ∇
```

```
       ∇ONE[⎕]∇
     ∇ ONE
[1]    'ENTER VALUES'
[2]    TEST
[3]    VV←TT
[4]    EQUA GAME
[5]    8 2 ⍕RES
     ∇
```

```
       ∇PO[⎕]∇
     ∇ Z←Y PO X
[1]    →(0=⌈/|X)/L1
[2]    Z←(Y,(Y-2)⌊0⌈⌈Y-4+10⍟⌈/|X)⍕X←,X
[3]    →0
[4]   L1:Z←(Y,0)⍕X←,X
     ∇
```

Instructions for Game Director

Enter the program into the computer work space of the experimenter. Execute program DIR. The program will request the game number. There are three games from which to choose. If the experimenter has requested game 2, type in the number 2. The program then requests the MIN; that is, the minimum level of each variable in the experimental space. These values are given below. In project game 2, there must be 10 numbers. The program then asks for the MAX; again, the 10 maximum levels must be entered. The program then requests the value of SIGMA. It is suggested that the number 1 be used. If the experimenter wants to make the project more difficult, the value of SIGMA can be made larger, e.g., 2.

Do not change the control variables.

After the above data have been entered, the game director must save the work space by using the command

)SAVE

and then signing off. The experimenter can now conduct experiments within the experimental space defined by the game director.

Note: the game director must tell the experimenter which game is being used. Under no circumstances should the experimenter be told the number of variables or the variance that was used.

Min and Max Values

Game 1

MIN	0	0	0	0	0	0	0	0	0.4	0	0	0	0	0
MAX	3.5	3	2	5	2.5	2	2.5	2	2	3	1.2	2	2	1.5

Game 2

MIN	0	0	0	0	0	0	0.4	0	0	0
MAX	3.5	3	2	2	2.5	2	2	1.2	2	1.5

Game 3

MIN	0	0	0	0	0	0	0	0	0	0	0	0
MAX	3	2	3	2	3	3	3	2	2	2	2	3

Response Equations

- Game 1

$$\begin{aligned}\text{Life} = {} & \{|15[2 - (A - 1)^2 - (F - 1)^2]|\}^{1/2} \\ & + 15B^2 e^{0.64 - B^2 - 10(B - G)^2} \\ & + 50Ce^{0.04 - C^2 - 10(6C - H)^2} \\ & + \tfrac{1}{2}(DE) \\ & + 5(|I - K|) \\ & + J^2 + M^2 - JM \\ & + 10e^{-100[(0.72L - 1.3)^2 + (N - 1.3)^2]} \\ & + 0(O, P, Q, R).\end{aligned}$$

MAX response = 70.33.

MIN response = 0.

• Game 2

$$\begin{aligned}\text{Life} = &\{|15[2 - (A - 1)^2 - (D - 1)^2]|\}^{1/2} \\ &+ 15B^2 e^{0.64 - B^2 - 10(B - E)^2} \\ &+ 50C e^{0.04 - C^2 - 10(6C - F)^2} \\ &+ 5(|G - H|) \\ &+ 10^{-100[(0.72I - 1.3)^2 + (J - 1.3)^2]} \\ &+ 0(K) + 0(L).\end{aligned}$$

MAX response = 58.

MIN response = 0.

• Game 3

$$\begin{aligned}\text{Life} = &\{|15[2 - (A - 1)^2 - (H - 1)^2]|\}^{1/2} \\ &+ 10E^2 e^{1 - E^2 - 10(E - L)^2} \\ &+ 50B e^{0.04 - B^2 - 10(6B - D)^2} \\ &+ I^2 + J^2 - 1 \\ &+ 9 - 3(|F + G - 3|) \\ &+ 4[1 + (C - 1.5)^2 - (K - 1)^2 \\ &- 0.2(C - 1.5)(K - 1)]\end{aligned}$$

MAX response = 61.3.

MIN response = −1.0.

24 Estimation of Variance

In previous chapters, various methods of estimating the variance from a set of experimental data were discussed. At this time, these methods are summarized, and one additional procedure is given for the case of saturated designs, where it is apparently impossible to obtain an estimate of the variance.

The first procedure for obtaining an estimate of the variance is to conduct an experiment in replicate, for example, make two samples of each treatment combination in a Hadamard matrix experiment. This is the best procedure and costs nothing if the additional samples are required in any event, to achieve a proper α and β risk. However, it is very costly and time consuming if the addition samples are not required for the proper α and β risk and if the Hadamard matrix is relatively large.

The next best procedure is to include center points in the design. Any number of points can be used; few if they are expensive and time consuming; many if they are easy to obtain and inexpensive. However, this procedure is limited to quantitative variables only. It has, of course, the advantage (which might be worth the cost) of also indicating whether there are any quadratic effects. Furthermore, the center points can be used as control samples if the experiment is conducted on several different days.

The third procedure, which is particularly advantageous with large Hadamard Matrix experiments, is to use the higher-order interactions contrasts as estimates of variance. This procedure has the particular advantage that no extra trials are required.

A fourth procedure, particularly useful for Resolution III designs, is to estimate the variance from all contrast columns which are not utilized for main effects. This procedure is efficient in that no extra trials are required, but it frequently gives estimates of σ^2, which are biased high since the contrast columns that are used also measure two-factor interactions, which can be large.

In the case of saturated designs, for example, 15 variables in 16 trials, it is impossible to obtain an "unbiased" estimate of the variance without conducting additional trials. However, it is possible to obtain a "good" estimate of the variance from the experiment without additional effort. The procedure is as follows for Resolution III Hadamard Matrix designs:

- Using the computer program, compute the mean effect and SS value for all contrast columns.
- Pick the 3 largest effects for the 16-trial design, the 4 largest effects in a 32-trail design, etc.
- Make up a three-way table in a 16-trial design, using the variables which produced the 3 largest effects. In the 32-trial design, make up a four-way table; for 64 trials make up a five-way table, etc.
- Enter the response in the proper blocks. There will be two responses in each block. In some cases, a four-way table is required with a 16 trial experiment, in order to have two responses in each block.
- Compute the estimated variance with one degree of freedom from the pair of data in each block.
- Compute the average estimate of variance using the estimates from all blocks. In the case of the 16-trial experiment, this will give an average estimate of σ^2, with eight degrees of freedom.
- Test the SS value of all the variable, including those used in the n-way table, for significance at the $\alpha = .08$ level with the F-test. For the 16-trial experiment, the F-value is looked up with one and eight degrees of freedom.
- If any variable other than those in the n-way table are significant, then all of the responses are corrected for this variable and a new n-way table compiled, and a new (smaller) estimate of the variance is obtained, using the procedure above.
- This procedure is repeated until no further variables are found to be significant at the $\alpha = .08$ level. The last estimate of variance is the "best" estimate of σ^2.
- If any of the variables in the original n-way table are shown to be insignificant at the $\alpha = .08$ level on the first F-test, the experimenter should be suspicious that the estimate of the variance is too low.

The above procedure was validated by testing a number of simulated projects with 14 variables in a Resolution III design of 16 trials using different known variances. The following examples are several runs of this research projects.

The true effects of the variables were:

A = −1.68	E = .08	I = .65	M = .05
B = .16	F = −.19	J = .05	N = 0
C = −12.90	G = 2.32	K = −.45	BD+DE = −3.98
D = .27	H = −.26	L = 0	

EXAMPLE 1

$\sigma^2 = 1.0$

The results of one of the computer runs were as follows:

Variable	Mean Effect	SS
A	−1.09	4.77
B	.49	.96
C	−12.80	655.04
D	−.14	.07
E	.01	0
F	−1.29	6.62
G	2.69	28.86
H	.32	.42
I	.24	.24
J	.37	.56
K	−.71	2.00
L	−3.98	63.50
M	.27	.29
N	−.86	2.99

The three variables with the largest effects are C, G, and L. Therefore, make up a three-way table with these three variables. Note: Since we know God's numbers, we know that L is not really significant, but rather the value attributed to L is really the sum of BD + DE interactions. This is a risk we were willing to take when we chose a Resolution III design. However, it will be shown that it has no effect on our estimate of σ^2.

	1		1	
	1	g	1	g
1	1 16 27.13 28.16 $S^2 = .53$	7 9 29.14 29.65 $S^2 = .12$	13 2 23.64 22.83 $S^2 = .33$	15 12 27.61 25.55 $S^2 = 2.12$
c	11 6 14.96 12.63 $S^2 = 2.71$	10 3 18.62 18.17 $S^2 = .10$	5 4 13.04 9.39 $S^2 = 6.66$	14 8 11.72 12.81 $S^2 = .59$

Average $S^2 = 1.65$; $\phi = 8$

The estimate of σ^2 with eight degrees of freedom is obtained as follows:

$$\frac{.53 + .12 + .33 + 2.12 + 2.71 + .10 + 6.66 + .59}{8} = 1.65$$

The next largest effect after C, G, and L is variable F.

$$F_F = \frac{6.61}{1.65} = 4.01$$

is compared to F* from table 10. Therefore, variable F is significant at the 92 percent level. The responses in the three-way table can be corrected for F by subtracting 1.29, the mean effect of F from all the results at low F. However, in this case it is apparent, if one checks the treatment combinations, that trials 1 and 16 are both at low F, and that trials 11 and 6 are both at high F. Likewise for every other block. Therefore, correction for F will not change the estimate of the variance.

The next largest effect is A.

$$F_A = \frac{4.77}{1.65} = 2.89$$

which value is compared to F*. Since F_A is less than F*, variable A is not significant at the 92 percent confidence level. Since all of the other variables have a smaller SS number than variable A, they will necessarily also have F values smaller than F*.

The best estimated variance is therefore 1.65, which is a fairly good estimate of the true variance of 1.0. In addition, of course, the experimenter would conclude that the four prime variables are C, F, G, and L. Two of these conclusions are right and two are wrong. However, as noted earlier, when a follow-up experiment is conducted with these four variables, the truth will be revealed and a still better product will be obtained.

EXAMPLE 2

$\sigma^2 = 9$

The results of a computer run with the same simulated projects as Example 1. were as follows:

Variable	Mean Effect	SS
A	−5.30	112.30
B	−1.35	7.32
C	−12.00	576.45
D	−1.58	9.93
E	−.29	.33
F	−.02	0
G	2.25	20.33
H	.45	.84
I	1.47	8.69
J	−.77	2.38
K	4.13	68.30
L	5.23	109.50
M	2.02	16.32
N	2.06	16.97

The following three-way table results.

		1	1
1	1	30.81 34.49 S^2 = 6.77	32.38 25.43 S^2 = 24.22
	a	31.70 30.47 S^2 = .76	21.09 24.55 S^2 = 24.55
c	1	23.95 17.82 S^2 = 18.79	16.92 22.29 S^2 = 14.42
	a	18.98 15.61 S^2 = 5.68	8.98 10.33 S^2 = .91

The new average estimate of σ^2 is 9.69. Using this estimate of σ^2, the next largest effect, which is variable G, is tested.

$$F_G = \frac{20.33}{9.69} = 2.10$$

which is not significant at the 92 percent confidence level. Therefore, the final estimate of the variance is:

$$S^2 = 9.69$$

which compares favorably with the true value of $\sigma^2 = 9.00$.

When significant variables are present, the estimate of the variance will be biased high up to $\sigma^2 = 9$ for the type of project and the size of the effects used in this study. If the variance is greater than 9, the estimates will be biased low. If there are no significant effects in the project, the variance will always be biased low, and as a result insignificant variables will appear to be significant.

Since, in most projects, the engineer will most certainly choose some variables that are significant and hopefully the procedure for production and testing of the various treatment combinations will be somewhat reasonable, the above procedure will give a "good," but biased, estimate of σ^2, which will be particularly useful for computing a proper sample size for the follow-up experiments.

EXERCISE

A resolution III experiment with 15 variables in 16 trials has been conducted with the following results:

aehikmno	70
abfijlno	56
abcgjkmo	55

abcdhkln	53
bcdeilmo	53
acdefjmn	69
bdefgkno	49
acefghlo	56
abdfghim	55
bceghijn	55
cdfhijko	51
adegijkl	56
befhjklm	52
cfgiklmn	65
dghjlmno	68
(1)	52

Obtain an estimate of the variance from the data and determine the significant variables.

25 Testing Distributions

It was pointed out in the first chapter that, with regard to designed experiments where one is concerned with making decisions about means, it is not necessary to be concerned with distributions. In some engineering problems, however, it is necessary to estimate the distribution within the population. A designed experiment might consist only of taking N samples of the population and analyzing this data. All of the rules previously given for designing a good experiment apply to this problem.

The procedure recommended is called the Kolmogorov-Smirnov Method. There are two types of problems:

a) $H_0{:}x$ is distributed, for example, normally with μ = a and σ^2 = b, that is, $x \sim N$(a,b)
   $H_a{:}x$ is not distributed normally with μ = a and σ^2 = b, that is, $x\chi N$(a,b)
b) $H_0{:}x$ is distributed the same as y. F(x) = F(y)
   $H_a{:}x$ is not distributed the same as y. F(x) $\neq$ F(y)

CASE 1

One Population

The quality control specification for the location of an electronic chip in a recess is extremely tight, ± .01 mil from the center. Calculations show that, in order to meet this specification, the distribution of the deviation of the center of the assembled chips from the right edge of the recess must be normal with μ = 0, σ = .005 mil.

A. State the problem.

$H_0{:}x \sim N$ (0, .000025)
$H_a{:}x \nsim N$ (0, .000025)

B. Choose α and D_n^*. α is the usual risk of accepting H_a when, in fact, H_0 is true. β is not used in these problems because of the difficulty in determing its value. D_n^* is the difference between the population and the sample distribution function at which α applies.

Let $\alpha = .10$
Let $D_n^* = .24$

C. Determine sample size.

Sample size is obtained from table 14.

In table 14, read across the row of α's to .10, then read down this column to .24 and across the row to the sample size. In this case, it is 25.

D. Make up 25 samples of the assembly and measure the deviation from the center toward the right edge (fig. 25-1).

Results

−.0043	.0022	−.0142	.0075	−.0035
.0058	−.0095	−.0030	.0065	−.0052
.0152	.0105	.0085	−.0048	.0084
−.0100	.0072	−.0120	.0092	−.0015
−.0125	.0095	−.0075	−.0010	.0025

E. Order the data and assign a cumulative probability at each data point. Each point has a probability of $\frac{1}{25} = .04$.

Results	Cumulative Probability
−.0142	.04
−.0125	.08
−.0120	.12
−.0100	.16
−.0095	.20
−.0075	.24
−.0048	.28
−.0043	.32
−.0035	.36
−.0030	.40
−.0015	.44
−.0010	.48
.0022	.52
.0025	.56
.0052	.62
.0058	.64
.0065	.68
.0072	.72
.0075	.76
.0084	.80
.0085	.84
.0092	.88
.0095	.92
.0105	.96
.0152	1.00

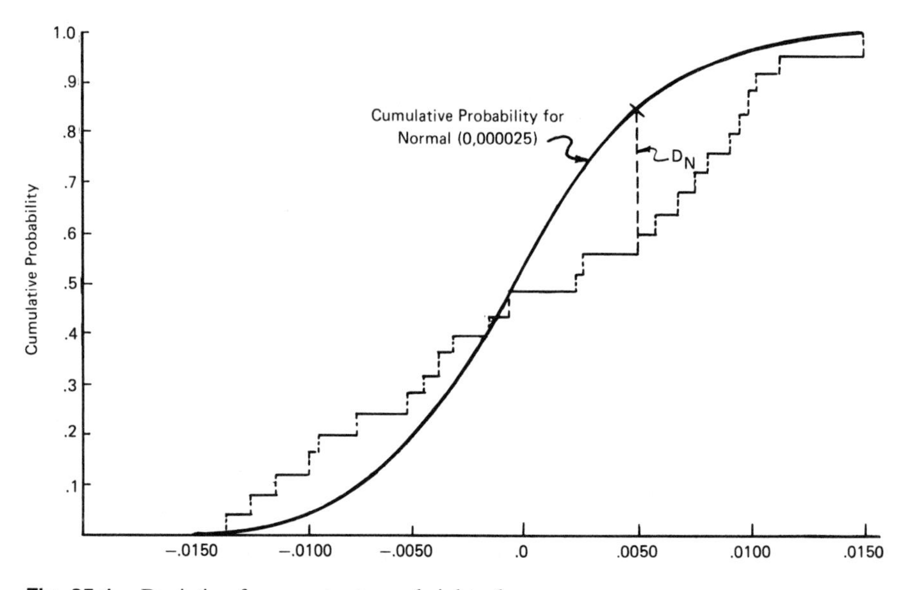

Fig. 25-1. Deviation from center toward right edge.

F. Plot the cumulative distribution curve of the population which is Normal (0, .000025), and superimpose the sample cumulative probability computed above.

G. Make the decision. If experimental D_n the largest difference, is greater than the tabled value of D_n for α risk, reject the null hypothesis.

The experimental D_n is .30. D_n^* for .10 risk is .24.

Therefore, reject the null hypothesis with at least 90 percent confidence. The population is not Normal (0, .000025). Observation of the data suggests the distribution is more likely uniform.

CASE 2

Two Populations

A company produces two different machines, each using a different technology, to do a certain job. The customer service cost on both products is excessive. Data is collected on the frequency of failure of 100 field machines of each type over a period of one year. The mean failure rate appears to be the same, but there is some question as to the distribution of the failures.

A. State the problem.

H_0: Distribution A = Distribution B
H_a: Distribution A $\neq$ Distribution B

B. Choose α.

$\alpha = .05$

C. Calculate criterion.

$$D_n^* = 1.36 \sqrt{\frac{n_1 + n_2}{n_1 n_2}} \text{ for } \alpha = .05$$

$$D_n^* = 1.63 \sqrt{\frac{n_1 + n_2}{n_1 n_2}} \text{ for } \alpha = .01$$

$$D_n^* = 1.36 \frac{100 + 100}{(100)(100)} = 0.027$$

D. Tabulate the data; the data of interest is the frequency of calls on any given machine.

Number of Machines with X Calls/Year

Calls/ Yr.	Type A Machines	Type B Machines	Cumulative A	Cumulative B
0	10	4	10	4
1	33	30	43	34
2	22	26	65	60
3	16	19	81	79
4	6	18	87	97
5	6	1	93	98
6	3	0	96	98
7	1	1	97	99
8	1	0	98	99
9	1	0	99	99
10	1	1	100	100

Total Calls: 231 233

Note that the number of calls is almost exactly the same in both machines. The largest cumulative difference is $87 - 97 = -10$, and the cumulative absolute relative frequency is $D_n = 10/100 = .1$, $D_n^* = 0.027$.

E. Make the decision.

Since $.1 > .027$, reject the null hypothesis with at least 95 percent confidence. The distribution of failure frequency is different for the two machines. Type B Machines have a larger number of very bad machines that require an excessive number of calls on the same machine. If the cause of the relatively few very bad machines of Type A can be found and eliminated, Machine A will be superior.

F. Present the data graphically if desired.

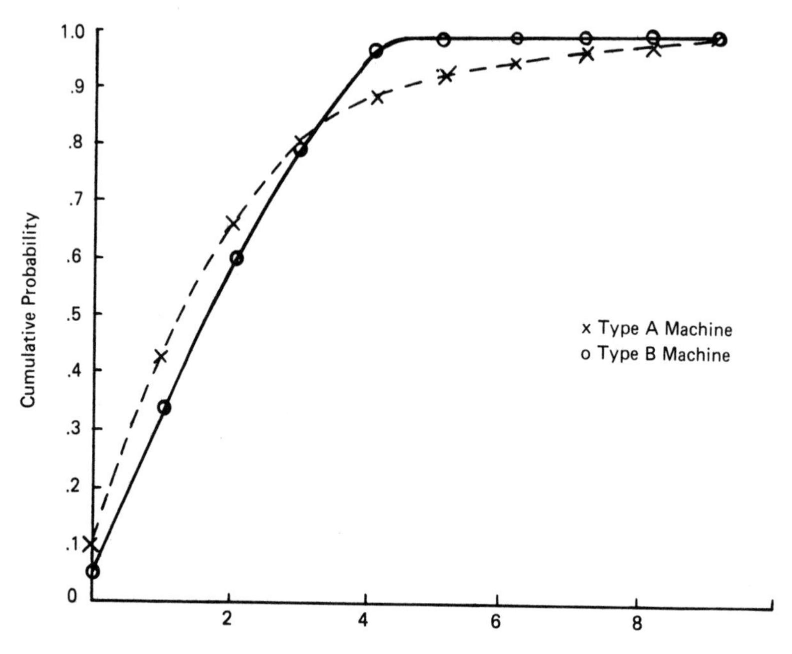

Fig. 25-2. Calls/Yr.

EXERCISE

A manufacturer of T.V. sets installed a counter into the on-off switching circuit of 20 sets and, after one year, read the number of times the set had been switched on. The following results were obtained.

178	424
220	427
275	439
340	448
365	461
370	482
374	498
382	540
395	670
422	820

The manufacturer had presumed the distribution to be uniformally distributed with a mean of 360/year and with a standard deviation of about 100. Does this data verify or refute the proposed distribution?

REFERENCE

Lindgren, B. W. 1969. *Statistical theory*; Section 6.4. New York: Macmillan Publishing Co.

Part Five

General References, Symbols, Tables, and Answers to Exercises

General References

There are any number of textbooks on basic statistics, experiment design, and related topics that could be of benefit to the new inductee to statistical experiment design. Most such texts, however, would contain much duplication of each other and of this book. (The book by Peter John, recommended below, is an exception to this statement.) A few texts are recommended that, I believe, are particularly good for certain topics, primarily for the novice who wishes to upgrade himself or herself to any desired level of skill.

In addition, all the references in *Technometrics* that are concerned with experiment design are listed. At this time, I would recommend to the novice (and maybe even the experienced experiment designer) that, after reading this book, they should obtain a set of *Technometrics* and, as a minimum, examine the titles of all the papers published there since 1959. It would also be useful if the readers of this book would also read those papers that are not on experiment design but that particularly apply to their types of problems, as well as the numerous papers on experiment design that will apply to their types of problems.

The bible on experiment design is *Statistical Design and Analysis of Experiments* by Peter W. M. John, published in 1971 by the Macmillan Publishing Co. That book is intended for graduate students in statistics and requires a knowledge of matrix algebra and calculus and a preliminary course on basic statistics. The statistical and mathematical foundation and proof of all the designs discussed in previous chapters are discussed in detail by John.

For a description of matrix algebra, there are many texts. One that this author recommends is *Matrices* by William Vann Parker and James Clifton Eaves, published by Ronald Press. A more modern text that covers a variety of other topics that are useful for design of experiments and probability, as well as matrix algebra, is *Finite Mathematics* by Wheeler and Peeples, published by Wadsworth Publishing Co.

For an introduction to the theory of statistics and probability, there are several recommended texts. *Introductory Statistical Analysis* by Anderson and Sclove, Houghton Mifflin Co., is a particularly readable text for the beginner. Another excellent text is *Introduction to Probability and Statistics* by Mendenhall, published by Wadsworth Publishing Co. For a more theoretical, advanced discussion of statistical theory, as applied to experiment design, one of the best texts is *An Introduction to the Theory and Applications of the Linear Statistical Model* by Graybill, published by Wadsworth Publishing Co. For pure statistical theory at an advanced level, the textbook is *Statistical Theory* by B. W. Lindgren, published by Macmillan Publishing Co.

For a treatment of experiment design between the level of this book and that of John's text, there are a number of excellent books. Most of them are particularly good for one or several chapters. *Experimental Statistics* by Mary Natrella, National Bureau of Standards Handbook 91, from the U.S. Government Printing Office, is particularly recommended for its clear and easily understood chapters on linear regression, polynomial regression, and multivariable regression. Her description of incomplete block designs (not discussed in this book) are easily understood by the novice. Likewise, she gives designs for up to 12 × 12 Latin squares. Her book also includes some excellent tables; e.g., the *t* distribution down to $\alpha = 0.40$. *Experimental Designs* by Cochran and Cox, John Wiley & Sons, is an excellent general text. *Experiment Design in Psychological Research* by Edwards, published by Holt, Rinehart and Winston, has a particularly good treatment of covariance analysis and the expectation of mean squares. Two of the more modern texts on experiment design in general are *Design of Experiments* by McLean and Anderson, published by Marcel Dekker, and *Parameter Estimation in Engineering and Science* by Beck and Arnold, published by John Wiley & Sons.

The most modern texts available by authors who are both theoretically sound, as well as practical, are *Statistics for Experimenters* by Box, Hunter, and Hunter, published by John Wiley & Sons; *Applications of Statistics to Industrial Experimentation* by Cuthbert Daniel, published by John Wiley & Sons; and *Experimental Design* by Johnson and Leone, published by John Wiley & Sons. The numerous examples in these books cover many of the fine points of good experimenting.

The best applied text on regression analysis, which also covers the theory of regression analysis, is *Applied Regression Analysis* by Draper and Smith, published by John Wiley & Sons. Another text that should be mentioned is *Evolutionary Operations: A Statistical Method for Process Improvement* by Box and Draper, also published by John Wiley & Sons.

For mechanical engineers, *Probabilistic Approach to Design* by Haugen, published by John Wiley & Sons, is an absolute must. This text is also recommended for its formulas for calculating both correlated and uncorrelated moments of sums, squares, products, etc., of functions.

Though this text deals only indirectly with probability, there are two texts that will be of benefit to most experimenters. *Probability and Random Processes* by Davenport, published by McGraw-Hill; and *Fundamentals of Applied Probability Theory* by Drake, published by McGraw-Hill.

It should be pointed out that all of the textbooks referred to above have many additional references to other texts that are particularly good on some aspect of experiment design.

There are a number of texts on multivariate analysis, but, for the practical experimenter, the best reference is a series of four papers by Kramer and Jensen titled "Fundamentals of Multivariate Analysis." These papers were published in the *Journal of Quality Technology*:

Volume 1, No. 2	April 1969
Volume 1, No. 3	July 1969
Volume 1, No. 4	October 1969
Volume 2, No. 1	January 1970

The following references from *Technometrics* are particularly recommended to readers of this book.

	Vol.	No.	Page
Response Surface Design for Three Factors at Three Levels	1	1	1
Partial Duplication of Factorial Experiments	1	1	63
Factorial Experiments in Life Testing	1	3	269
Experimental Designs to Adjust for Time Trends	2	1	67
Partial Duplication of Response Surface Designs	2	2	185

	Vol.	No.	Page
Some New Three Level Designs for the Study of Quantitative Variables	2	4	455
On Methods of Constructing Sets of Mutually Orthogonal Latin Squares Using a Computer	2	4	507
An Application of a Balanced Incomplete Design	3	1	51
On Methods of Constructing Sets of Mutually Orthogonal Squares Using a Computer, II	3	1	111
The 2^{k-p} Fractional Factorial Designs	3	3	311
Partial Confounding in Fractional Replicates	3	3	353
Finding New Fractions of Fractional Factorial Designs	3	3	359
A Study of Group Screening Methods	3	3	371
Missing Values in Response Surface Designs	3	3	389
*Orthogonal Main-Effects for Asymmetrical Factorial Experiments	4	1	21
Symmetrical and Asymmetrical Fractional Factorial Plans	4	1	47
Estimation of Sample Size	4	1	59
Life Test Sampling Plans for Normal and Log Normal Distributions	4	2	151
Some Acceptance Sampling Plans Based on the Theory of Runs	4	2	177
Three Factor Additive Designs More General Than the Latin Square	4	2	187
Group Screening With More Than Two Stages	4	2	209
Third Order Rotatable Designs in Three Factors	4	2	219
A Useful Method for Model Building	4	3	301
Four Factor Additive Designs More General Than the Greco-Latin Square	4	3	361
An Optimal Sequential Accelerated Life Test	4	3	367
Optimal Accelerated Life Designs for Estimation	4	3	381
Sequential Application of Simplex Designs in Optimisation and Evolutionary Operations	4	4	441
Simplex Lattice Designs for Multicomponent Systems	4	4	463
Some Systematic Supersaturated Designs	4	4	489
Partially Duplicated Fractional Factorial Designs	5	1	71
The Effect of Errors in the Factor Levels and Experiment Design	5	1	247
A Quick Method for Choosing a Transformation	5	3	317
Progressively Censored Samples in Life Testing	5	3	327
A Comparison of Three Different Procedures for Estimating Variance Components	5	4	421
The Design of Screening Tests	5	4	481
Sequential Factorial Estimation	6	1	41
Serial Designs for Routine Quality Control and Experimentation	6	1	77
Some Two-Level Factorial Plans with Split Plot Confounding	6	3	253
Methods for Estimating the Composition of a Three Component Liquid Mixture	6	4	343
Designs for Sequential Application of Factors	6	4	365
A Procedure for Constructing Incomplete Block Designs	6	4	389
The Experimental Study of Physical Mechanisms	7	1	23
Estimating the Fraction of Acceptable Product	7	1	43
Designs for Discriminating Between Two Rival Models	7	3	307
Balanced Sets of Balanced Incomplete Block Designs of Block Size Three	7	4	553
Missing Values in Factorial Experiments	7	4	649
Selection of Variables for Fitting Equations to Data	8	1	27
Factorial 2^{p-q} Plans Robust Against Linear and Quadratic Trends	8	2	259
Evolutionary Operation: A Review	8	3	389
Extreme Vertices Design of Mixture Experiments	8	3	447
Discussion of "Extreme Vertices Design of Mixture Experiments"	8	3	455
Augmented 2^{n-1} Designs	8	3	469
A Review of Response Surface Methodology: A Literature Survey	8	4	571
Distribution-Free Life Test Sampling Plans	8	4	591
Use and Abuse of Regression	8	4	625
An Addleman 2^{17-9} Resolution V Plan	8	4	705
A General Approach to the Estimation of Variance Components	9	1	93
Some New Families of Partially Balanced Designs of the Latin Square Type and Related Designs	9	2	229
Systematic Methods for Analyzing 2^n3^m Factorial Experiments with Applications	9	2	245
Sampling Mixtures of Particles	9	3	365

	Vol.	No.	Page
Three Dimensional Models of Extreme Vertices Designs for Four Component Mixtures	9	3	472
Comparison of Designs and Estimation Procedure for Estimating Parameters in a Two-Stage Nested Process	9	4	499
Sequential Designs for Spherical Weight Functions	9	4	517
Sequential Analysis, Direct Method	10	1	125
Response Surface Designs for Factors at Two and Three Levels	10	1	177
The Design of Experiments for Parameter Estimation	10	2	271
Non-Orthogonal Designs of Even Resolution	10	2	291
Factor Changes and Linear Trends in Eight-Run Two Level Factorial Designs	10	2	301
Asymmetrical Rotatable Designs and Orthogonal Transformations	10	2	313
Sequential Testing of Sample Size	10	2	331
Saturated Sequential Factorial Designs	10	3	535
Some Further Remarks Concerning "A General Approach to the Estimation of Variance Components"	10	3	551
Orthogonal Main Effects $2^n 3^m$ Designs and Two-Factor Interaction Aliasing	10	3	559
Response Surface Designs for Mixture Problems	10	4	739
Optimal Experimental Designs for Estimating the Independent Variables in Regression	10	4	811
Transformations: Some Examples Revisited	11	1	23
Computer Aided Design of Experiments	11	1	137
Some General Remarks on Consulting in Statistics	11	2	241
The Statistical Consultant in a Scientific Laboratory	11	2	247
Results on Factorial Designs of Resolution IV for the 2^n and $2^n 3^m$ Series	11	3	431
Minimum Bias Estimation and Experimental Design for Response Surfaces	11	3	461
Sequence of Two-Level Factorial Plans	11	3	477
Balanced Incomplete Block Designs with Sets of Identical Blocks	11	3	613
Constrained Maximization and the Design of Experiments	11	3	616
Orthogonal Main Effect Plans Permitting Estimation of All Two-Factor Interactions for the $2^n 3^m$ Factorial Series of Designs	11	4	747
A Method of Estimating Variance Components in Unbalanced Designs	12	2	207
Three Stage Nested Designs for Estimating Variance Components	12	3	487
Simplified Experimental Design	13	1	19
Small Incomplete Factorial Designs for Two- and Three-Level Factors	13	2	243
Balanced Optimal 2^m Fractional Factorial Designs of Resolution V, $m \leqq 6$	13	2	257
Linear Combinations in Designing Experiments	13	3	575
Factorial Designs, the $[X\text{-}X]$ Criterion and Some Related Matters	13	4	731
An Application of Variational Methods to Experiment Design	13	4	771
Randomization and Experimentation	14	1	13
Designs for Estimating the Slope of a Second Order Linear Model	14	2	341
Design of Over-Stress Life Test Experiments When Failure Times Have the Two-Parameter Weibull Distributions	14	2	437
A Boyesian Sequential Life Test	14	2	453
Alternative Multi-Level Continuous Sampling Plans	14	3	645
"Repairing" Response Surface Designs	14	3	767
Response Surface Techniques for Dual Response Systems	15	2	301
Secondary Criteria in the Selection of Minimum Bias Designs in Two Variables	15	2	319
Experiments with mixtures: A Review	15	3	437
Some Run Orders Requiring a Minimum Number of Factor Level Changes for the 2^4 and 2^5 Main Effect Plans	16	1	31
Computer Construction of D-Optional First Order Designs	16	2	211
A Sequential Screening Procedure	16	2	229
Economical Second Order Designs Based on Irregular Fractions of the 3^n Factorial	16	3	375
Design of Experiments for Subsets of Parameters	16	3	425
Optimum Composite Designs	16	4	561
Nearly Optimal Allocation of Experimental Units Using Observed Covariate Values	16	4	589
On Minimum-Point Second Order Designs	16	4	613

	Vol.	No.	Page
Some Comments on Designs for Cox's Mixture Polynomial	17	1	25
A Comparison of Designs and Estimators for the Two-Stage Nested Random Model	17	1	37
Experimental Designs for Quadratic Models in Constrained Mixture Spaces	17	2	149
Sample Size Calculation in Exponential Life Testing	17	2	229
Design of Experiments for the Identification of Linear Dynamic Systems	17	3	303
Screening Designs for Poly-Factor Experimentation	17	4	487
A Comparison of Four Designs for Estimating the Variances of Random Effects in a Two-Way Classification	17	4	495
Screening Concepts and Designs for Experiments with Mixtures	18	1	19
Design and Analysis of Factorial Experiments Via Interactive Computing in APL	18	2	135
Designing Experiments When Run Order Is Important	18	3	249
Applications of the Direct Method in Sequential Analysis	18	3	301
Which Response Surface Design Is Best	18	4	411
Hybrid Designs for Quadratic Response Surfaces	18	4	419
An Algorithm for Deriving Optimal Block Designs	18	4	451
Restricted Exploration of Mixtures by Symmetric-Simplex Design	19	1	47
Optimal Designs for Two Non-Interactive Treatments	19	1	53
The Design of Incomplete Multinomial Samples: The Nested Case	19	1	59
Screening Based on Normal Variables	19	1	65
Accounting for Heteroscedesticity in Experiment Design	19	2	109
Designs in Three and Four Components for Mixture Models with Inverse Terms	19	2	117
Experimental Design for Sensitivity Testing: The Weibull Model	19	4	405
Comparison of Box-Draper and D-Optimum Designs for Experiments with Mixtures	19	4	441
Comparison of Simplex Designs for Quadratic Mixture Models	19	4	445
Sequential Designs for Replacements in Exponential Life Tests	19	4	455
A Note on Orthogonal Main Effect Plans	19	4	511
A Review of Experimental Design Procedures for Regression Model Discrimination	20	1	15
Experimental Designs for Fitting Segmented Polynomial Regression Models	20	2	151
A New Balanced Design for Eighteen Varieties	20	2	155
Comparison of Designs to Estimate Variance Components in Two-Way Classification Model	20	2	159
Experiments with Mixtures; An Update and Bibliography	21	1	95
Missing Points in 2^N and 2^{N-K} Factorial Designs	21	2	225
Orthogonal Main Effect Plans for Asymmetrical Factorials	21	2	269
Computer Augmentation of Experimental Designs to Maximize $\|X'X\|$	21	3	321
Near Minimal Resolution IV Designs for 5^N Factorial Experiment	21	3	331
An Improved Sign Test for Experiments in which Neutral Responses are Possible	21	4	525
Minimum Aberrations 2^{k-p} Designs	22	4	601
Incomplete Block Designs for Comparing treatments with a Control: General Theory	23	1	45
Sequentially Generated Second-Order Quasi-*D*-Optimal Designs for Experiments with Mixture and Process Variables	23	3	233
Measuring Component Effects in Constrained Mixture Experiments	24	1	29
Model Robust, Linear-Optimal Designs	24	1	49
First order "Interruptable" Designs	24	1	55
Center Points in Second-Order Response Surface Designs	24	2	127
Orthogonal Main-Effect Plans for Agymetrical Factorials	24	2	135
Multiple Comparisons for Orthogonal Contrasts: Examples and Tables	24	3	213
The Performance of Two-Stage Group Screening in Factor-Screening Experiments	24	4	325
Design of Experiments for Comparing Treatments with a Control: Tables of Optimal Allocation of Observations	25	1	87

	Vol.	No.	Page
Defining Consistant Constraint Regions in Mixture Experiments	25	1	97
Orthogonal Resolution IV Designs for Some Asymmetrical Factorials	25	2	197
Calculating Centroids in Constrained Mixture Experiments	25	3	279
Two-Level Multifactor Designs for Detecting the Presence of Interventions	25	4	345
Selecting Check Points for Testing of Fit in Response Surface Models	25	4	357
A New Algorithm for Extreme Vertices Designs for Linear Mixture Models	25	4	367
Experimental Design in the Complex World	26	1	19
Experimental Design: Review and Comment	26	2	71–128
Construction of Orthogonal Main Effect Plans Using Hadamard Matrices	27	1	37
Models for Mixture Experiments When the Response Depends on the Total Amount	27	3	219
An Analysis for Unreplicated Fractional Factorials	28	1	11
Dispersion Effects from Fractional Factorial Designs	28	1	19
The Geometry of Constrained Mixture Experiments	28	2	95
Testing in Industrial Experiments With Ordered Categorical Data	28	4	283–328
A Class of Multifactor Design for Estimating the Shapes of Response Surfaces	29	4	449
Signal-to-Noise Ratios, Performance Criteria and Transformations	30	1	1–40
Testing for Interactions Using the Analysis of Means	30	1	53

Symbols

N Number of trials required or obtained at one level of a variable in a matrix experiment.

$\bar{X}$ Sample mean.

S^2 Sample variance.

S Sample standard deviation.

X_i The ith sample observation.

μ The population mean.

σ^2 The population variance.

σ The population standard deviation.

$\frac{\sigma}{\sqrt{N}}$ The standard error of the mean.

H_0 The null hypothesis.

H_a The alternative hypothesis.

α The risk of making an error of the first kind; that is, stating that the population mean is different than some number when, in fact, this statement is false.

β The risk of making an error of the second kind; that is, stating that the population mean is the same as some number when, in fact, this statement is false.

p Computed probability.

U_α Standard normal deviate associated with type 1 error for means.

U_β Standard normal deviate associated with type 2 error for means.

$\bar{X}^*$ Decision criterion for population means.

$|\bar{X}_A - \bar{X}_B|^*$ Decision criterion for the difference of population means.

t_α The t deviate associated with type 1 error.

t_β The t deviate associated with type 2 error.

ϕ Degrees of freedom.

χ^2 The chi square deviate associated with α error for variances.

$^*S^2$ Decision criterion for population variances.

F^* Decision criterion for comparing two population variances.

F The ratio of two variances.

δ The difference between two means, which is important from an engineering viewpoint.

SS Sum of squares.

TSS Total sum of squares.

MS Mean square.

$\bar{Y}$ Sample mean of the response Y.

$\hat{\mu}$ Estimate of population mean.

$\hat{\sigma}^2$ Estimate of the population variance.

λ Expected value of the mean and variance in the Poisson distribution function.

ρ Population probability or population failure rate.

Σ Summation in mathematical operations. Also used as the variance-covariance matrix in multivariate analysis.

$\times$ Multiplication in mathematical operations.

$+$ High level of variable in matrix experiment designs.

$-$ Low level of variable in matrix experiment designs.

0 Middle level of variable in matrix experiment designs.

μ_0 A fixed number against which some population mean is compared.

σ_0^2 A fixed number against which some population variance is compared.

T Number of trials in a Hadamard matrix experiment.

$>$ Mathematical symbol for "greater than."

$<$ Mathematical symbol for "less than."

$\geqq$ Mathematical symbol for "equal to or greater than."

$\leqq$ Mathematical symbol for "equal to or less than."

$\ngtr$ Mathematical symbol for "not greater than."

$\nless$ Mathematical symbol for "not less than."

D_i^* Decision criterion for sequential random-strategy experiments.

$\| \|$	Mathematical symbol for "absolute value of."
∞	Infinity.
$\sim$	Distributed as; for example, $X \sim N\ (0,\ 10)$ means X is distributed normally with mean 0 and variance 10.
b_0	The intercept of a line in a regression equation.
b_1	The coefficient of the linear effect of the first variable in a regression equation.
b_{11}	The coefficient of the quadratic (squared) term for the first variable in a regression equation.
b_{12}	The coefficient of the cross-product term of the first and second variables in a regression equation.
γ	Percentage of good space within the total experimental space.
Ψ	The value of the star points in central composite rotatable designs.
$\prod_{i=1}^{n}$	Mathematical symbol for multiplication of n terms.

Tables

Table 1 Probability points of the normal distribution: single-sided; σ^2 known.

ρ (α or β)	U
0.001	3.090
0.005	2.576
0.010	2.326
0.015	2.170
0.020	2.054
0.025	1.960
0.050	1.645
0.100	1.282
0.150	1.036
0.200	0.842
0.300	0.524
0.400	0.253
0.500	0.000
0.600	−0.253

Table 2 Probability points of the normal distribution: double-sided; σ^2 known.

p (α only)	U
0.001	3.291
0.005	2.807
0.010	2.576
0.015	2.432
0.020	2.326
0.025	2.241
0.050	1.960
0.100	1.645
0.150	1.440
0.200	1.282
0.300	1.036
0.400	0.842
0.500	0.675
0.600	0.524

Table 3 Probability points of t-distribution: single-sided; σ^2 unknown.

	p						
ϕ	0.005	0.01	0.025	0.05	0.10	0.20	0.30
1	63.66	31.82	12.71	6.31	3.08	1.38	0.73
2	9.93	6.97	4.30	2.92	1.89	1.06	0.62
3	5.84	4.54	3.18	2.35	1.64	0.98	0.58
4	4.60	3.75	2.78	2.13	1.53	0.94	0.57
5	4.03	3.37	2.57	2.02	1.48	0.92	0.56
6	3.71	3.14	2.45	1.94	1.44	0.91	0.56
7	3.50	3.00	2.37	1.90	1.42	0.90	0.55
8	3.36	2.90	2.31	1.86	1.40	0.90	0.55
9	3.25	2.82	2.26	1.83	1.38	0.89	0.54
10	3.17	2.76	2.23	1.81	1.37	0.89	0.54
15	2.95	2.60	2.13	1.75	1.34	0.87	0.54
20	2.85	2.53	2.09	1.73	1.33	0.86	0.53
25	2.79	2.49	2.06	1.71	1.32	0.86	0.53
30	2.75	2.46	2.04	1.70	1.31	0.85	0.53
60	2.66	2.39	2.00	1.67	1.30	0.85	0.53
120	2.62	2.36	1.98	1.66	1.29	0.85	0.53
∞	2.58	2.33	1.96	1.65	1.28	0.84	0.52

Table 4 Probability points of t-distribution: double-sided; σ^2 unknown.

	p						
ϕ	0.005	0.01	0.02	0.05	0.10	0.20	0.30
1	127.00	63.70	31.82	12.71	6.31	3.08	1.96
2	14.10	9.93	6.97	4.30	2.92	1.89	1.39
3	7.45	5.84	4.54	3.18	2.35	1.64	1.25
4	5.60	4.60	3.75	2.78	2.13	1.53	1.19
5	4.77	4.03	3.37	2.57	2.02	1.48	1.16
10	3.58	3.17	2.76	2.23	1.81	1.37	1.09
15	3.29	2.95	2.60	2.13	1.75	1.34	1.07
20	3.15	2.85	2.53	2.09	1.73	1.33	1.06
25	3.08	2.79	2.49	2.06	1.71	1.32	1.06
30	3.03	2.75	2.46	2.04	1.70	1.31	1.05
60	2.91	2.66	2.39	2.00	1.67	1.30	1.05
120	2.86	2.62	2.36	1.98	1.66	1.29	1.05
∞	2.81	2.58	2.33	1.96	1.65	1.28	1.04

Table 5 Sample size for testing a decrease in variance using the χ^2 test.

ϕ	$\alpha = 0.01$		$\alpha = 0.05$		$\alpha = 0.10$	
	$\beta = 0.01$	$\beta = 0.05$	$\beta = 0.01$	$\beta = 0.05$	$\beta = 0.01$	$\beta = 0.05$
1	0.00002	0.00004	0.0006	0.0010	0.0024	0.0041
2	0.0022	0.0034	0.0111	0.0171	0.0229	0.0352
3	0.0101	0.0147	0.0310	0.0450	0.0515	0.0748
4	0.0224	0.0313	0.0535	0.0749	0.0812	0.1121
5	0.0367	0.0501	0.0759	0.1035	0.1067	0.1455
6	0.0519	0.0693	0.0973	0.1299	0.1311	0.1750
7	0.0671	0.0881	0.1173	0.1541	0.1534	0.2014
8	0.0820	0.1062	0.1360	0.1762	0.1737	0.2250
9	0.0963	0.1234	0.1535	0.1965	0.1924	0.2464
10	0.1102	0.1397	0.1698	0.2152	0.2096	0.2657
15	0.1710	0.2092	0.2375	0.2905	0.2795	0.3419
20	0.2199	0.2630	0.2889	0.3454	0.3312	0.3962
25	0.2610	0.3050	0.3284	0.3850	0.3700	0.4340
30	0.2939	0.3416	0.3634	0.4225	0.4047	0.4706
40	0.3479	0.3975	0.4161	0.4755	0.4562	0.5211
50	0.3880	0.4360	0.4550	0.5110	0.4920	0.5550
60	0.4241	0.4739	0.4888	0.5461	0.5258	0.5875
120	0.5467	0.5931	0.6020	0.6527	0.6329	0.6863
∞	1.0000	1.0000	1.0000	1.0000	1.0000	1.0000

Example: $\alpha = 0.05$

$\beta = 0.05$

$$R = \frac{\sigma_0^2 - \delta}{\sigma_0^2} = \frac{5}{20} = 0.25 \qquad \therefore \phi = 12 \qquad N = \phi + 1 = 13$$

Table 6 Sample size for testing an increase in variance using the χ^2 test.

ϕ	$\alpha = 0.01$			$\alpha = 0.05$		
	$\beta = 0.01$	$\beta = 0.05$	$\beta = 0.10$	$\beta = 0.01$	$\beta = 0.05$	$\beta = 0.10$
1	42240	1687	420	24450	977	243
2	458	90	44	298	58	28
3	99	32	19	68	22	13
4	45	19	12	32	13	8.9
5	27	13	9.4	20	9.7	6.9
6	19	10	7.6	14	7.7	5.7
7	15	8.5	6.5	11	6.5	5.0
8	12	7.3	5.8	9.4	5.7	4.4
9	10.4	6.5	5.2	8.1	5.1	4.1
10	9.1	5.9	4.8	7.2	4.6	3.8
15	5.8	4.2	3.6	4.8	3.4	2.9
20	4.5	3.5	3.0	3.8	2.9	2.5
25	3.8	3.0	2.7	3.3	2.6	2.3
30	3.4	2.8	2.5	2.9	2.4	2.1
40	2.9	2.4	2.2	2.5	2.1	1.9
50	2.6	2.2	2.0	2.3	1.9	1.8
60	2.4	2.0	1.9	2.1	1.8	1.7
120	1.8	1.7	1.1	1.7	1.5	1.4
∞	1.0	1.0	1.0	1.0	1.0	1.0

Example: $\alpha = 0.01$

$\beta = 0.05$

$$R = \frac{\sigma_0^2 + \delta}{\sigma_0^2} = \frac{200}{20} = 10 \qquad \therefore \phi = 6 \qquad N = 6 + 1 = 7$$

Table 7 Probability points of χ^2 distribution; H_a: $\sigma_1^2 < \sigma_0^2$.

ϕ	$\chi^2_{0.005}$	$\chi^2_{0.01}$	$\chi^2_{0.025}$	$\chi^2_{0.05}$	$\chi^2_{0.10}$
1	0.00004	0.0002	0.001	0.004	0.016
2	0.01	0.02	0.05	0.10	0.21
3	0.07	0.12	0.22	0.35	0.58
4	0.21	0.30	0.48	0.71	1.06
5	0.41	0.55	0.831	1.15	1.61
6	0.68	0.87	1.24	1.64	2.20
7	0.99	1.24	1.69	2.17	2.83
8	1.34	1.65	2.18	2.73	3.49
9	1.73	2.09	2.70	3.33	4.17
10	2.16	2.56	3.25	3.94	4.87
11	2.60	3.05	3.82	4.57	5.58
12	3.07	3.57	4.40	5.23	6.30
13	3.57	4.11	5.01	5.89	7.04
14	4.07	4.66	5.63	6.57	7.79
15	4.60	5.23	6.26	7.26	8.55
30	13.79	14.95	16.79	18.49	20.60
40	20.71	22.16	24.43	26.51	29.05
60	35.53	37.48	40.48	43.19	46.46
120	83.85	86.92	91.58	95.70	100.62

Table 8 Probability points of χ^2 distribution; H_a:$\sigma_1^2 > \sigma_0^2$.

ϕ	$\chi^2_{0.10}$	$\chi^2_{0.05}$	$\chi^2_{0.025}$	$\chi^2_{0.01}$	$\chi^2_{0.005}$
1	2.71	3.84	5.02	6.63	7.88
2	4.61	5.99	7.38	9.21	10.60
3	6.25	7.81	9.35	11.34	12.84
4	7.78	9.49	11.14	13.28	14.86
5	9.24	11.07	12.83	15.09	16.75
6	10.64	12.59	14.45	16.81	18.55
7	12.02	14.07	16.01	18.48	20.28
8	13.36	15.51	17.53	20.09	21.96
9	14.68	16.92	19.02	21.67	23.59
10	15.99	18.31	20.48	23.21	25.19
11	17.28	19.68	21.92	24.73	26.76
12	18.55	21.03	23.34	26.22	28.30
13	19.81	22.36	24.74	27.69	29.82
14	21.06	23.68	26.12	29.14	31.32
15	22.31	25.00	27.49	30.58	32.80
20	28.41	31.41	34.17	37.57	40.00
30	40.26	43.77	46.98	50.89	53.67
40	51.81	55.76	59.34	63.69	66.77
60	74.40	79.08	83.30	88.38	91.95
120	140.23	146.57	152.21	158.95	163.64

Table 9 Sample size for testing a difference in variance using the F-test.

ϕ	$\alpha = 0.01$, $\beta = 0.01$	$\alpha = 0.01$, $\beta = 0.05$	$\alpha = 0.01$, $\beta = 0.10$	$\alpha = 0.05$, $\beta = 0.01$	$\alpha = 0.05$, $\beta = 0.05$	$\alpha = 0.05$, $\beta = 0.10$
1	—	—	—	—	—	6436
2	9801	1881	891	1881	361	171
3	868	273	159	273	86	50
4	255	102	66	102	41	26
5	120	55	38	55	26	17
6	72	36	26	36	18	13
7	49	26	19	26	14	11
8	36	21	16	21	12	8.9
9	29	17	13	17	10	7.6
10	24	14	11	14	8.9	6.9
15	12	8.5	6.9	8.5	5.8	4.7
20	8.6	6.2	5.3	6.2	4.5	3.8
25	6.8	5.1	4.4	5.1	3.8	3.3
30	5.7	4.4	3.8	4.4	3.4	3.0
40	4.5	3.6	3.2	3.6	2.9	2.5
50	3.8	3.1	2.8	3.1	2.6	2.3
60	3.4	2.8	2.6	2.8	2.4	2.1
120	2.4	2.1	1.9	2.1	1.8	1.7
∞	1.0	1.0	1.0	1.0	1.0	1.0

Example:

$\alpha = 0.01$

$\beta = 0.10$

$$R = \frac{\sigma_1^2}{\sigma_2^2} \text{ or } \frac{\sigma_2^2}{\sigma_1^2} = 11 \qquad \therefore \phi_1 = \phi_2 = 10 \quad \text{and} \quad N_1 = N_2 = 11$$

Table 10* Probability points of the variance ratio (F distribution).

Probability point	ϕ_1	Numerator 1	2	3	4	5	6	7	8	9	10	12	15	20	24	30	40	60	120	∞
0.1	1	39.9	49.5	53.6	55.8	57.2	58.2	58.9	59.4	59.9	60.2	60.7	61.2	61.7	62.0	62.3	62.5	62.8	63.1	63.3
0.05		161	199	216	225	230	234	237	239	241	242	244	246	248	249	250	251	252	253	254
0.01		4052	4999	5403	5625	5764	5859	5928	5982	6022	6056	6106	6157	6209	6235	6261	6287	6313	6339	6366
0.1	2	8.53	9.00	9.16	9.24	9.29	9.33	9.35	9.37	9.38	9.39	9.41	9.42	9.44	9.45	9.46	9.47	9.47	9.48	9.49
0.05		18.5	19.0	19.2	19.2	19.3	19.3	19.4	19.4	19.4	19.4	19.4	19.4	19.4	19.5	19.5	19.5	19.5	19.5	19.5
0.01		98.5	99.0	99.2	99.2	99.3	99.3	99.4	99.4	99.4	99.4	99.4	99.4	99.4	99.5	99.5	99.5	99.5	99.5	99.5
0.1	3	5.54	5.46	5.39	5.34	5.31	5.28	5.27	5.25	5.24	5.23	5.22	5.20	5.18	5.18	5.17	5.16	5.15	5.14	5.13
0.05		10.1	9.55	9.28	9.12	9.01	8.94	8.89	8.85	8.81	8.79	8.74	8.70	8.66	8.64	8.62	8.59	8.57	8.55	8.53
0.01		34.1	30.8	29.5	28.7	28.2	27.9	27.7	27.5	27.3	27.2	27.1	26.9	26.7	26.6	26.5	26.4	26.3	26.2	26.1
0.1	4	4.54	4.32	4.19	4.11	4.05	4.01	3.98	3.95	3.94	3.92	3.90	3.87	3.84	3.83	3.82	3.80	3.79	3.78	3.76
0.05		7.71	6.94	6.59	6.39	6.26	6.16	6.09	6.04	6.00	5.96	5.91	5.86	5.80	5.77	5.75	5.72	5.69	5.66	5.63
0.01		21.2	18.0	16.7	16.0	15.5	15.2	15.0	14.8	14.7	14.5	14.4	14.2	14.0	13.9	13.8	13.7	13.7	13.6	13.5
0.1	5	4.06	3.78	3.62	3.52	3.45	3.40	3.37	3.34	3.32	3.30	3.27	3.24	3.21	3.19	3.17	3.16	3.14	3.12	3.10
0.05		6.61	5.79	5.41	5.19	5.05	4.95	4.88	4.82	4.77	4.74	4.68	4.62	4.56	4.53	4.50	4.46	4.43	4.40	4.36
0.01		16.3	13.3	12.1	11.4	11.0	10.7	10.5	10.3	10.2	10.1	9.89	9.72	9.55	9.47	9.38	9.29	9.20	9.11	9.02
0.1	6	3.78	3.46	3.29	3.18	3.11	3.05	3.01	2.98	2.96	2.94	2.90	2.87	2.84	2.82	2.80	2.78	2.76	2.74	2.72
0.05		5.99	5.14	4.76	4.53	4.39	4.28	4.21	4.15	4.10	4.06	4.00	3.94	3.87	3.84	3.81	3.77	3.74	3.70	3.67
0.01		13.7	10.9	9.78	9.15	8.75	8.47	8.26	8.10	7.98	7.87	7.72	7.56	7.40	7.31	7.23	7.14	7.06	6.97	6.88
0.1	7	3.59	3.26	3.07	2.96	2.88	2.83	2.78	2.75	2.72	2.70	2.67	2.63	2.59	2.58	2.56	2.54	2.51	2.49	2.47
0.05		5.59	4.74	4.35	4.12	3.97	3.87	3.79	3.73	3.68	3.64	3.57	3.51	3.44	3.41	3.38	3.34	3.30	3.27	3.23
0.01		12.2	9.55	8.45	7.85	7.46	7.19	6.99	6.84	6.72	6.62	6.47	6.31	6.16	6.07	5.99	5.91	5.82	5.74	5.65
0.1	8	3.46	3.11	2.92	2.81	2.73	2.67	2.62	2.59	2.56	2.54	2.50	2.46	2.42	2.40	2.38	2.36	2.34	2.32	2.29
0.05		5.32	4.46	4.07	3.84	3.69	3.58	3.50	3.44	3.39	3.35	3.28	3.22	3.15	3.12	3.08	3.04	3.01	2.97	2.93
0.01		11.3	8.65	7.59	7.01	6.63	6.37	6.18	6.03	5.91	5.81	5.67	5.52	5.36	5.28	5.20	5.12	5.03	4.95	4.86
0.1	9	3.36	3.01	2.81	2.69	2.61	2.55	2.51	2.47	2.44	2.42	2.38	2.34	2.30	2.28	2.25	2.23	2.21	2.18	2.16
0.05		5.12	4.26	3.86	3.63	3.48	3.37	3.29	3.23	3.18	3.14	3.07	3.01	2.94	2.90	2.86	2.83	2.79	2.75	2.71
0.01		10.6	8.02	6.99	6.42	6.06	5.80	5.61	5.47	5.35	5.26	5.11	4.96	4.81	4.73	4.65	4.57	4.48	4.40	4.31
0.1	10	3.28	2.92	2.73	2.61	2.52	2.46	2.41	2.38	2.35	2.32	2.28	2.24	2.20	2.18	2.16	2.13	2.11	2.08	2.06
0.05		4.96	4.10	3.71	3.48	3.33	3.22	3.14	3.07	3.02	2.98	2.91	2.84	2.77	2.74	2.70	2.66	2.62	2.58	2.54
0.01		10.0	7.56	6.55	5.99	5.64	5.39	5.20	5.06	4.94	4.85	4.71	4.56	4.41	4.33	4.25	4.17	4.08	4.00	3.91
0.1	11	3.23	2.86	2.66	2.54	2.45	2.39	2.34	2.30	2.27	2.25	2.21	2.17	2.12	2.10	2.08	2.05	2.03	2.00	1.97
0.05		4.84	3.98	3.59	3.36	3.20	3.09	3.01	2.95	2.90	2.85	2.79	2.72	2.65	2.61	2.57	2.53	2.49	2.45	2.40
0.01		9.65	7.21	6.22	5.67	5.32	5.07	4.89	4.74	4.63	4.54	4.40	4.25	4.10	4.02	3.94	3.86	3.78	3.69	3.60
0.1	12	3.18	2.81	2.61	2.48	2.39	2.33	2.28	2.24	2.21	2.19	2.15	2.10	2.06	2.04	2.01	1.99	1.96	1.93	1.90
0.05		4.75	3.89	3.49	3.26	3.11	3.00	2.91	2.85	2.80	2.75	2.69	2.62	2.54	2.51	2.47	2.43	2.38	2.34	2.30
0.01		9.33	6.93	5.95	5.41	5.06	4.82	4.64	4.50	4.39	4.30	4.16	4.01	3.86	3.78	3.70	3.62	3.54	3.45	3.36
0.1	13	3.14	2.76	2.56	2.43	2.35	2.28	2.23	2.20	2.16	2.14	2.10	2.05	2.01	1.98	1.96	1.93	1.90	1.88	1.85
0.05		4.67	3.81	3.41	3.18	3.03	2.92	2.83	2.77	2.71	2.67	2.60	2.53	2.46	2.42	2.38	2.34	2.30	2.25	2.21
0.01		9.07	6.70	5.74	5.21	4.86	4.62	4.44	4.30	4.19	4.10	3.96	3.82	3.66	3.59	3.51	3.43	3.34	3.25	3.17
0.1	14	3.10	2.73	2.52	2.39	2.31	2.24	2.19	2.15	2.12	2.10	2.05	2.01	1.96	1.94	1.91	1.89	1.86	1.83	1.80
0.05		4.60	3.74	3.34	3.11	2.96	2.85	2.76	2.70	2.65	2.60	2.53	2.46	2.39	2.35	2.31	2.27	2.22	2.18	2.13
0.01		8.86	6.51	5.56	5.04	4.69	4.46	4.28	4.14	4.03	3.94	3.80	3.66	3.51	3.43	3.35	3.27	3.18	3.09	3.00
0.1	15	3.07	2.70	2.49	2.36	2.27	2.21	2.16	2.12	2.09	2.06	2.02	1.97	1.92	1.90	1.87	1.85	1.82	1.79	1.76
0.05		4.54	3.68	3.29	3.06	2.90	2.79	2.71	2.64	2.59	2.54	2.48	2.40	2.33	2.29	2.25	2.20	2.16	2.11	2.07
0.01		8.68	6.36	5.42	4.89	4.56	4.32	4.14	4.00	3.89	3.80	3.67	3.52	3.37	3.29	3.21	3.13	3.05	2.96	2.87
0.1	16	3.05	2.67	2.46	2.33	2.24	2.18	2.13	2.09	2.06	2.03	1.99	1.94	1.89	1.87	1.84	1.81	1.78	1.75	1.72
0.05		4.49	3.63	3.24	3.01	2.85	2.74	2.66	2.59	2.54	2.49	2.42	2.35	2.28	2.24	2.19	2.15	2.11	2.06	2.01
0.01		8.53	6.23	5.29	4.77	4.44	4.20	4.03	3.89	3.78	3.69	3.55	3.41	3.26	3.18	3.10	3.02	2.93	2.84	2.75
0.1	17	3.03	2.64	2.44	2.31	2.22	2.15	2.10	2.06	2.03	2.00	1.96	1.91	1.86	1.84	1.81	1.78	1.75	1.72	1.69
0.05		4.45	3.59	3.20	2.96	2.81	2.70	2.61	2.55	2.49	2.45	2.38	2.31	2.23	2.19	2.15	2.10	2.06	2.01	1.96
0.01		8.40	6.11	5.18	4.67	4.34	4.10	3.93	3.79	3.68	3.59	3.46	3.31	3.16	3.08	3.00	2.92	2.83	2.75	2.65
0.1	18	3.01	2.62	2.42	2.29	2.20	2.13	2.08	2.04	2.00	1.98	1.93	1.89	1.84	1.81	1.78	1.75	1.72	1.69	1.66
0.05		4.41	3.55	3.16	2.93	2.77	2.66	2.58	2.51	2.46	2.41	2.34	2.27	2.19	2.15	2.11	2.06	2.02	1.97	1.92
0.01		8.29	6.01	5.09	4.58	4.25	4.01	3.84	3.71	3.60	3.51	3.37	3.23	3.08	3.00	2.92	2.84	2.75	2.66	2.57

* This table is taken from Table V of Fisher and Yates: *Statistical Tables for Biological, Agricultural and Medical Research*, published by Longman Group, Ltd., London (previously published by Oliver and Boyd, Edinburgh) by permission of the authors and publishers.

Table 10 (continued)

Probability point	ϕ_2	Numerator 1	2	3	4	5	6	7	8	9	10	12	15	20	24	30	40	60	120	∞
0.1	19	2.99	2.61	2.40	2.27	2.18	2.11	2.06	2.02	1.98	1.96	1.91	1.86	1.81	1.79	1.76	1.73	1.70	1.67	1.63
0.05		4.38	3.52	3.13	2.90	2.74	2.63	2.54	2.48	2.42	2.38	2.31	2.23	2.16	2.11	2.07	2.03	1.98	1.93	1.88
0.01		8.18	5.93	5.01	4.50	4.17	3.94	3.77	3.63	3.52	3.43	3.30	3.15	3.00	2.92	2.84	2.76	2.67	2.58	2.49
0.1	20	2.97	2.59	2.38	2.25	2.16	2.09	2.04	2.00	1.96	1.94	1.89	1.84	1.79	1.77	1.74	1.71	1.68	1.64	1.61
0.05		4.35	3.49	3.10	2.87	2.71	2.60	2.51	2.45	2.39	2.35	2.28	2.20	2.12	2.08	2.04	1.99	1.95	1.90	1.84
0.01		8.10	5.85	4.94	4.43	4.10	3.87	3.70	3.56	3.46	3.37	3.23	3.09	2.94	2.86	2.78	2.69	2.61	2.52	2.42
0.1	21	2.96	2.57	2.36	2.23	2.14	2.08	2.02	1.98	1.95	1.92	1.87	1.83	1.78	1.75	1.72	1.69	1.66	1.62	1.59
0.05		4.32	3.47	3.07	2.84	2.68	2.57	2.49	2.42	2.37	2.32	2.25	2.18	2.10	2.05	2.01	1.96	1.92	1.87	1.81
0.01		8.02	5.78	4.87	4.37	4.04	3.81	3.64	3.51	3.40	3.31	3.17	3.03	2.88	2.80	2.72	2.64	2.55	2.46	2.36
0.1	22	2.95	2.56	2.35	2.22	2.13	2.06	2.01	1.97	1.93	1.90	1.86	1.81	1.76	1.73	1.70	1.67	1.64	1.60	1.57
0.05		4.30	3.44	3.05	2.82	2.66	2.55	2.46	2.40	2.34	2.30	2.23	2.15	2.07	2.03	1.98	1.94	1.89	1.84	1.78
0.01		7.95	5.72	4.82	4.31	3.99	3.76	3.59	3.45	3.35	3.26	3.12	2.98	2.83	2.75	2.67	2.58	2.50	2.40	2.31
0.1	23	2.94	2.55	2.34	2.21	2.11	2.05	1.99	1.95	1.92	1.89	1.85	1.80	1.74	1.72	1.69	1.66	1.62	1.59	1.55
0.05		4.28	3.42	3.03	2.80	2.64	2.53	2.44	2.37	2.32	2.27	2.20	2.13	2.05	2.00	1.96	1.91	1.86	1.81	1.76
0.01		7.88	5.66	4.76	4.26	3.94	3.71	3.54	3.41	3.30	3.21	3.07	2.93	2.78	2.70	2.62	2.54	2.45	2.35	2.26
0.1	24	2.93	2.54	2.33	2.19	2.10	2.04	1.98	1.94	1.91	1.88	1.83	1.78	1.73	1.70	1.67	1.64	1.61	1.57	1.53
0.05		4.26	3.40	3.01	2.78	2.62	2.51	2.42	2.36	2.30	2.25	2.18	2.11	2.03	1.98	1.94	1.89	1.84	1.79	1.73
0.01		7.82	5.61	4.72	4.22	3.90	3.67	3.50	3.36	3.26	3.17	3.03	2.89	2.74	2.66	2.58	2.49	2.40	2.31	2.21
0.1	25	2.92	2.53	2.32	2.18	2.09	2.02	1.97	1.93	1.89	1.87	1.82	1.77	1.72	1.69	1.66	1.63	1.59	1.56	1.52
0.05		4.24	3.39	2.99	2.76	2.60	2.49	2.40	2.34	2.28	2.24	2.16	2.09	2.01	1.96	1.92	1.87	1.82	1.77	1.71
0.01		7.77	5.57	4.68	4.18	3.86	3.63	3.46	3.32	3.22	3.13	2.99	2.85	2.70	2.62	2.54	2.45	2.36	2.27	2.17
0.1	26	2.91	2.52	2.31	2.17	2.08	2.01	1.96	1.92	1.88	1.86	1.81	1.76	1.71	1.68	1.65	1.61	1.58	1.54	1.50
0.05		4.23	3.37	2.98	2.74	2.59	2.47	2.39	2.32	2.27	2.22	2.15	2.07	1.99	1.95	1.90	1.85	1.80	1.75	1.69
0.01		7.72	5.53	4.64	4.14	3.82	3.59	3.42	3.29	3.18	3.09	2.96	2.82	2.66	2.58	2.50	2.42	2.33	2.23	2.13
0.1	27	2.90	2.51	2.30	2.17	2.07	2.00	1.95	1.91	1.87	1.85	1.80	1.75	1.70	1.67	1.64	1.60	1.57	1.53	1.49
0.05		4.21	3.35	2.96	2.73	2.57	2.46	2.37	2.31	2.25	2.20	2.13	2.06	1.97	1.93	1.88	1.84	1.79	1.73	1.67
0.01		7.68	5.49	4.60	4.11	3.78	3.56	3.39	3.26	3.15	3.06	2.93	2.78	2.63	2.55	2.47	2.38	2.29	2.20	2.10
0.1	28	2.89	2.50	2.29	2.16	2.06	2.00	1.94	1.90	1.87	1.84	1.79	1.74	1.69	1.66	1.63	1.59	1.56	1.52	1.48
0.05		4.20	3.34	2.95	2.71	2.56	2.45	2.36	2.29	2.24	2.19	2.12	2.04	1.96	1.91	1.87	1.82	1.77	1.71	1.65
0.01		7.64	5.45	4.57	4.07	3.75	3.53	3.36	3.23	3.12	3.03	2.90	2.75	2.60	2.52	2.44	2.35	2.26	2.17	2.06
0.1	29	2.89	2.50	2.28	2.15	2.06	1.99	1.93	1.89	1.86	1.83	1.78	1.73	1.68	1.65	1.62	1.58	1.55	1.51	1.47
0.05		4.18	3.33	2.93	2.70	2.55	2.43	2.35	2.28	2.22	2.18	2.10	2.03	1.94	1.90	1.85	1.81	1.75	1.70	1.64
0.01		7.60	5.42	4.54	4.04	3.73	3.50	3.33	3.20	3.09	3.00	2.87	2.73	2.57	2.49	2.41	2.33	2.23	2.14	2.03
0.1	30	2.88	2.49	2.28	2.14	2.05	1.98	1.93	1.88	1.85	1.82	1.77	1.72	1.67	1.64	1.61	1.57	1.54	1.50	1.46
0.05		4.17	3.32	2.92	2.69	2.53	2.42	2.33	2.27	2.21	2.16	2.09	2.01	1.93	1.89	1.84	1.79	1.74	1.68	1.62
0.01		7.56	5.39	4.51	4.02	3.70	3.47	3.30	3.17	3.07	2.98	2.84	2.70	2.55	2.47	2.39	2.30	2.21	2.11	2.01
0.1	40	2.84	2.44	2.23	2.09	2.00	1.93	1.87	1.83	1.79	1.76	1.71	1.66	1.61	1.57	1.54	1.51	1.47	1.42	1.38
0.05		4.08	3.23	2.84	2.61	2.45	2.34	2.25	2.18	2.12	2.08	2.00	1.92	1.84	1.79	1.74	1.69	1.64	1.58	1.51
0.01		7.31	5.18	4.31	3.83	3.51	3.29	3.12	2.99	2.89	2.80	2.66	2.52	2.37	2.29	2.20	2.11	2.02	1.92	1.80
0.1	60	2.79	2.39	2.18	2.04	1.95	1.87	1.82	1.77	1.74	1.71	1.66	1.60	1.54	1.51	1.48	1.44	1.40	1.35	1.29
0.05		4.00	3.15	2.76	2.53	2.37	2.25	2.17	2.10	2.04	1.99	1.92	1.84	1.75	1.70	1.65	1.59	1.53	1.47	1.39
0.01		7.08	4.98	4.13	3.65	3.34	3.12	2.95	2.82	2.72	2.63	2.50	2.35	2.20	2.12	2.03	1.94	1.84	1.73	1.60
0.1	120	2.75	2.35	2.13	1.99	1.90	1.82	1.77	1.72	1.68	1.65	1.60	1.54	1.48	1.45	1.41	1.37	1.32	1.26	1.19
0.05		3.92	3.07	2.68	2.45	2.29	2.18	2.09	2.02	1.96	1.91	1.83	1.75	1.66	1.61	1.55	1.50	1.43	1.35	1.25
0.01		6.85	4.79	3.95	3.48	3.17	2.96	2.79	2.66	2.56	2.47	2.34	2.19	2.03	1.95	1.86	1.76	1.66	1.53	1.38
0.1	∞	2.71	2.30	2.08	1.94	1.85	1.77	1.72	1.67	1.63	1.60	1.55	1.49	1.42	1.38	1.34	1.30	1.24	1.17	1.00
0.05		3.84	3.00	2.60	2.37	2.21	2.10	2.01	1.94	1.88	1.83	1.75	1.67	1.57	1.52	1.46	1.39	1.32	1.22	1.00
0.01		6.63	4.61	3.78	3.32	3.02	2.80	2.64	2.51	2.41	2.32	2.18	2.04	1.88	1.79	1.70	1.59	1.47	1.32	1.00

Table 11 Tolerance factors for normal distributions.[1]
Factors K such that the probability is $(1 - \alpha)$ that at least a proportion $1 - \gamma$ of the distribution will be included between $\bar{X} \pm KS$, where $\bar{X}$ and S are estimators of the mean and the standard deviation computed from a random sample of size N.

γ / N	$(1-\alpha) = 0.75$					$(1-\alpha) = 0.90$				
	0.25	0.10	0.05	0.01	0.001	0.25	0.10	0.05	0.01	0.001
2	4.498	6.301	7.414	9.531	11.920	11.407	15.978	18.800	24.167	30.227
3	2.501	3.538	4.187	5.431	6.844	4.132	5.847	6.919	8.974	11.309
4	2.035	2.892	3.431	4.471	5.657	2.932	4.166	4.943	6.440	8.149
5	1.825	2.599	3.088	4.033	5.117	2.454	3.494	4.152	5.423	6.879
6	1.704	2.429	2.889	3.779	4.802	2.196	3.131	3.723	4.870	6.188
7	1.624	2.318	2.757	3.611	4.593	2.034	2.902	3.452	4.521	5.750
8	1.568	2.238	2.663	3.491	4.444	1.921	2.743	3.264	4.278	5.446
9	1.525	2.178	2.593	3.400	4.330	1.839	2.626	3.125	4.098	5.220
10	1.492	2.131	2.537	3.328	4.241	1.775	2.535	3.018	3.959	5.046
11	1.465	2.093	2.493	3.271	4.169	1.724	2.463	2.933	3.849	4.906
12	1.443	2.062	2.456	3.223	4.110	1.683	2.404	2.863	3.758	4.792
13	1.425	2.036	2.424	3.183	4.059	1.648	2.355	2.805	3.682	4.697
14	1.409	2.013	2.398	3.148	4.016	1.619	2.314	2.756	3.618	4.615
15	1.395	1.994	2.375	3.118	3.979	1.594	2.278	2.713	3.562	4.545
16	1.383	1.977	2.355	3.092	3.946	1.572	2.246	2.676	3.514	4.484
17	1.372	1.962	2.337	3.069	3.917	1.552	2.219	2.643	3.471	4.430
18	1.363	1.948	2.321	3.048	3.891	1.535	2.194	2.614	3.433	4.382
19	1.355	1.936	2.307	3.030	3.867	1.520	2.172	2.588	3.399	4.339
20	1.347	1.925	2.294	3.013	3.846	1.506	2.152	2.564	3.368	4.300
21	1.340	1.915	2.282	2.998	3.827	1.493	2.135	2.543	3.340	4.264
22	1.334	1.906	2.271	2.984	3.809	1.482	2.118	2.524	3.315	4.232
23	1.328	1.898	2.261	2.971	3.793	1.471	2.103	2.506	3.292	4.203
24	1.322	1.891	2.252	2.959	3.778	1.462	2.089	2.489	3.270	4.176
25	1.317	1.883	2.244	2.948	3.764	1.453	2.077	2.474	3.251	4.151
26	1.313	1.877	2.236	2.938	3.751	1.444	2.065	2.460	3.232	4.127
27	1.309	1.871	2.229	2.929	3.740	1.437	2.054	2.447	3.215	4.106
28	1.305	1.865	2.222	2.920	3.728	1.430	2.044	2.435	3.199	4.085
29	1.301	1.860	2.216	2.911	3.718	1.423	2.034	2.424	3.184	4.066
30	1.297	1.855	2.210	2.904	3.708	1.417	2.025	2.413	3.170	4.049
31	1.294	1.850	2.204	2.896	3.699	1.411	2.017	2.403	3.157	4.032
32	1.291	1.846	2.199	2.890	3.690	1.405	2.009	2.393	3.145	4.016
33	1.288	1.842	2.194	2.883	3.682	1.400	2.001	2.385	3.133	4.001
34	1.285	1.838	2.189	2.877	3.674	1.395	1.994	2.376	3.122	3.987
35	1.283	1.834	2.185	2.871	3.667	1.390	1.988	2.368	3.112	3.974
36	1.280	1.830	2.181	2.866	3.660	1.386	1.981	2.361	3.102	3.961
37	1.278	1.827	2.177	2.860	3.653	1.381	1.975	2.353	3.092	3.949
38	1.275	1.824	2.173	2.855	3.647	1.377	1.969	2.346	3.083	3.938
39	1.273	1.821	2.169	2.850	3.641	1.374	1.964	2.340	3.075	3.927
40	1.271	1.818	2.166	2.846	3.635	1.370	1.959	2.334	3.066	3.917
41	1.269	1.815	2.162	2.841	3.629	1.366	1.954	2.328	3.059	3.907
42	1.267	1.812	2.159	2.837	3.624	1.363	1.949	2.322	3.051	3.897
43	1.266	1.810	2.156	2.833	3.619	1.360	1.944	2.316	3.044	3.888
44	1.264	1.807	2.153	2.829	3.614	1.357	1.940	2.311	3.037	3.879
45	1.262	1.805	2.150	2.826	3.609	1.354	1.935	2.306	3.030	3.871
46	1.261	1.802	2.148	2.822	3.605	1.351	1.931	2.301	3.024	3.863
47	1.259	1.800	2.145	2.819	3.600	1.348	1.927	2.297	3.018	3.855
48	1.258	1.798	2.143	2.815	3.596	1.345	1.924	2.292	3.012	3.847
49	1.256	1.796	2.140	2.812	3.592	1.343	1.920	2.288	3.006	3.840
50	1.255	1.794	2.138	2.809	3.588	1.340	1.916	2.284	3.001	3.833

[1] Reproduced from C. Eisenhart, M. W. Hastay, and W. A. Wallis, *Techniques of Statistical Analysis*, Chapter 2. Used with permission of McGraw-Hill Book Company, Inc., New York, 1947.

Table 11 (continued).

N \ γ	(1 − α) = 0.95					(1 − α) = 0.99				
	0.25	0.10	0.05	0.01	0.001	0.25	0.10	0.05	0.01	
2	22.858	32.019	37.674	48.430	60.573	114.363	160.193	188.491	242.300	303.054
3	5.922	8.380	9.916	12.861	16.208	13.378	18.930	22.401	29.055	36.616
4	3.779	5.369	6.370	8.299	10.502	6.614	9.398	11.150	14.527	18.383
5	3.002	4.275	5.079	6.634	8.415	4.643	6.612	7.855	10.260	13.015
6	2.604	3.712	4.414	5.775	7.337	3.743	5.337	6.345	8.301	10.548
7	2.361	3.369	4.007	5.248	6.676	3.233	4.613	5.488	7.187	9.142
8	2.197	3.136	3.732	4.891	6.226	2.905	4.147	4.936	6.468	8.234
9	2.078	2.967	3.532	4.631	5.899	2.677	3.822	4.550	5.966	7.600
10	1.987	2.839	3.379	4.433	5.649	2.508	3.582	4.265	5.594	7.129
11	1.916	2.737	3.259	4.277	5.452	2.378	3.397	4.045	5.308	6.766
12	1.858	2.655	3.162	4.150	5.291	2.274	3.250	3.870	5.079	6.477
13	1.810	2.587	3.081	4.044	5.158	2.190	3.130	3.727	4.893	6.240
14	1.770	2.529	3.012	3.955	5.045	2.120	3.029	3.608	4.737	6.043
15	1.735	2.480	2.954	3.878	4.949	2.060	2.945	3.507	4 605	5.876
16	1.705	2.437	2.903	3.812	4.865	2.009	2.872	3.421	4.492	5.732
17	1.679	2.400	2.858	3.754	4.791	1.965	2.808	3.345	4.393	5.607
18	1.655	2.366	2.819	3.702	4.725	1.926	2.753	3.279	4.307	5.497
19	1.635	2.337	2.784	3.656	4.667	1.891	2.703	3.221	4.230	5.399
20	1.616	2.310	2.752	3.615	4.614	1.860	2.659	3.168	4.161	5.312
21	1.599	2.286	2.723	3.577	4.567	1.833	2.620	3.121	4.100	5.234
22	1.584	2.264	2.697	3.543	4.523	1·808	2.584	3.078	4.044	5.163
23	1.570	2.244	2.673	3.512	4.484	1.785	2.551	3.040	3.993	5.098
24	1.557	2.225	2.651	3.483	4.447	1.764	2.522	3.004	3.947	5.039
25	1.545	2.208	2.631	3.457	4.413	1.745	2.494	2.972	3.904	4.985
26	1.534	2.193	2.612	3.432	4.382	1.727	2.469	2.941	3.865	4.935
27	1.523	2.178	2.595	3.409	4.353	1.711	2.446	2.914	3.828	4.888
28	1.514	2.164	2.579	3.388	4.326	1.695	2.424	2.888	3.794	4.845
29	1.505	2.152	2.554	3.368	4.301	1.681	2.404	2.864	3.763	4.805
30	1.497	2.140	2.549	3.350	4.278	1.668	2.385	2.841	3.733	4.768
31	1.489	2.129	2.536	3.332	4.256	1.656	2.367	2.820	3.706	4.732
32	1.481	2.118	2.524	3.316	4.235	1.644	2.351	2.801	3.680	4.699
33	1.475	2.108	2.512	3.300	4.215	1.633	2.335	2.782	3.655	4.668
34	1.468	2.099	2.501	3.286	4.197	1.623	2.320	2.764	3.632	4.639
35	1.462	2.090	2.490	3.272	4.179	1.613	2.306	2.748	3.611	4.611
36	1.455	2.081	2.479	3.258	4.161	1.604	2.293	2.732	3.590	4.585
37	1.450	2.073	2.470	3.246	4.146	1.595	2.281	2.717	3.571	4.560
38	1.446	2.068	2.464	3.237	4.134	1.587	2.269	2.703	3.552	4.537
39	1.441	2.060	2.455	3.226	4.120	1.579	2.257	2.690	3.534	4.514
40	1.435	2.052	2.445	3.213	4.104	1.571	2.247	2.677	3.518	4.493
41	1.430	2.045	2.437	3.202	4.090	1.564	2.236	2.665	3.502	4.472
42	1.426	2.039	2.429	3.192	4.077	1.557	2.227	2.653	3.486	4.453
43	1.422	2.033	2.422	3.183	4.065	1.551	2.217	2.642	3.472	4.434
44	1.418	2.027	2.415	3.173	4.053	1.545	2.208	2.631	3.458	4.416
45	1.414	2.021	2.408	3.165	4.042	1.539	2.200	2.621	3.444	4.399
46	1.410	2.016	2.402	3.156	4.031	1.533	2.192	2.611	3.431	4.383
47	1.406	2.011	2.396	3.148	4.021	1.527	2.184	2.602	3.419	4.367
48	1.403	2.006	2.390	3.140	4.011	1.522	2.176	2.593	3.407	4.352
49	1.399	2.001	2.384	3.133	4.002	1.517	2.169	2.584	3.396	4.337
50	1.396	1.969	2.379	3.126	3.993	1.512	2.162	2.576	3.385	4.323

Table 12 One-sided tolerance factors for normal distributions. *Factors K such that the probability is* $(1 - \alpha)$ *that at least a proportion* $1 - \gamma$ *of the distribution will be less than* $\bar{X} + KS$ *(or greater than* $\bar{X} - KS$*), where* $\bar{X}$ *and S are estimators of the mean and the standard deviation computed from a random sample of size N.*

γ	$(1-\alpha) = 0.75$					$(1-\alpha) = 0.90$				
N	0.25	0.10	0.05	0.01	0.001	0.25	0.10	0.05	0.01	0.001
3	1.464	2.501	3.152	4.396	5.805	2.602	4.258	5.310	7.340	9.651
4	1.256	2.134	2.680	3.726	4.910	1.972	3.187	3.957	5.437	7.128
5	1.152	1.961	2.463	3.421	4.507	1.698	2.742	3.400	4.666	6.112
6	1.087	1.860	2.336	3.243	4.273	1.540	2.494	3.091	4.242	5.556
7	1.043	1.791	2.250	3.126	4.118	1.435	2.333	2.894	3.972	5.201
8	1.010	1.740	2.190	3.042	4.008	1.360	2.219	2.755	3.783	4.955
9	0.984	1.702	2.141	2.977	3.924	1.302	2.133	2.649	3.641	4.772
10	0.964	1.671	2.103	2.927	3.858	1.257	2.065	2.568	3.532	4.629
11	0.947	1.646	2.073	2.885	3.804	1.219	2.012	2.503	3.444	4.515
12	0.933	1.624	2.048	2.851	3.760	1.188	1.966	2.448	3.371	4.420
13	0.919	1.606	2.026	2.822	3.722	1.162	1.928	2.403	3.310	4.341
14	0.909	1.591	2.007	2.796	3.690	1.139	1.895	2.363	3.257	4.274
15	0.899	1.577	1.991	2.776	3.661	1.119	1.866	2.329	3.212	4.215
16	0.891	1.566	1.977	2.756	3.637	1.101	1.842	2.299	3.172	4.164
17	0.883	1.554	1.964	2.739	3.615	1.085	1.820	2.272	3.136	4.118
18	0.876	1.544	1.951	2.723	3.595	1.071	1.800	2.249	3.106	4.078
19	0.870	1.536	1.942	2.710	3.577	1.058	1.781	2.228	3.078	4.041
20	0.865	1.528	1.933	2.697	3.561	1.046	1.765	2.208	3.052	4.009
21	0.859	1.520	1.923	2.686	3.545	1.035	1.750	2.190	3.028	3.979
22	0.854	1.514	1.916	2.675	3.532	1.025	1.736	2.174	3.007	3.952
23	0.849	1.508	1.907	2.665	3.520	1.016	1.724	2.159	2.987	3.927
24	0.845	1.502	1.901	2.656	3.509	1.007	1.712	2.145	2.969	3.904
25	0.842	1.496	1.895	2.647	3.497	0.999	1.702	2.132	2.952	3.882
30	0.825	1.475	1.869	2.613	3.454	0.966	1.657	2.080	2.884	3.794
35	0.812	1.458	1.849	2.588	3.421	0.942	1.623	2.041	2.833	3.730
40	0.803	1.445	1.834	2.568	3.395	0.923	1.598	2.010	2.793	3.679
45	0.795	1.435	1.821	2.552	3.375	0.908	1.577	1.986	2.762	3.638
50	0.788	1.426	1.811	2.538	3.358	0.894	1.560	1.965	2.735	3.604

Reproduced from Chapter 2 of C. Eisenhart, M. W. Hastay, and W. A. Wallis, *Techniques of Statistical Analysis*, 1947. Used with permission of McGraw-Hill Publishing Company, Inc., New York.

Table 12 (continued).

N \ γ	$(1-\alpha)=0.95$					$(1-\alpha)=0.99$				
	0.25	0.10	0.05	0.01	0.001	0.25	0.10	0.05	0.01	0.001
3	3.804	6.158	7.655	10.552	13.857					
4	2.619	4.163	5.145	7.042	9.215					
5	2.149	3.407	4.202	5.741	7.501					
6	1.895	3.006	3.707	5.062	6.612	2.849	4.408	5.409	7.334	9.540
7	1.732	2.755	3.399	4.641	6.061	2.490	3.856	4.730	6.411	8.348
8	1.617	2.582	3.188	4.353	5.686	2.252	3.496	4.287	5.811	7.566
9	1.532	2.454	3.031	4.143	5.414	2.085	3.242	3.971	5.389	7.014
10	1.465	2.355	2.911	3.981	5.203	1.954	3.048	3.739	5.075	6.603
11	1.411	2.275	2.815	3.852	5.036	1.854	2.897	3.557	4.828	6.284
12	1.366	2.210	2.736	3.747	4.900	1.771	2.773	3.410	4.633	6.032
13	1.329	2.155	2.670	3.659	4.787	1.702	2.677	3.290	4.472	5.826
14	1.296	2.108	2.614	3.585	4.690	1.645	2.592	3.189	4.336	5.651
15	1.268	2.068	2.566	3.520	4.607	1.596	2.521	3.102	4.224	5.507
16	1.242	2.032	2.523	3.463	4.534	1.553	2.458	3.028	4.124	5.374
17	1.220	2.001	2.486	3.415	4.471	1.514	2.405	2.962	4.038	5.268
18	1.200	1.974	2.453	3.370	4.415	1.481	2.357	2.906	3.961	5.167
19	1.183	1.949	2.423	3.331	4.364	1.450	2.315	2.855	3.893	5.078
20	1.167	1.926	2.396	3.295	4.319	1.424	2.275	2.807	3.832	5.003
21	1.152	1.905	2.371	3.262	4.276	1.397	2.241	2.768	3.776	4.932
22	1.138	1.887	2.350	3.233	4.238	1.376	2.208	2.729	3.727	4.866
23	1.126	1.869	2.329	3.206	4.204	1.355	2.179	2.693	3.680	4.806
24	1.114	1.853	2.309	3.181	4.171	1.336	2.154	2.663	3.638	4.755
25	1.103	1.838	2.292	3.158	4.143	1.319	2.129	2.632	3.601	4.706
30	1.059	1.778	2.220	3.064	4.022	1.249	2.029	2.516	3.446	4.508
35	1.025	1.732	2.166	2.994	3.934	1.195	1.957	2.431	3.334	4.364
40	0.999	1.697	2.126	2.941	3.866	1.154	1.902	2.365	3.250	4.255
45	0.978	1.669	2.092	2.897	3.811	1.122	1.857	2.313	3.181	4.168
50	0.961	1.646	2.065	2.863	3.766	1.096	1.821	2.296	3.124	4.096

Table 13 Barnard's sequential t test. Boundary values U_0 and U_1

$D = 0{\cdot}10$

	α = 0·01						α = 0·05						α = 0·20					
	β = 0·01		β = 0·05		β = 0·20		β = 0·01		β = 0·05		β = 0·20		β = 0·01		β = 0·05		β = 0·20	
n	U_0	U_1	U_0	U_1	U_0	U_1	U_0	U_1	U_0	U_1	U_0	U_1	U_0	U_1	U_0	U_1	U_0	U_1
2																		
4																		
6																		
8																	[−5·14]	[4·97]
10					[−5·24]						[−5·09]			[5·13]		[5·01]	[−4·49]	[4·49]
15					[−4·11]						[−4·00]			[4·27]		[4·17]	−3·53	3·75
20			[−6·79]		−3·46				[−6·68]		−3·36	[6·28]		3·77	[−6·27]	3·68	−2·96	3·31
25			[−5·96]		−3·02			[6·07]	[−5·87]	[6·00]	−2·94	[5·67]		3·42	[−5·50]	3·34	−2·58	3·00
30			−5·36		−2·70			[5·62]	−5·27	[5·55]	−2·63	5·25		3·18	−4·94	3·10	−2·30	2·79
35			−4·89		−2·45		[−7·70]	5·26	−4·81	5·19	−2·38	4·91	[−7·39]	2·98	−4·51	2·92	−2·08	2·63
40	[−7·20]	[7·41]	−4·51	[7·35]	−2·25	[7·09]	[−7·13]	4·97	−4·44	4·91	−2·18	4·64	[−6·84]	2·83	−4·16	2·77	−1·90	2·50
45	[−6·71]	[7·04]	−4·20	[6·98]	−2·08	[6·73]	−6·65	4·73	−4·14	4·67	−2·01	4·42	−6·38	2·71	−3·88	2·65	−1·76	2·39
50	−6·33	6·72	−3·94	6·67	−1·93	6·44	−6·27	4·53	−3·88	4·47	−1·87	4·23	−6·02	2·60	−3·63	2·55	−1·63	2·31
60	−5·64	6·19	−3·51	6·15	−1·69	5·93	−5·59	4·20	−3·47	4·15	−1·64	3·93	−5·36	2·44	−3·25	2·39	−1·42	2·17
70	−5·15	5·81	−3·21	5·77	−1·52	5·57	−5·10	3·95	−3·16	3·90	−1·47	3·70	−4·89	2·31	−2·95	2·26	−1·26	2·06
80	−4·76	5·50	−2·93	5·46	−1·36	5·27	−4·71	3·75	−2·88	3·70	−1·31	3·51	−4·51	2·21	−2·69	2·17	−1·12	1·98
90	−4·43	5·25	−2·71	5·21	−1·23	5·03	−4·38	3·59	−2·66	3·55	−1·18	3·37	−4·20	2·14	−2·48	2·10	−1·00	1·92
100	−4·14	5·04	−2·52	5·00	−1·11	4·83	−4·10	3·46	−2·47	3·41	−1·07	3·24	−3·92	2·08	−2·30	2·04	−0·90	1·87
150	−3·16	4·33	−1·84	4·30	−0·69	4·16	−3·12	3·03	−1·80	2·99	−0·66	2·85	−2·98	1·90	−1·64	1·87	−0·52	1·73
200	−2·56	3·93	−1·42	3·90	−0·42	3·78	−2·53	2·80	−1·38	2·77	−0·40	2·65	−2·40	1·83	−1·25	1·80	−0·27	1·68
	n_0	n_1	n_0	n_1	n_0	n_1	n_0	n_1	n_0	n_1	n_0	n_1	n_0	n_1	n_0	n_1	n_0	n_1
	46	48	30	47	16	46	45	31	29	31	16	29	44	17	28	17	14	15
	$\bar{n}_0$	$\bar{n}_1$	$\bar{n}_0$	$\bar{n}_1$	$\bar{n}_0$	$\bar{n}_1$	$\bar{n}_0$	$\bar{n}_1$	$\bar{n}_0$	$\bar{n}_1$	$\bar{n}_0$	$\bar{n}_1$	$\bar{n}_0$	$\bar{n}_1$	$\bar{n}_0$*	$\bar{n}_1$	$\bar{n}_0$	$\bar{n}_1$
	1000	1000	600	900	400	700	900	600	600	600	300	400	700	300	400	300	200	200

$D = 0{\cdot}25$

	α = 0·01						α = 0·05						α = 0·20					
	β = 0·01		β = 0·05		β = 0·20		β = 0·01		β = 0·05		β = 0·20		β = 0·01		β = 0·05		β = 0·20	
n	U_0	U_1	U_0	U_1	U_0	U_1	U_0	U_1	U_0	U_1	U_0	U_1	U_0	U_1	U_0	U_1	U_0	U_1
2																	[−5·20]	
4					[−3·48]						[−3·37]			[3·30]		[3·23]	[−2·94]	[2·93]
6					[−2·56]						[−2·48]			[2·83]		[2·77]	−2·16	[2·51]
8			[−4·39]		−2·06			[4·29]	[−4·32]	[4·24]	−1·99	[4·03]		2·55	[−4·02]	2·49	−1·76	2·27
10	[−6·26]	[5·68]	[−3·73]	[5·64]	−1·73	[5·46]	[−6·20]	[3·95]	[−3·67]	[3·91]	−1·67	[3·71]	[−5·91]	2·36	[−3·41]	2·32	−1·43	2·11
15	[−4·67]	[4·91]	−2·76	[4·87]	−1·20	[4·72]	[−4·62]	3·43	−2·72	3·39	−1·16	3·23	[−4·41]	2·10	−2·52	2·06	−0·98	1·89
20	−3·79	4·44	−2·21	4·41	−0·89	4·29	−3·75	3·13	−2·17	3·10	−0·85	2·95	−3·58	1·96	−2·00	1·92	−0·70	1·78
25	−3·21	4·12	−1·82	4·09	−0·67	3·97	−3·17	2·93	−1·77	2·90	−0·63	2·77	−3·02	1·88	−1·64	1·85	−0·49	1·71
30	−2·78	3·90	−1·54	3·87	−0·49	3·75	−2·75	2·80	−1·50	2·77	−0·46	2·65	−2·61	1·83	−1·37	1·80	−0·33	1·68
35	−2·45	3·72	−1·31	3·70	−0·35	3·59	−2·42	2·70	−1·28	2·67	−0·32	2·56	−2·30	1·80	−1·16	1·77	−0·20	1·66
40	−2·18	3·59	−1·12	3·57	−0·22	3·46	−2·15	2·63	−1·09	2·60	−0·20	2·50	−2·03	1·78	−0·98	1·76	−0·09	1·65
45	−1·95	3·48	−0·96	3·46	−0·12	3·36	−1·92	2·57	−0·93	2·55	−0·09	2·45	−1·81	1·77	−0·83	1·75	0·01	1·65
50	−1·75	3·40	−0·81	3·37	−0·02	3·28	−1·73	2·53	−0·79	2·51	0·00	2·41	−1·63	1·77	−0·69	1·75	0·10	1·65
60	−1·43	3·26	−0·58	3·24	0·14	3·15	−1·41	2·46	−0·56	2·44	0·16	2·36	−1·32	1·78	−0·47	1·75	0·25	1·67
70	−1·17	3·17	−0·39	3·15	0·28	3·07	−1·15	2·43	−0·37	2·41	0·30	2·33	−1·07	1·79	−0·28	1·77	0·38	1·69
80	−0·95	3·11	−0·22	3·09	0·40	3·01	−0·93	2·41	−0·20	2·39	0·42	2·32	−0·85	1·81	−0·12	1·80	0·49	1·72
90	−0·76	3·06	−0·07	3·04	0·51	2·97	−0·74	2·41	−0·06	2·39	0·52	2·32	−0·67	1·84	0·02	1·83	0·60	1·76
100	−0·59	3·03	0·05	3·01	0·60	2·95	−0·58	2·41	0·07	2·39	0·62	2·32	−0·51	1·87	0·14	1·86	0·69	1·79
150	0·03	2·98	0·55	2·97	1·00	2·92	0·04	2·47	0·57	2·46	1·01	2·41	0·10	2·03	0·62	2·02	1·07	1·97
200	0·46	3·03	0·92	3·01	1·31	2·97	0·48	2·58	0·93	2·57	1·32	2·52	0·52	2·20	0·98	2·19	1·37	2·14
	n_0	n_1	n_0	n_1	n_0	n_1	n_0	n_1	n_0	n_1	n_0	n_1	n_0	n_1	n_0	n_1	n_0	n_1
	18	20	12	20	7	20	18	14	12	13	7	13	17	8	11	7	6	7
	$\bar{n}_0$	$\bar{n}_1$	$\bar{n}_0$	$\bar{n}_1$	$\bar{n}_0$	$\bar{n}_1$	$\bar{n}_0$	$\bar{n}_1$	$\bar{n}_0$	$\bar{n}_1$	$\bar{n}_0$	$\bar{n}_1$	$\bar{n}_0$	$\bar{n}_1$	$\bar{n}_0$	$\bar{n}_1$	$\bar{n}_0$	$\bar{n}_1$
	150	150	100	150	50	125	150	100	100	100	50	75	125	50	75	50	50	50

n_0 and n_1 refer to the smallest number of observations for which a decision is possible when $\mu = \mu_0$ and $\mu = \mu_0 + D\sigma$ respectively.
$\bar{n}_0$ and $\bar{n}_1$ are the approximate average sample numbers when $\mu = \mu_0$ and $\mu = \mu_0 + D\sigma$ respectively.
Values of U shown in square brackets are included to assist interpolation and the drawing of boundaries and must not be used in making the test.

Table 13 (continued)

$D = 0·50$

	α = 0·01					
	β = 0·01		β = 0·05		β = 0·20	
n	U_0	U_1	U_0	U_1	U_0	U_1
2			[−7·14]		[−2·66]	
4			[−3·19]		−1·24	
6	[−3·96]		−2·11		−0·75	
8	[−2·98]	[3·49]	−1·55	[3·46]	−0·45	[3·37]
10	−2·38	[3·32]	−1·18	[3·30]	−0·23	[3·21]
15	−1·50	3·07	−0·59	3·05	0·14	2·97
20	−0·98	2·96	−0·22	2·94	0·40	2·87
25	−0·60	2·87	0·06	2·86	0·61	2·80
30	−0·31	2·85	0·28	2·84	0·78	2·79
35	−0·07	2·85	0·47	2·84	0·93	2·79
40	0·13	2·86	0·64	2·85	1·06	2·80
45	0·31	2·88	0·78	2·87	1·18	2·83
50	0·47	2·91	0·91	2·90	1·29	2·85
60	0·74	2·97	1·15	2·96	1·49	2·92
70	0·98	3·04	1·35	3·03	1·67	3·00
80	1·19	3·12	1·53	3·11	1·83	3·07
90	1·38	3·20	1·70	3·19	1·99	3·16
100	1·55	3·28	1·86	3·27	2·12	3·24
150	2·26	3·67	2·51	3·67	2·71	3·64
200	2·81	4·04	3·03	4·04	3·21	4·01
	n_0	n_1	n_0	n_1	n_0	n_1
	9	11	6	11	4	11
	$\bar{n}_0$	$\bar{n}_1$	$\bar{n}_0$	$\bar{n}_1$	$\bar{n}_0$	$\bar{n}_1$
	45	45	30	45	15	35

	α = 0·05					
	β = 0·01		β = 0·05		β = 0·20	
n	U_0	U_1	U_0	U_1	U_0	U_1
2			[−6·96]		[−2·56]	
4		[3·03]	[−3·13]	[3·01]	−1·20	[2·89]
6	[−3·91]	[2·75]	−2·07	[2·73]	−0·71	[2·62]
8	[−2·94]	2·59	−1·51	2·56	−0·41	2·47
10	−2·35	2·49	−1·15	2·46	−0·20	2·37
15	−1·47	2·36	−0·57	2·34	0·17	2·27
20	−0·96	2·33	−0·21	2·31	0·42	2·24
25	−0·59	2·31	0·07	2·30	0·62	2·24
30	−0·30	2·34	0·29	2·32	0·79	2·27
35	−0·06	2·37	0·48	2·36	0·94	2·31
40	0·14	2·41	0·65	2·40	1·07	2·35
45	0·32	2·45	0·79	2·44	1·19	2·39
50	0·48	2·50	0·92	2·49	1·30	2·44
60	0·75	2·59	1·16	2·58	1·50	2·54
70	0·99	2·69	1·36	2·68	1·68	2·64
80	1·20	2·79	1·54	2·78	1·84	2·74
90	1·39	2·88	1·71	2·88	2·00	2·84
100	1·56	2·98	1·87	2·97	2·13	2·94
150	2·26	3·42	2·51	3·42	2·72	3·39
200	2·81	3·83	3·03	3·83	3·22	3·80
	n_0	n_1	n_0	n_1	n_0	n_1
	9	8	6	7	4	7
	$\bar{n}_0$	$\bar{n}_1$	$\bar{n}_0$	$\bar{n}_1$	$\bar{n}_0$	$\bar{n}_1$
	45	30	30	30	15	20

	α = 0·20					
	β = 0·01		β = 0·05		β = 0·20	
n	U_0	U_1	U_0	U_1	U_0	U_1
2		[2·38]	[−6·28]	[2·34]	[−2·16]	[2·15]
4	[−5·65]	1·99	[−2·86]	1·94	−1·00	1·79
6	[−3·70]	1·82	−1·89	1·79	−0·56	1·67
8	−2·78	1·75	−1·37	1·73	−0·29	1·62
10	−2·21	1·73	−1·03	1·70	−0·09	1·60
15	−1·38	1·73	−0·48	1·70	0·25	1·62
20	−0·87	1·77	−0·12	1·75	0·50	1·68
25	−0·51	1·82	0·14	1·81	0·69	1·74
30	−0·23	1·89	0·36	1·87	0·85	1·81
35	0·00	1·95	0·54	1·94	0·99	1·88
40	0·19	2·02	0·70	2·00	1·12	1·95
45	0·37	2·08	0·84	2·07	1·24	2·02
50	0·53	2·14	0·97	2·13	1·35	2·09
60	0·80	2·27	1·20	2·26	1·54	2·22
70	1·03	2·39	1·40	2·38	1·72	2·34
80	1·23	2·50	1·58	2·49	1·88	2·46
90	1·42	2·61	1·75	2·60	2·03	2·57
100	1·59	2·72	1·90	2·71	2·16	2·68
150	2·29	3·21	2·54	3·20	2·75	3·18
200	2·84	3·64	3·05	3·63	3·24	3·61
	n_0	n_1	n_0	n_1	n_0	n_1
	8	4	6	4	3	4
	$\bar{n}_0$	$\bar{n}_1$	$\bar{n}_0$	$\bar{n}_1$	$\bar{n}_0$	$\bar{n}_1$
	35	15	20	15	10	10

$D = 0·75$

	α = 0·01					
	β = 0·01		β = 0·05		β = 0·20	
n	U_0	U_1	U_0	U_1	U_0	U_1
2			[−3·96]		−1·29	
4	[−3·28]	[2·99]	−1·53	[2·97]	−0·35	[2·91]
6	−1·93	[2·83]	−0·78	[2·82]	0·05	[2·74]
8	−1·24	2·73	−0·37	2·72	0·30	2·66
10	−0·81	2·69	−0·08	2·68	0·50	2·62
15	−0·14	2·68	0·42	2·68	0·87	2·63
20	0·31	2·73	0·77	2·72	1·15	2·68
25	0·64	2·81	1·04	2·80	1·38	2·76
30	0·91	2·89	1·27	2·88	1·58	2·85
35	1·14	2·97	1·48	2·97	1·76	2·93
40	1·35	3·06	1·66	3·05	1·92	3·02
45	1·54	3·15	1·83	3·14	2·08	3·12
50	1·71	3·24	1·98	3·24	2·22	3·21
60	2·02	3·41	2·27	3·41	2·48	3·38
70	2·29	3·58	2·52	3·58	2·72	3·55
80	2·53	3·75	2·75	3·74	2·94	3·72
90	2·75	3·90	2·96	3·89	3·13	3·87
100	2·97	4·05	3·15	4·05	3·33	4·03
150	3·87	4·75	4·00	4·75	4·16	4·72
200	4·61	5·33	4·75	5·33	4·85	5·33
	n_0	n_1	n_0	n_1	n_0	n_1
	6	8	4	8	2	8
	$\bar{n}_0$	$\bar{n}_1$	$\bar{n}_0$	$\bar{n}_1$	$\bar{n}_0$	$\bar{n}_1$
	25	25	20	25	15	20

	α = 0·05					
	β = 0·01		β = 0·05		β = 0·20	
n	U_0	U_1	U_0	U_1	U_0	U_1
2		[2·61]	[−3·90]	[2·60]	−1·23	[2·50]
4	[−3·21]	[2·31]	−1·49	[2·30]	−0·32	[2·22]
6	−1·90	2·22	−0·76	2·20	0·07	2·13
8	−1·21	2·18	−0·35	2·16	0·32	2·10
10	−0·79	2·18	−0·07	2·16	0·51	2·11
15	−0·12	2·24	0·44	2·23	0·88	2·18
20	0·33	2·34	0·78	2·33	1·16	2·29
25	0·65	2·45	1·05	2·44	1·39	2·40
30	0·92	2·56	1·28	2·55	1·59	2·51
35	1·15	2·66	1·49	2·66	1·77	2·63
40	1·36	2·77	1·67	2·76	1·92	2·73
45	1·54	2·88	1·84	2·87	2·09	2·84
50	1·71	2·98	1·99	2·97	2·22	2·95
60	2·02	3·17	2·27	3·17	2·48	3·14
70	2·29	3·36	2·52	3·35	2·72	3·33
80	2·54	3·54	2·76	3·53	2·94	3·51
90	2·77	3·71	2·97	3·71	3·13	3·68
100	2·98	3·87	3·17	3·87	3·33	3·84
150	3·87	4·59	4·00	4·59	4·16	4·56
200	4·61	5·23	4·75	5·23	4·85	5·20
	n_0	n_1	n_0	n_1	n_0	n_1
	6	5	4	5	2	5
	$\bar{n}_0$	$\bar{n}_1$	$\bar{n}_0$	$\bar{n}_1$	$\bar{n}_0$	$\bar{n}_1$
	25	20	20	20	10	15

	α = 0·20					
	β = 0·01		β = 0·05		β = 0·20	
n	U_0	U_1	U_0	U_1	U_0	U_1
2		[1·77]	[−3·48]	[1·73]	−0·99	[1·62]
4	[−3·01]	1·63	−1·32	1·61	−0·19	1·52
6	−1·76	1·62	−0·63	1·61	0·17	1·53
8	−1·11	1·65	−0·27	1·63	0·40	1·56
10	−0·71	1·70	0·02	1·69	0·58	1·63
15	−0·05	1·84	0·48	1·83	0·93	1·78
20	0·38	1·99	0·82	1·98	1·20	1·93
25	0·69	2·13	1·09	2·12	1·43	2·08
30	0·96	2·27	1·32	2·26	1·63	2·22
35	1·19	2·39	1·52	2·39	1·80	2·35
40	1·39	2·52	1·70	2·51	1·96	2·48
45	1·58	2·64	1·87	2·63	2·12	2·60
50	1·74	2·75	2·02	2·75	2·26	2·72
60	2·05	2·97	2·30	2·97	2·51	2·93
70	2·32	3·17	2·55	3·16	2·75	3·14
80	2·57	3·36	2·78	3·35	2·97	3·33
90	2·79	3·52	2·99	3·53	3·17	3·51
100	3·00	3·71	3·20	3·70	3·36	3·68
150	3·87	4·45	4·03	4·45	4·19	4·43
200	4·61	5·12	4·78	5·12	4·88	5·10
	n_0	n_1	n_0	n_1	n_0	n_1
	6	3	4	3	2	3
	$\bar{n}_0$	$\bar{n}_1$	$\bar{n}_0$	$\bar{n}_1$	$\bar{n}_0$	$\bar{n}_1$
	20	10	15	10	10	10

n_0 and n_1 refer to the smallest number of observations for which a decision is possible when $\mu = \mu_0$ and $\mu = \mu_0 + D\sigma$ respectively.
$\bar{n}_0$ and $\bar{n}_1$ are the approximate average sample numbers when $\mu = \mu_0$ and $\mu = \mu_0 + D\sigma$ respectively.
Values of U shown in square brackets are included to assist interpolation and the drawing of boundaries and must not be used in making the test.

Table 13 (continued)

$D = 1{\cdot}00$

	α = 0·01						α = 0·05						α = 0·20					
	β = 0·01		β = 0·05		β = 0·20		β = 0·01		β = 0·05		β = 0·20		β = 0·01		β = 0·05		β = 0·20	
n	U_0	U_1	U_0	U_1	U_0	U_1	U_0	U_1	U_0	U_1	U_0	U_1	U_0	U_1	U_0	U_1	U_0	U_1
2	[−5·80]		[−2·21]		−0·52	[2·64]	[−5·66]	[2·15]	[−2·14]	[2·13]	−0·49	[2·06]	[−5·16]	[1·56]	[−1·89]	[1·54]	−0·33	[1·46]
4	−1·68	[2·53]	−0·55	[2·52]	0·21	[2·48]	−1·65	[2·04]	−0·53	[2·03]	0·23	1·97	−1·51	1·54	−0·43	1·53	0·31	1·49
6	−0·73	[2·49]	0·01	[2·48]	0·56	2·44	−0·71	2·05	0·03	2·04	0·58	1·99	−0·62	1·65	0·10	1·64	0·64	1·58
8	−0·23	2·50	0·35	2·49	0·81	2·45	−0·22	2·10	0·37	2·09	0·82	2·05	−0·15	1·74	0·43	1·73	0·88	1·68
10	0·13	2·53	0·62	2·52	1·02	2·49	0·14	2·17	0·63	2·16	1·03	2·12	0·20	1·84	0·68	1·83	1·07	1·79
15	0·73	2·66	1·10	2·65	1·41	2·62	0·74	2·35	1·11	2·34	1·42	2·31	0·78	2·08	1·15	2·07	1·46	2·04
20	1·15	2·80	1·46	2·79	1·72	2·76	1·16	2·53	1·47	2·52	1·73	2·49	1·19	2·30	1·50	2·30	1·76	2·27
25	1·47	2·96	1·75	2·95	1·99	2·93	1·48	2·71	1·76	2·70	2·00	2·67	1·51	2·49	1·79	2·48	2·02	2·46
30	1·76	3·12	2·01	3·11	2·22	3·09	1·76	2·88	2·02	2·88	2·22	2·85	1·79	2·68	2·05	2·68	2·25	2·65
35	2·01	3·27	2·24	3·26	2·43	3·24	2·02	3·05	2·24	3·05	2·43	3·02	2·04	2·86	2·26	2·86	2·45	2·83
40	2·23	3·41	2·44	3·40	2·62	3·38	2·24	3·21	2·45	3·21	2·63	3·19	2·26	3·04	2·47	3·03	2·65	3·01
45	2·43	3·54	2·63	3·54	2·81	3·52	2·44	3·36	2·64	3·36	2·81	3·34	2·46	3·20	2·66	3·19	2·83	3·17
50	2·62	3·68	2·82	3·67	2·98	3·65	2·63	3·50	2·82	3·50	2·99	3·48	2·65	3·35	2·84	3·34	3·01	3·32
60	2·98	3·94	3·16	3·94	3·30	3·92	2·99	3·77	3·16	3·77	3·30	3·75	3·01	3·63	3·18	3·63	3·32	3·61
70	3·30	4·18	3·45	4·18	3·59	4·16	3·30	4·03	3·45	4·03	3·59	4·01	3·32	3·90	3·47	3·89	3·61	3·88
80	3·58	4·41	3·73	4·41	3·86	4·39	3·59	4·27	3·73	4·27	3·86	4·25	3·60	4·15	3·75	4·14	3·87	4·13
90	3·85	4·62	3·99	4·62	4·11	4·61	3·86	4·50	3·99	4·49	4·11	4·48	3·87	4·38	4·01	4·38	4·13	4·36
100	4·10	4·83	4·23	4·82	4·34	4·81	4·11	4·70	4·24	4·70	4·35	4·69	4·12	4·60	4·25	4·59	4·36	4·58
150	5·16	5·76	5·26	5·76	5·35	5·75	5·16	5·66	5·27	5·65	5·36	5·64	5·17	5·56	5·28	5·56	5·37	5·55
200	6·05	6·57	6·15	6·57	6·23	6·56	6·05	6·49	6·15	6·48	6·23	6·47	6·06	6·41	6·16	6·41	6·24	6·40
	n_0	n_1	n_0	n_1	n_0	n_1	n_0	n_1	n_0	n_1	n_0	n_1	n_0	n_1	n_0	n_1	n_0	n_1
	4	7	3	7	2	6	4	5	3	5	2	4	4	3	3	3	2	3
	$\bar{n}_0$	$\bar{n}_1$	$\bar{n}_0$	$\bar{n}_1$	$\bar{n}_0$	$\bar{n}_1$	$\bar{n}_0$	$\bar{n}_1$	$\bar{n}_0$	$\bar{n}_1$	$\bar{n}_0$	$\bar{n}_1$	$\bar{n}_0$	$\bar{n}_1$	$\bar{n}_0$	$\bar{n}_1$	$\bar{n}_0$	$\bar{n}_1$
	15	15	10	15	5	10	10	10	10	10	5	7	10	5	5	5	5	5

$D = 1{\cdot}50$

	α = 0·01						α = 0·05						α = 0·20					
	β = 0·01		β = 0·05		β = 0·20		β = 0·01		β = 0·05		β = 0·20		β = 0·01		β = 0·05		β = 0·20	
n	U_0	U_1	U_0	U_1	U_0	U_1	U_0	U_1	U_0	U_1	U_0	U_1	U_0	U_1	U_0	U_1	U_0	U_1
2	[−1·89]	[2·06]	−0·44	[2·05]	0·32	[2·02]	[−1·88]	[1·70]	−0·47	[1·69]	0·33	[1·64]	[−1·65]	1·35	−0·39	1·33	0·40	1·28
4	−0·03	[2·13]	0·49	[2·12]	0·85	[2·09]	−0·02	1·85	0·51	1·84	0·86	1·81	0·04	1·56	0·55	1·56	0·90	1·53
6	0·56	2·26	0·90	2·25	1·19	2·22	0·57	2·02	0·91	2·01	1·20	1·99	0·61	1·78	0·95	1·77	1·24	1·75
8	0·93	2·39	1·22	2·38	1·46	2·36	0·94	2·19	1·23	2·18	1·47	2·16	0·98	1·97	1·26	1·97	1·50	1·95
10	1·23	2·53	1·48	2·52	1·70	2·50	1·24	2·35	1·49	2·34	1·70	2·32	1·27	2·15	1·52	2·15	1·73	2·13
15	1·80	2·87	2·00	2·86	2·17	2·84	1·81	2·71	2·01	2·70	2·18	2·68	1·84	2·55	2·03	2·55	2·20	2·53
20	2·24	3·18	2·42	3·17	2·56	3·15	2·25	3·03	2·42	3·02	2·57	3·01	2·28	2·89	2·44	2·89	2·59	2·87
25	2·62	3·46	2·78	3·46	2·91	3·44	2·63	3·32	2·78	3·32	2·91	3·30	2·66	3·19	2·80	3·19	2·92	3·18
30	2·95	3·72	3·09	3·72	3·21	3·70	2·96	3·59	3·09	3·59	3·21	3·57	2·99	3·47	3·11	3·47	3·23	3·46
35	3·25	3·96	3·38	3·96	3·49	3·94	3·26	3·84	3·38	3·84	3·49	3·82	3·29	3·73	3·40	3·73	3·50	3·72
40	3·52	4·19	3·64	4·19	3·74	4·17	3·53	4·07	3·64	4·07	3·75	4·06	3·56	3·97	3·66	3·97	3·76	3·96
45	3·78	4·40	3·89	4·40	3·99	4·38	3·78	4·30	3·89	4·29	3·99	4·28	3·81	4·20	3·90	4·20	4·00	4·19
50	4·02	4·61	4·12	4·61	4·21	4·60	4·02	4·51	4·12	4·50	4·21	4·49	4·05	4·42	4·14	4·42	4·23	4·41
	n_0	n_1	n_0	n_1	n_0	n_1	n_0	n_1	n_0	n_1	n_0	n_1	n_0	n_1	n_0	n_1	n_0	n_1
	3	5	2	5	2	5	3	4	2	4	2	4	3	2	2	2	2	2

$\bar{n}_0$ and $\bar{n}_1$ are in each case less than 10.

n_0 and n_1 refer to the smallest number of observations for which a decision is possible when $\mu = \mu_0$ and $\mu = \mu_0 + D\sigma$ respectively.
$\bar{n}_0$ and $\bar{n}_1$ are the approximate average sample numbers when $\mu = \mu_0$ and $\mu = \mu_0 + D\sigma$ respectively.
Values of U shown in square brackets are included to assist interpolation and the drawing of boundaries and must not be used in making the test.

Table 13 (continued)

$D = 2{\cdot}00$

	$\alpha = 0{\cdot}01$						$\alpha = 0{\cdot}05$						$\alpha = 0{\cdot}20$					
	$\beta = 0{\cdot}01$		$\beta = 0{\cdot}05$		$\beta = 0{\cdot}20$		$\beta = 0{\cdot}01$		$\beta = 0{\cdot}05$		$\beta = 0{\cdot}20$		$\beta = 0{\cdot}01$		$\beta = 0{\cdot}05$		$\beta = 0{\cdot}20$	
n	U_0	U_1	U_0	U_1	U_0	U_1	U_0	U_1	U_0	U_1	U_0	U_1	U_0	U_1	U_0	U_1	U_0	U_1
2	−0·26	[1·79]	0·36	[1·79]	0·73	[1·76]	−0·24	[1·57]	0·37	[1·56]	0·74	[1·54]	−0·16	1·36	0·42	1·35	0·78	1·32
4	0·75	2·00	1·02	2·00	1·23	1·98	0·74	1·83	1·03	1·82	1·23	1·80	0·77	1·66	1·06	1·66	1·27	1·64
6	1·22	2·22	1·42	2·22	1·59	2·20	1·22	2·07	1·43	2·06	1·59	2·05	1·25	1·93	1·45	1·93	1·61	1·91
8	1·56	2·42	1·73	2·42	1·87	2·41	1·56	2·29	1·74	2·29	1·87	2·27	1·59	2·17	1·75	2·16	1·89	2·15
10	1·85	2·62	2·00	2·61	2·12	2·60	1·85	2·49	2·00	2·49	2·12	2·48	1·87	2·38	2·01	2·38	2·14	2·36
15	2·42	3·03	2·54	3·03	2·64	3·02	2·42	2·94	2·54	2·94	2·64	2·93	2·44	2·85	2·55	2·84	2·65	2·83
20	2·87	3·41	2·97	3·41	3·06	3·40	2·87	3·32	2·97	3·32	3·06	3·31	2·88	3·25	2·98	3·24	3·07	3·23
25	3·28	3·76	3·36	3·75	3·43	3·74	3·28	3·67	3·36	3·67	3·43	3·66	3·29	3·60	3·37	3·60	3·44	3·59
30	3·63	4·07	3·71	4·07	3·78	4·06	3·63	3·99	3·71	3·99	3·78	3·98	3·64	3·93	3·72	3·93	3·79	3·92
35	3·96	4·36	4·03	4·36	4·09	4·35	3·96	4·29	4·03	4·29	4·09	4·28	3·97	4·23	4·04	4·23	4·10	4·22
40	4·25	4·63	4·32	4·63	4·38	4·62	4·25	4·57	4·32	4·57	4·38	4·56	4·26	4·51	4·33	4·51	4·39	4·50
45	4·54	4·89	4·60	4·89	4·65	4·88	4·54	4·83	4·60	4·83	4·65	4·82	4·54	4·78	4·61	4·77	4·66	4·77
50	4·80	5·14	4·86	5·14	4·91	5·13	4·80	5·08	4·86	5·08	4·91	5·07	4·80	5·03	4·87	5·03	4·92	5·02
	n_0	n_1	n_0	n_1	n_0	n_1	n_0	n_1	n_0	n_1	n_0	n_1	n_0	n_1	n_0	n_1	n_0	n_1
	2	4	2	4	2	4	2	3	2	3	2	3	2	2	2	2	2	2

$\bar{n}_0$ and $\bar{n}_1$ are in each case less than 10.

$D = 3{\cdot}00$

	$\alpha = 0{\cdot}01$						$\alpha = 0{\cdot}05$						$\alpha = 0{\cdot}20$					
	$\beta = 0{\cdot}01$		$\beta = 0{\cdot}05$		$\beta = 0{\cdot}20$		$\beta = 0{\cdot}01$		$\beta = 0{\cdot}05$		$\beta = 0{\cdot}20$		$\beta = 0{\cdot}01$		$\beta = 0{\cdot}05$		$\beta = 0{\cdot}20$	
n	U_0	U_1	U_0	U_1	U_0	U_1	U_0	U_1	U_0	U_1	U_0	U_1	U_0	U_1	U_0	U_1	U_0	U_1
2	0·77	[1·57]	0·95	[1·57]	1·09	[1·56]	0·77	[1·46]	0·95	[1·46]	1·09	[1·44]	0·79	1·36	0·97	1·35	1·11	1·34
4	1·39	1·94	1·50	1·94	1·59	1·93	1·40	1·86	1·50	1·85	1·59	1·85	1·41	1·78	1·51	1·78	1·60	1·77
6	1·81	2·26	1·90	2·26	1·97	2·25	1·82	2·19	1·90	2·19	1·97	2·17	1·82	2·12	1·91	2·12	1·98	2·11
8	2·15	2·54	2·22	2·53	2·28	2·53	2·16	2·47	2·22	2·47	2·28	2·46	2·16	2·41	2·23	2·41	2·29	2·40
10	2·44	2·78	2·50	2·78	2·56	2·78	2·44	2·73	2·50	2·73	2·56	2·72	2·45	2·68	2·51	2·68	2·57	2·67
15	3·05	3·34	3·10	3·34	3·15	3·33	3·06	3·29	3·10	3·29	3·15	3·28	3·06	3·25	3·11	3·25	3·15	3·24
	n_0	n_1	n_0	n_1	n_0	n_1	n_0	n_1	n_0	n_1	n_0	n_1	n_0	n_1	n_0	n_1	n_0	n_1
	2	4	2	4	2	4	2	3	2	3	2	3	2	2	2	2	2	2

$\bar{n}_0$ and $\bar{n}_1$ are in each case less than 5.

n_0 and n_1 refer to the smallest number of observations for which a decision is possible when $\mu = \mu_0$ and $\mu = \mu_0 + D\sigma$ respectively.
$\bar{n}_0$ and $\bar{n}_1$ are the approximate average sample numbers when $\mu = \mu_0$ and $\mu = \mu_0 + D\sigma$ respectively.
Values of U shown in square brackets are included to assist interpolation and the drawing of boundaries and must not be used in making the test.

Note: this table was designed and computed by Dr. S. Rushton, Imperial College of Science and Technology, under the direction of Professor G. A. Barnard and with financial assistance from Imperial Chemical Industries, Ltd. Adopted with permission of Professor Barnard and Imperial Chemical Industries, Ltd.

Table 14 Kolmogorov-Smirnov probabilities

	Level of Significance				
Sample Size	.20	.15	.10	.05	.01
1	.90	.93	.95	.98	1.00
2	.68	.73	.78	.84	.93
3	.57	.60	.64	.71	.83
4	.49	.53	.56	.62	.73
5	.45	.47	.51	.56	.67
6	.41	.44	.47	.52	.62
7	.38	.41	.44	.49	.58
8	.36	.38	.41	.46	.54
9	.34	.36	.39	.43	.51
10	.32	.34	.37	.41	.49
11	.31	.33	.35	.39	.47
12	.30	.31	.34	.38	.45
13	.28	.30	.33	.36	.43
14	.27	.29	.31	.35	.42
15	.27	.28	.30	.34	.40
16	.26	.27	.30	.33	.39
17	.25	.27	.29	.32	.38
18	.24	.26	.28	.31	.37
19	.24	.25	.27	.30	.36
20	.23	.25	.26	.29	.35
>20	$\frac{1.07}{\sqrt{N}}$	$\frac{1.14}{\sqrt{N}}$	$\frac{1.22}{\sqrt{N}}$	$\frac{1.36}{\sqrt{N}}$	$\frac{1.63}{\sqrt{N}}$

Answers to Exercises

CHAPTER 1

1. The histogram should be close to normally distributed, with the most frequent occurrence being 5 heads and the least frequent being 0 and 10 heads.
2. With a single, uniform. With $\bar{X}$ of six dice, normal.
3. $\bar{X} = 5.85 \quad S^2 = 0.105 \quad S = 0.324 \quad \phi = 9$

CHAPTER 2

1. $N = 7 \quad X = 31.45$
2. Yes
3. Vendor A, about 91 percent; Vendor B about 96.4 percent
4. F T T T F T F T F

CHAPTER 3

1. $\bar{X}^*$ should be 7.8 mW, and δ should be chosen to be 0.6 mW. α and β can be any value but probably should be equal to or less than 0.05. Probably α should be less than β. If $\alpha = 0.01$, $\beta = 0.05$, and $\delta = 0.6$, $N = 63$ and $\bar{X}^* = 7.85$ mW. If the sample mean is $\bar{X} = 7.9$ mW, the additional chip should be added to the circuit.
   The 95 percent confidence interval is 7.6 to 8.2.
2. $N = 11$; $H_0: \mu_A = \mu_B$, $H_a: \mu_A \neq \mu_B$.
   Material A is not significantly different than material B at the 95 percent confidence level. Because the proper sample has not been tested, you cannot state that the materials are equal with 90 percent confidence. Additional samples should be obtained and tested if possible. Otherwise, buy material B.
3. Choose α to be larger, for example, 0.20; choose β to be smaller, for example, 0.05; choose δ to be 1 or less. With $\alpha = 0.20$, $\beta = 0.05$, and $\delta = 1$, $N = 28$ and $|\bar{X}_{30} - \bar{X}_{40}|^* = 0.27$. The engineering decision should be to use the 30° angle.

CHAPTER 4

1. A decrease of 1 lumen in the standard deviation will result in a break-even financial situation for the supplier. The change in standard deviation should therefore be twice this decrease. The present standard deviation is 5.66. Therefore, the new standard deviation should be $5.66 - 2 = 3.66$. Therefore, δ should be $32 - (3.66)^2 = 32 - 13.40 = 18.60$.

$R = \frac{32 - 18.6}{32} = 0.42$, $N = 30$, and $*S^2 = 20.4$. Since $S^2 = 18.4$, the process should be changed. Additional changes in the process should be studied, because the product cost will still be excessive due to excess scrap.

2. With at least 99 percent confidence, $\sigma_B^2 < \sigma_A^2$.

CHAPTER 5

1. $N = 44$. A decision to accept H_a will be made on the eleventh trial. The 99 percent confidence interval is 278.2 to 293.8. Starting with the bottom number, H_a will be accepted after 10 trials.

2. Tenth trial.

CHAPTER 6

1.

A	B	C	Trial	Expected result
+	−	−	1	25 + 5 = 30
+	+	−	2	25 + 5 − 2 = 28
+	+	+	3	25 + 5 − 2 + 4 + 3 = 35
−	+	+	4	25 − 2 + 4 = 27
+	−	+	5	25 + 5 + 4 + 3 = 37
−	+	−	6	25 − 2 = 23
−	−	+	7	25 + 4 = 29
−	−	−	8	25 = 25

2. $\bar{X}_{\text{highA}} - \bar{X}_{\text{lowA}} = 5.25$.
$\bar{X}_{\text{highB}} - \bar{X}_{\text{lowB}} = 8.75$.
AB interaction = 2.75.
Estimate of variance is the same by both methods: $S^2 = 1.625$.

3. Conclusions:
High A is better than low A.
High B is better than low B.
Propose a follow-up experiment with higher A and higher B. Possibly include variable D at a lower level.

4. 5,002,846 **5.** 4.8 × 100 sq. ft.

CHAPTER 7

1. I = ABCDEF
I = ACDEG
I = BFG

B is confounded with FG.
F is confounded with BG.
G is confounded with BF.
Use figure 9-1 for labeling all columns.

2. I = ABCDF
I = BCDEG
I = AEFG

This is a better design. It is Resolution IV, whereas the design for question 1 is Resolution III.

3. Resolution IV: third order and higher; BD and for AE are significant.

4. A = 4.125, B = 4.375, AB = 5.125, all others zero.

CHAPTER 8

1. a. Same as on page 147
   b. The interaction of low A and low B reduces the number of jams.
   c. The interaction of low C and high D reduces the number of jams. In addition, the conclusion of (b) is confirmed.
   d. No additional information is obtained. Nothing will be gained by testing the treatment combinations of block i. The ad treatment combination will be about 10, the ac treatment combination will be about 15, the b treatment combination will be about 19, and the bcd treatment combination will be about 20.

2. Blocks iii and iv have the identity I = −BCD. Multiply every possible combination of A, B, C, and D with the identity. Only A, AB, AC, and AD will not be confounded with another main factor or two-factor interaction.

3.
$$\begin{array}{lrl} \text{ii} & I = & -ABE + CDF - ABCDEF \\ \text{iii} & +\,I = & +ABE - CDF - ABCDEF \\ \hline & 2I = & -2ABCDEF \\ & I = & -ABCDEF \end{array}$$

$$\begin{array}{lrl} \text{iii} & I = & +ABE - CDF - ABCDEF \\ \text{iv} & +\,I = & -ABE - CDF + ABCDEF \\ \hline & 2I = & -2CDF \\ & I = & -CDF \end{array}$$

CHAPTER 9

The first part of the design shows that high A, high B, and high C have fewer uncharged bits. There is also a BG interaction.

The second part of the design shows that high E, high F, and high G have fewer uncharged bits.

CHAPTER 11

1. N = 32 parts at each treatment combination (double-sided). Criterion = 13.06.
2. N = 197 sq. ft/treatment combination (double-sided). Criterion = 5.66.

CHAPTER 12

2. a. Order 44. b. Order 64. c. Order 192. d. Order 128.

CHAPTER 16

1.

	P_1		P_2		P_3		P_4	
	L_1	L_2	L_1	L_2	L_1	L_2	L_1	L_2
S_1								
S_2								
S_3								

$\phi_P = 3; \phi_L = 1, \phi_S = 2.$
$\phi_{LP} = 3; \phi_{PS} = 6, \phi_{LS} = 2.$
$\phi_{LPS} = 6.$

2. Rubber E and adhesive 2 or rubber A with adhesive 3 are the best systems. Design 1 is better than design 2.

3. MS(documents) = 9.4.
MS(typist) = 635.
MS(days) = 1.17.

a. Yes. **b.** No. **c.** No.

CHAPTER 17

1. The experiment should consist of a central composite rotatable design (CCRD) with the 0.020 wire and a CCRD with 0.030 wire.

0-level temperature = 250°.	0-level voltage = 125.
+1-level temperature = 285°.	+1-level voltage = 143.
$+\psi$-level temperature = 300°.	$+\psi$-level voltage = 150.

2. The fuse design will not perform its required function at any wire thickness. With the 0.03 wire, the operating time at, for example, 270 and 135 volts will be less than 4000 seconds. If the wire is thicker than 0.03, the operating time at, for example, 210° and 160 volts will be more than 3000 seconds before blowing.

3.

A_1	C	D	E
0	−1	−1	+1
+1	−1	−1	−1
+1	+1	−1	+1
+1	+1	+1	−1
0	+1	+1	+1
0	+1	+1	+1
0	−1	+1	−1
0	+1	−1	−1
+1	−1	+1	+1
0	+1	−1	−1
−1	+1	+1	−1
0	−1	+1	−1
0	−1	−1	+1
−1	+1	−1	+1
−1	−1	+1	+1
−1	−1	−1	−1

4. The middle level of variable A, the high level of variable C, and the low level of variable E will give the highest results.

CHAPTER 18

1.

A	B	C	Concentration
100%	0%	0%	1%
0%	100%	0%	1%
0%	0%	100%	1%
50%	50%	0%	1%
50%	0%	50%	1%
0%	50%	50%	1%
33.3%	33.3%	33.3%	1%

Repeat above combinations of A, B, and C with 3 percent concentration.

2. One proposal might be:

A	B	C
59	12	39
59	18	33
53	18	29
57	16	27

CHAPTER 19

Stop at the seventh trial.

CHAPTER 20

1. Use contrast columns 12 and 14, for example, to construct blocks.

Equipment	Treatment combinations
1	ab, c, d, abcd
2	a, acd, bd, bc
3	bcd, ac, ad, b
4	abc, abd, cd, (1)

2. Set variable E = ABCD, and block on ABC; interaction DE will be confounded with blocks.
   Technician 1: ae, abce, abcd, bd, cd, ad, be, ce.
   Technician 2: ab, bcde, acde, ac, abde, bc, de, (1).

3. $S^2 = 1.67$. There is no day-to-day bias. There is curvature. A result of 30 could be expected at higher A and higher D, about +2 level.

CHAPTER 21

$MS(\text{testing}) = 0.0010$. $\hat{\sigma}_0^2 = 0.0010$.
$MS(\text{sampling}) = 0.0323$. $\hat{\sigma}_1^2 = 0.0157$.
$MS(\text{lots}) = 0.1038$. $\hat{\sigma}_2^2 = 0.0119$.

The variability between lots and within lots is too large, but the testing procedure is outstandingly good.

If each batch is sampled eight times and each part is tested once, the variance of the batch mean will be 0.0022. The 90 percent confidence interval of the mean of each batch will be ± 0.09.

Use table 13 to determine the acceptance limits.

CHAPTER 24

1. $S^2 = 5.855$ $\phi = 12$

Index